Magnetic Resonance in Food Science
Challenges in a Changing World

Magnetic Resonance in Food Science
Challenges in a Changing World

Edited by

María Guðjónsdóttir
Matvælarannsóknir Íslands, Reykjavík, Iceland

Peter Belton
School of Chemical Sciences and Pharmacy, University of East Anglia, Norwich, UK

Graham Webb
c/o Royal Society of Chemistry, Burlington House , Piccadilly, London, UK

RSCPublishing

The proceedings of the 9th International Conference on the Applications of Magnetic Resonance in Food Science: Challenges in a Changing World held in Reykjavik, Iceland on 15-17 September 2008.

Special Publication No. 319

ISBN: 978-0-85404-117-6

A catalogue record for this book is available from the British Library

Published by The Royal Society of Chemistry,
Thomas Graham House, Science Park, Milton Road,
Cambridge CB4 0WF, UK

Registered Charity Number 207890

For further information see our web site at www.rsc.org

Preface

This volume of Magnetic Resonance in Food Science is the result of the very interesting 9th International Conference on the Applications of Magnetic Resonance in Food Science, which was held in the Nordic House in Reykjavik, Iceland on September 15th–17th 2008. The conference had 80 participants from 20 countries from all over the world, coming from countries as far as Brazil, Japan and South Africa. This wide origin of the participants led to a great variety in the published material at the conference. The conference included 33 oral contributions, of which 29 are presented in these proceedings, as well as 33 poster presentations. The conference was divided into four sessions: New techniques; Food systems and processing; ESR and other techniques; and Fish and Meat.

As the chairman of this conference I must declare my sincere thanks to my co-editors, Peter Belton and Graham Webb, for giving me the chance of taking part in the organization of this event. The work has been a pleasure from start to finish and the value of getting in contact with some of the most prestigious scientists within the applications of Magnetic Resonance in Food Science is of endless value for a new scientist in the field as myself. I also hope that the conference will lead to increased cooperation between the participants and other readers of these proceedings in various research projects to come.

This volume of Magnetic Resonance in Food Science is dedicated to my new born son. May he do well in his future challenges in a changing world.

María Guðjónsdóttir

Reykjavík, Iceland

Contents

Food Systems and Processing

ESR and Other Techniques

Contents

Fish and Meat

New Techniques

HIGH RESOLUTION NMR ANALYSIS OF COMPLEX MIXTURES

Adrian J. Charlton

Central Science Laboratory, Department for Environment, Food & Rural Affairs, Sand Hutton, York, YO41 1LZ, UK.

1 INTRODUCTION

Advances in analytical techniques and reduced computation times have facilitated broad ranging characterisation of complex chemical and biochemical systems. Resolution and sensitivity enhancements have ensured a key role for high-resolution liquid state nuclear magnetic resonance (NMR) spectroscopy for the determination of compositional variation in complex matrices including foods. A range of studies utilising NMR spectroscopy and advanced data exploration techniques have illustrated the applicability of these techniques to issues facing the food and agricultural sectors.

In the era of the discerning consumer, attitudes toward food choice have changed markedly from the need to provide basic nutrition to the desire to make informed choices relating to food intake. Consumer choices in relation to food intake are often made on the basis of the perceived health benefits that food borne components may impart, where scientific methods to substantiate these claims are often lacking. Whilst the organoleptic properties of food are doubtless a major factor for continued food consumption, it is also clear that initial choices are often made on the basis of promotional labelling.[1] Information that is present on food labels acts as a comparative index by which initial consumer choices are made. For example, regional produce may be associated with a distinctive flavour. Similarly, production processes including organic, corn-fed or free range are associated with superior quality produce or morality probing implications.

Informed consumer preferences are thus leading to a greater range of choices and a concurrent increase in the number of claims that are made by food producers, manufacturers and the broadcast media. These choices can only be considered as informed when public awareness of current scientific knowledge is prevalent. In the scientific communities an extensive phase of evidence gathering is being undertaken in relation to food composition. This is largely being driven by a desire to fully understand the composition of food and its implications with respect to consumer choice and public health. Large bodies of data are being collected and interpreted against specific claims relating to food authenticity, quality, safety and nutrition.

An insight is provided here into the current state-of-the-art for the compositional analysis of molecules in food utilising high-resolution NMR spectroscopy in conjunction with multivariate data analysis techniques. The recent Human Metabolome Project (http://www.metabolomics.ca) has identified 2,500 metabolites, 1,200 drugs, and 3,500 food components that can be found in the human body. Whilst numerical incidence is not a robust measure of significance, it is clear from this information that it would be impossible to completely understand human physiology without a detailed understanding of factors effecting food composition.

2 METHODS

2.1 Nuclear magnetic resonance (NMR) spectroscopy

The application of nuclear magnetic resonance spectroscopy within the food sector has, until recently, focussed primarily on the use of time domain (TD) techniques. These enable the quantitative measurement of bulk properties such as water and fat content in whole foods. The measurement relies on the intrinsic relaxation properties of the proton nucleus when a radio frequency pulse is applied to a sample placed in a magnetic field. The differential between the relaxation properties of major food components allows the proportion of these components to be estimated by reference to a calibration series. This form of NMR spectroscopy is routinely applied for quality and composition checks and is often undertaken *in situ* as the instrumentation is both inexpensive and robust.

In more recent times, higher magnetic field strengths have enabled the measurement of specific resonance frequencies relating to individual components of food and it is the high-resolution (HR) form of NMR spectroscopy that has been adopted as one of the key techniques in a wide range of studies relating to food composition. HR NMR spectroscopy uses the same principles of pulse excitation that is used for TD NMR studies. However, it is the precession frequency of a given nucleus within a magnetic field that is the primary source of information. This frequency is recorded in the time domain as the free induction decay (FID) and the Fourier transformation is used to generate what is recognised as the frequency domain NMR spectrum. The spectrum is characterised by a number of resonance peaks which are plotted as spectral intensities and frequencies (chemical shifts) usually displayed in parts per million (ppm) of the NMR carrier frequency. Assuming the correct instrumental set-up there is a linear correlation between the number of nuclei that represent a resonance peak and its intensity. Therefore NMR measurements can largely be considered as quantitative. High-resolution NMR spectra of both liquid and solid food (or extracts thereof) can be recorded with minimal sample preparation. The ubiquity of the NMR response from the small organic molecules that are found in food make HR NMR spectroscopy an excellent method for compositional screening and molecular characterisation.

The sensitivity of the NMR measurement is a function of the time used for data acquisition with a high number of repetitions leading to a greater signal to noise ratio. This feature of NMR spectroscopy is one of the principle reasons for its increasing adoption for the determination of food composition, illustrating the high degree of analytical reproducibility that is offered.

The large amount of interpretable data that is obtainable by NMR spectroscopy when it is applied to complex mixtures is perhaps the main advantage that the technique offers over other spectroscopic techniques. NMR spectra can be routinely interpreted to determine the range of compounds that are present in complex mixtures particularly when two-dimensional (2D) NMR techniques are employed. Limitations of the technique are often quoted as being poor sensitivity and resolution. Both of these limitations are progressively being overcome by increasing the available magnetic field strengths (currently the maximum is 22.3 tesla [950MHz ^{1}H]), and even greater strides towards high sensitivity measurements are being made by the development of polarisation techniques and the advent of cryogenically cooled probes.

2.2 Multivariate Analysis

Examination of the wide diversity of small molecules in complex mixtures requires multivariate methods that are able to reduce the high dimensionality of analytical datasets to fewer characteristic dimensions. Multivariate methods to extract information from large datasets are becoming increasingly widespread. One of the most common chemometric methods is principal components analysis (PCA). This method performs a coordinate transformation on multivariate data so that they are represented as a number of principal component scores on new coordinate axis (principal components). The initial principal components capture the most significant sources of variation and this decreases for each subsequent component. PCA is an unsupervised method in which no prior information about experimental groupings is used in the transformation. It therefore avoids the need for the extensive cross-validation required in supervised methods, but can be less efficient in finding differences between experimental groupings when there is a large degree of natural variation.

Soft independent modelling of class analogy (SIMCA) is a supervised multivariate statistical method. SIMCA is used to compute scores and residuals from or within a component, plane or hyperplane of a principal components analysis. Critical distances from (residuals) and within (scores) the plane of the model are then used to determine thresholds for class membership. SIMCA models allow predictions to be made, using test data, about the membership (or otherwise) of the modelled class. In judging class membership two types of outlier are defined: moderate and strong outliers. Strong outliers have a great leverage on a PCA and are thus detected by the PC scores. Threshold values for strong outliers can be set by the determination of a probability ellipsoid and this is often calculated using the Hotelling's T^2 statistic, which is a multivariate generalisation of Student's t-test. Moderate outliers are detected by consideration of the model residuals. This is the distance from an observation to the plane of the SIMCA model and is also known as the residual error or the residual standard deviation (RSD).

Partial least squares (PLS) regression is a multivariate method that seeks to find linear combinations of the variables that result in the best separation between specific experimental groupings. In broad terms the method can be considered to be mathematically equivalent to PCA where vector rotation is performed until maximum categorical variance is captured as component planes. As *a-priori* knowledge of group membership is required when developing a PLS model it is a supervised method, and requires cross-validation to avoid overfitting the data. PLS is used in combination with an appropriate discriminant analysis technique to determine the probably of group membership. A PLS calculation also generates a measure of the contribution that each variable in the calculation makes to the discrimination of the tested groups. This is called the variable importance of projection (VIP) and is used to interpret the PLS calculation in the context of the non-transformed data.

Artificial intelligence has also been applied to analytical data sets, including artificial neural networks, genetic algorithms and genetic programming (GP). Genetic algorithms and genetic programming are computational learning techniques based on Darwin's theory of evolution and are popular for solving optimization problems. A population of possible solutions to a problem is randomly created and this is considered the first generation. New generations are achieved through mutation and crossover (sexual reproduction) evolving optimal solutions. The fitness of each solution is evaluated using a fitness function, which for classification problems can be simply the number of correct classifications achieved. Where genetic algorithms create a string of numbers to represent solutions, genetic programming creates computer programs (symbolic expressions). These computer programs are often represented as trees and consist of mathematical operators and data variables (Figure 1). A program that provides an optimised solution can therefore be related to the data from which it was generated.

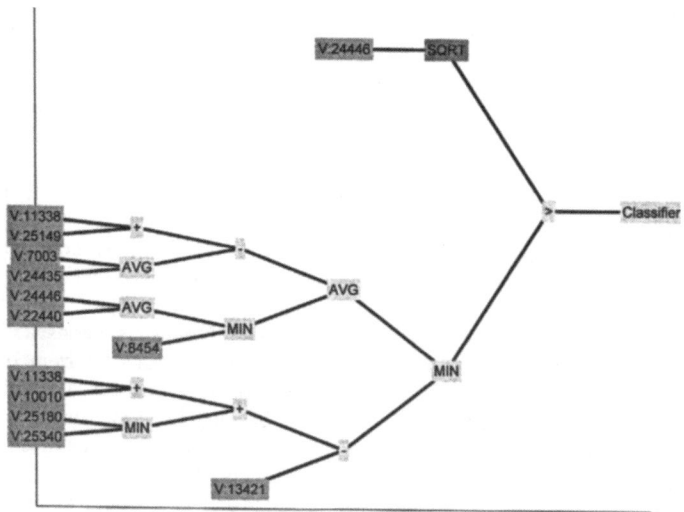

Figure 1 A typical genetic programming tree. V: = variable, SQRT = square root, + = summation, - = subtraction, min = minimum and AVG = average.

3 RESULTS AND DISCUSSION

3.1 Incident Detection

In 1997 the Laboratory Environmental Analysis Proficiency (LEAP) Emergency Scheme was initiated[2]. The primary function of this scheme is to establish the reliability of water testing laboratories to determine the contents of a simulated potable water contamination sample. In the event of a potable water contamination incident, water testing laboratories may be called upon to rapidly identify contaminants that are present in the water with little information about the likely source of contamination. Samples are analysed for inorganic and organic chemical contamination.

Determination of organic contamination of the potable water supply is usually performed using techniques based on the application of high performance liquid chromatography (HPLC) or gas chromatography (GC). The choice of chromatographic method is determined by the suite of analytes that is to be detected and therefore these approaches are inherently targeted to detect specific compounds. Ideally, a non-targetted technique, such as NMR spectroscopy should be used, to minimise the probability that a contaminant will not be detected. NMR spectroscopy is commonly perceived to be a relatively insensitive technique, however developments in the field of NMR are constantly leading to improved sensitivity and include improvements such as increases in the maximum available magnetic field strength and cryoprobes. In the absence of high salt concentrations, the sensitivity obtained by using the cryoprobe is approximately 3 to 4 fold greater than that obtained by using the equivalent standard probe and this corresponds to a reduction of 9 to 16 fold in experimental time. Further improvements in the signal to noise ratio of NMR measurements can be made by effective instrument set-up. This includes the choice of pulse sequence, correct calibration of required pulses and effective choice of probehead. Recent advances in the sensitivity of NMR spectroscopy have facilitated the determination and characterisation of molecules present in solution at much lower concentrations than prior to these developments.

Pre-concentration of a sample using rotary evaporation has also been shown to improve detection limits for NMR measurement of non-volatile analytes in water. However, one of the principal advantages of NMR spectroscopy for the detection of unknown contaminants in potable water is that the analysis can be performed without significant sample preparation. Table 1 summarises the analysis of three consecutive LEAP organic contamination emergency samples using cryoprobe nuclear magnetic resonance (NMR)[1] spectroscopy.

An efficient method for detecting malicious and accidental contamination of foods has been developed using a combined ^{1}H nuclear magnetic resonance (NMR) and chemometrics approach[3]. The method has been demonstrated using a commercially available carbonated soft drink, as being capable of identifying atypical products and to identify contaminant resonances. Soft independent modelling of class analogy (SIMCA) was used to compare ^{1}H NMR profiles of genuine products (obtained from the manufacturer) against retail products spiked in the laboratory with impurities. The benefits of using feature selection for extracting contaminant NMR frequencies were also assessed. Using example impurities (paraquat, *p*-cresol and glyphosate) NMR spectra were analysed using multivariate methods resulting in detection limits of approximately 0.075 mM, 0.2 mM, and 0.06 mM for *p*-cresol, paraquat and glyphosate, respectively. These detection limits, assuming soft beverage consumption of 500 mL and an adult weight to be 70 kg, were shown to be approximately 100 fold lower than the minimum lethal dose for paraquat. The methodology is used to assess the composition of complex matrices for the presence of contaminating molecules without *a priori* knowledge of the nature of potential contaminants. The ability to detect if a sample does not fit into the expected profile without recourse to multiple targeted analyses is a valuable tool for incident detection and forensic applications.

3.2 Food Authenticity and Traceability

NMR spectroscopy has been extensively utilised for the verification of the provenance of food commodities. Particular examples include the determination of the manufacturing origin of instant coffee[4] and more recently the use of artificial intelligence techniques to elucidate biomarkers of the botanical and geographical origin of honey[5].

Principal components analysis followed by linear discriminant analysis of the NMR spectra from 98 instant spray dried coffees, obtained from 3 different producers, correctly attributed 99% of the samples to their manufacturer. Blind testing of the PCA model with a further 36 samples of instant coffee resulted in a 100% success rate in identifying the samples from the 3 manufacturers. Coffees from one manufacturer were also assigned into 2 groups using these techniques

	Analyte	Concentration (μgL^{-1})	^{1}H Chemical shift	Observed without pre-concentration	Observed with pre-concentration	Minimum experimental time used to observe (Hr)
LEAP Water Sample 10	Picric Acid	504	8.987	Yes	Yes	0.2[b]
	Amitrole	50	7.792	Yes	Yes	2.0[b]
	Dicamba	50	7.400 7.237 3.909	Yes	Yes	0.5[b]
	Pyridine	200	8.541 7.915 7.492	Yes	No	2.0
	Mevinphos	80	5.853 3.923 3.901 3.753 2.382	Yes	Yes	1.0[b]
LEAP Water Sample 11	Aldicarb	75	7.740 2.820 2.812 2.010 1.470	Yes	Yes	0.3[b]
	Aniline	339	7.261 6.9091[a] 6.8943[a] 6.8797[a] 6.864[a]	Yes	No	1.3
	Glyphosate	202	3.731 3.026 3.002	Yes	No	2.0
	MCPA	11	7.257 7.193 6.757 2.246	No	Yes	4.0[b]
	Phosphamidon	70	3.961 3.865 3.480 2.247 2.088 1.20	No	Yes	0.3[b]
LEAP Water Sample 12	2,4-D	75	7.533 7.311 6.899 4.571	Yes	Yes	0.3[b]
	2,6-Dibromo-phenol	5	N/D	No	No	>30Hr
	Pentachloro-phenol	20	N/D	No	No	N/A

Table 1 Summary of results obtained from ^{1}H NMR spectroscopic analyses of potable water and the concentration of the organic contaminants present in the three LEAP water samples tested.

---

[a] Two overlapping multiplets.

[b] Acquisition time used after pre-concentration.

and the compound 5-(hydroxymethyl)-2-furaldehyde was identified as the primary marker of differentiation.

EU legislation (Council Directive 2001/110/EC) requires that the geographical origin of honey be presented on the packaging. Honeys from specified botanical sources will often command a premium price due to their organoleptic or pharmacoactive properties. To prevent the fraudulent marketing of honey, analytical techniques are required to confirm its origin.

HR ^{1}H-NMR spectroscopy and multivariate analysis techniques were used to classify honey into two groups by geographical origin. Honey from Corsica (Miel de Corse) was used as an example of a protected designation of origin (PDO) product. Mathematical models were constructed to determine the feasibility of distinguishing between honey from Corsica and that from other geographical locations in Europe, using ^{1}H-NMR spectroscopy. Honey from 10 different regions within 5 countries was analysed. ^{1}H-NMR spectra were used as input variables for PLS-LDA and genetic programming. Models were generated using three methods, PLS-LDA, 2-stage GP and a combination of PLS and GP (PLS-GP). The PLS-GP model used variables selected by PLS for subsequent genetic programming calculations. All models were generated using Venetian blind cross-validation. Overall classification rates for the discrimination of Corsican and non-Corsican honey of 76.4, 94.5 and 96.2 % were determined using PLS-LDA, 2-stage GP and PLS-GP, respectively. The variables utilised by PLS-GP were related to their ^{1}H-NMR chemical shifts and this led to the identification of trigonelline in honey for the first time. Biomarkers of sweet chestnut, Corsican spring Maquis and Arbousier (strawberry-tree) honeys were identified. Kynurenic acid was found to be a biomarker of sweet chestnut honey. α-isophorone and 2,5-dihydroxyphenylacetic acid were confirmed as markers of strawberry-tree honey. Additional compounds specific to strawberry-tree and Corsican spring Maquis honey were partially characterised.

3.3 Food Safety

Constitutive levels of metabolites readily respond to perturbations in environmental conditions, breeding processes as well as direct molecular bioengineering approaches. NMR profiling, typically in conjunction with multivariate analyses, is particularly adept at identifying such changes. In other words, NMR-based metabolite profiling is well suited to monitor and quantify the degree of metabolic impact induced by genetics, environment or bioengineering. This clearly has implications for the determination of the quality and safety of food crops.

HR NMR spectroscopy has been applied to pea (*Pisum sativum*) extracts in order to define the impact of environment and genetic diversity on baseline metabolite profiles. The plants included 20 diverse lines of *Pisum*, 20 recombinant inbred lines (RILs) derived from a wide cross (between *P. sativum* cv. Ethiopia and *P. sativum* cv. Cennia), four independent transgenic lines derived from *Agrobacterium*-mediated transformation of *Pisum sativum* (cv. Puget) using a construct composed of a trypsin inhibitor (TI) gene promoter-*GUS* fusion and a *bar* selectable marker gene, and control lines which were azygous segregants identified at T_2 or T_3 generations for every transformed line. This allowed eight data groupings (identified in parenthesis below) and defined as follows:

i)   RILs grown in 2003 (RI-03)
ii) RILs grown in 2004 (RI-04)

iii)   cultivar Puget plants sown at monthly intervals (Mnth) under glasshouse
         conditions in an   experiment designed to investigate the effect of growing
         season on metabolite profiles.

iv)   cultivar Puget plants grown in a range of environments (Envi) including
         controlled environment room, glasshouse, polytunnel and outdoor plots.

v)   diverse germplasm lines harvested in 2003 (GP03).

vi)   drought-stressed transgenic and azygous plants (Drought) grown in one season.

vii)   well-watered transgenic plants (W/T) grown in one season.

viii)   well-watered azygous plants (W/N) grown in one season.

Leaf samples were harvested from all plants at the onset of flowering and seeds were harvested at maturity, following desiccation. NMR analysis on these samples followed a previously established protocol. The NMR profiles were subjected to a number of statistical tests, including; principal components analysis (PCA), partial least squares discriminant analysis (PLS-DA), analysis of variance (ANOVA) and Student's t-test. These spectra were subjected to statistical analysis to determine the regions of the NMR spectra that exhibited the greatest variation and also to identify profiles that could be correlated with genotype, environment or transgenesis. PCA identified the watering regime as the factor that affected the leaf composition most significantly. Clear separation between the well-watered plants and those that had been subjected to drought was seen when the first two principal component scores were plotted. It was also apparent that the genotype of the plants significantly affected the metabolome. Many of the individual genotypes from the diverse germplasm formed tight and distinct data clusters when subjected to PCA. Using PLS-LDA, it was possible to classify the leaf extracts as derived from transgenic or azygous plants and this classification was more successful when the plants were well-watered than when they were drought-stressed. It was not possible to separate the transgenic plants from the azygous controls when both the drought-stressed and well-watered plants were treated as one group. This strongly suggests that any impact on the metabolome that is conferred by transgenesis may be masked by larger environmental effects, such as drought.

Figure 2 shows the NMR intensities at approximately 2.33 ppm, determined for the eight groups. The results of the t-test performed on the well-watered transgenic (W/T) and azygous control (W/N) groups show that the metabolite that gave rise to this resonance is present at an elevated level in the transgenic plants ($p < 0.0001$). It is also clear that the wide range of genetic diversity for this metabolite, observed in both the RILs (RI-03, RI-04) and diverse germplasm (GP03), exceeds the elevated range determined for the transgenic plants. In the case of this particular metabolite, the difference between the transgenic and azygous control groups was statistically the most significant single point difference in the pea leaf metabolome, as observed by NMR. However, when the range of concentrations of this metabolite, as affected by environmental factors and genotype, is considered, it is clear that the relevance of this significant difference in the context of safety evaluation of genetically modified (GM) foods is minimal[6].

4 CONCLUSIONS

Rapid determination of biomarkers of product origin and contamination has illustrated the capability of modern data analysis techniques in combination with advanced NMR systems. Characterisation of biological systems has also been performed in support of genetic improvement and for the detection and characterisation of the effects of genetic modification. These application-focussed studies have led to the development of the fundamental knowledge base that underpins food and agricultural sciences. Examples include the determination of the effect on the pea metabolome of biotic and abiotic stressors, including the interaction between genetic and environmental factors and their relation to product composition and quality.

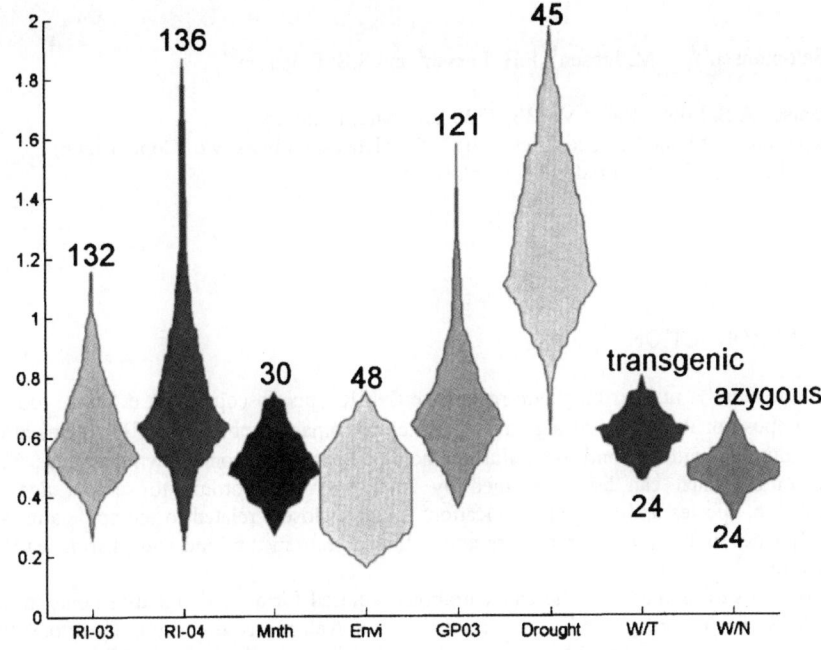

Figure 2 The range of NMR intensities observed for a metabolite resonating at 2.33ppm. The range of metabolite concentrations for each treatment is shown by the height of the coloured plot for each group. The width of the coloured areas represents the proportion of the total number of observations (in each group) having a particular NMR intensity. The number of plants contributing to every group is indicated above or below the group. The group names are defined in the text.

5 REFERENCES

1. Gregory, N.G. Consumer concerns about food. *Outlook on Agriculture* **2000**, *29* (4), 251-257.
2. Charlton A.J., Donarski, J.A., Jones, S.A., May, B.D., & Thompson, K.C., 2006, *J. Environ. Monitor.* **8**, 1106.
3. Charlton, A.J., Robb, P., Donarski, J.A. & Godward, J., 2008, *Anal. Chim. Acta.* **618**, 196.
4. Charlton, A.J., Farrington, W.H. & Brereton, P. (2002) *J. Agri. Food Chem.* 50, 3098-3103.
5. Donarski, J.A., Jones, S.A. & Charlton, A.J., 2008, *J. Agric. Food Chem.*, **56**, 5451.
6. Dixon, R.A., Gang, D.R., Charlton, A.J., Fiehn, O., Kuiper, H.A., Reynolds, T.L., Tjeerdema, R.S., Jeffery, E.H., German, J.B., Ridley, W. & Seiber, J.N., 2006, *J. Agric. Food Chem.* **54**, 8984.

THE QUANTITATIVE IMPACT OF WATER SUPPRESSION ON NMR SPECTRA FOR COMPOSITIONAL ANALYSIS OF ALGINATES

T. Salomonsen[1,2], H.M. Jensen[1], F.H. Larsen[2] and S.B. Engelsen[2]

[1] Danisco A/S, Edwin Rahrs Vej 38, 8220 Brabrand, Denmark
[2] Department of Food Science, Faculty of Life Sciences, University of Copenhagen, Rolighedsvej 30, 1958 Frederiksberg C, Denmark

1 INTRODUCTION

Solution-state ^{1}H nuclear magnetic resonance (NMR) spectroscopy is an effective tool in the compositional analysis of alginates,[1,2] which are binary copolymers of (1-4) linked β-D-mannuronic acid (M) and α-L-guluronic acid (G) extracted from brown seaweeds. The M/G ratio, which can be determined by NMR, varies according to season, age of population, species and geographic location[3-5] and is closely related to the application of the alginates as thickeners, stabilisers and gelling agents in the food and pharmaceutical industries.[6-8]

Solutions of alginates, in the concentrations required for a good signal-to-noise in the ^{1}H NMR spectra, are generally too viscous to give well-resolved spectra. Therefore, the viscosity is typically reduced by lowering the molecular weight prior to the NMR analysis. In addition, the spectra are recorded at 80-90 °C in order to reduce the viscosity further and to shift the water resonance away from the anomeric region, which contains the signals used in the calculations of the M/G ratio.

The magnitude of the water resonance is normally several orders of magnitude higher than the other signals in the spectra due to residual water in the alginates. Thus, it can be an advantage to apply water suppression, which allows one to recover signals that overlap the broad baseline of the intense water resonance. However, water suppression can possibly affect the intensity of other signals in the spectra, which may lead to significant errors in the quantitative analysis.

In a previous NMR study of alginates, water suppression was found to alter the intensities of the signals of interest[9] whereas other studies concluded that water suppression had no effect on the intensities of the signals near the water resonance.[10] In this study we set out to investigate how some of the common water suppression techniques influence the ^{1}H NMR spectra of alginates in solutions and thereby the calculated M/G ratios.

2 MATERIALS AND METHODS

2.1 The Samples

A total of 40 different commercial sodium alginates were kindly provided by Danisco A/S (Brabrand, Denmark). The average molecular weights (M_w) of the alginates were reduced from ~300 kDa to ~30 kDa by partial acid hydrolysis[1] in order to reduce the viscosity of the alginate solutions for the NMR analysis. The samples were prepared for NMR analysis by dissolving the hydrolysed alginates in D_2O (1% (w/v)) followed by neutralisation (pH 7). 3-(trimethylsilyl)propionic acid-d_4 sodium salt (TSP-d4) was added as chemical shift reference. The solutions (550 μl) were filtrated through Whatman 13 mm syringe filters (pore size 0.45 μm) into 5 mm NMR tubes.

2.2 NMR Data Acquisition

^{1}H NMR spectra of the samples were collected on a Bruker Avance[TM] 400 spectrometer (Bruker Biospin Gmbh, Rheinstetten, Germany) operating at 400.13 MHz for protons and equipped with a 5 mm broad band inverse probe. Four unique NMR spectra of each sample were acquired using the pulse sequences schematically depicted in Figure 1. Spectra without water suppression were recorded with a single 90° pulse (Figure 1a) and spectra with water suppression were recorded using three different water suppression techniques: conventional presaturation (Figure 1b), presaturation followed by a 90° composite pulse (Figure 1c) and NOESY-presaturation (Figure 1d). The experiments will be denoted *zg*, *zgpr*, *zgcppr* and *noesypr1d* (Bruker notation), respectively. In the *noesypr1d* experiment, saturation was also employed during the NOESY mixing period, which was set to 150 ms.

All spectra were acquired using a relaxation delay of 2 s, a spectral width of 8278 Hz, an acquisition time of 3.96 s and a sample temperature of 90 °C. All acquisitions were initiated with two dummy scans followed by 16 scans. All *zg* spectra were recorded with one receiver gain and all water-suppressed spectra with another (4.5 times higher).

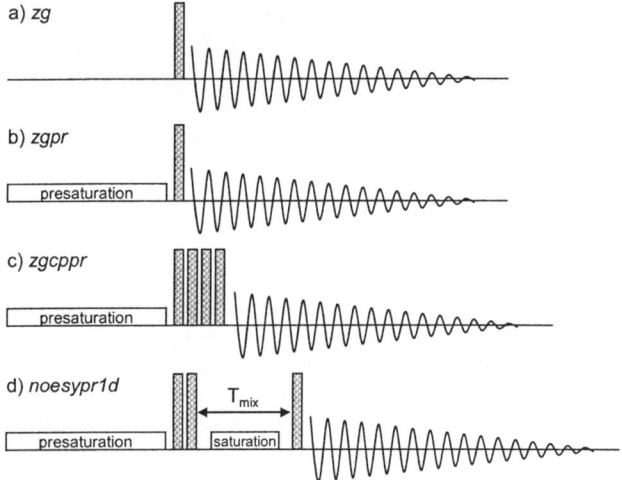

Figure 1 *Schematic drawing of the four different pulse sequences used in the study*

Magnetic Resonance in Food Science

The FIDs were zero filled to 64K data points and apodized by a 1.0 Hz exponential line broadening before Fourier transformation. The resulting spectra were individually phased and baseline corrected using the Bruker TOPSPIN 1.3 software (Bruker Biospin Gmbh, Rheinstetten, Germany). All spectra were referenced to TSP-d4 at 0.000 ppm.

2.3 FT-Raman Spectroscopy

Raman spectra were collected on a Perkin Elmer System NIR FT-Raman interferometer (Perkin Elmer Instruments, Waltham, Massachusetts, USA) using 1064 nm laser excitation with a power of 200 mW as described by Salomonsen et al.[11] The range 650-1800 cm^{-1} was used in the data analysis.

2.4 M/G Ratio Calculations

The M/G ratio was calculated using the relative areas of the signals in the anomeric region as described by Grasdalen et al.[1,2] The integration limits applied in the calculations of the areas of the signals denoted A, B and C in the anomeric region (Figure 2) were 4.96-5.18 ppm (H-1 of G at 5.07 ppm), 4.57-4.82 ppm (H-5 of G at 4.76 and 4.72 ppm, H-1 of M at 4.70 and 4.68) and 4.38-4.55 ppm (H-5 of G at 4.46 ppm), respectively. Assignments were made according to Grasdalen et al.[2]

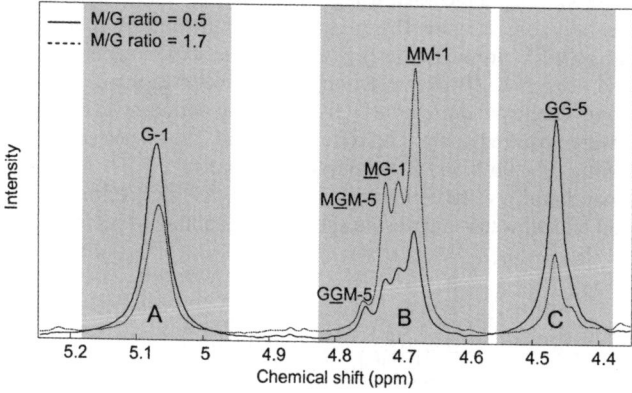

Figure 2 *^{1}H NMR spectra of two hydrolysed alginates in solution (1 % (w/v)). The M/G ratios are calculated from the relative areas of signal A, B and C. The integration limits are indicated by the grey boxes.*

2.5 Partial Least Squares Regression

Linear models between the multivariate Raman spectra and the M/G ratio values calculated from the four sets of NMR spectra were established using partial least squares regression (PLSR)[12]. The Raman spectra were transformed with extended inverted scatter correction (EISC)[13,14] and mean centred prior to PLSR model development. The PLSR models were validated using segmented cross-validation (6 segments). The PLSR analyses were performed using LatentiX version 2.0 (Latent5, Copenhagen, Denmark, http://www.latentix.com), whereas EISC pre-processing was performed in MatLab 7.4 (Mathworks, Natick, Massachusetts, USA) using the command line EMSC/EISC toolbox (http://www.models.life.ku.dk).

3 RESULTS AND DISCUSSION

The M/G ratio of alginate is traditionally determined from the relative intensities of the signals in the anomeric region in the ¹H NMR spectra of mildly hydrolysed alginate in solution. It is therefore highly relevant to establish if water suppression influence the quantitative analysis. We have evaluated the impact of three commonly used water suppression techniques, namely conventional presaturation (*zgpr*), presaturation followed by a composite pulse (*zgcppr*) and NOESY-presaturation (*noesypr1d*). The carbohydrate region (3.4-5.3 ppm) of the spectra of an alginate sample acquired without water suppression (*zg*) and with the three different water suppression techniques is shown in Figure 3. The relative areas of the three signals in the anomeric region as well as the M/G ratio calculated from these are indicated at each spectrum.

The intensity of the water signals (4.18 ppm) in all three water suppression experiments is reduced significantly compared to the intensity of the signal in the *zg* experiment. A 98% reduction is observed in the *zgpr* and *zgcppr* experiment and an almost 100% reduction is observed in the *noesypr1d* experiment. Thus, the *noesypr1d* experiment gives the most efficient suppression. The residual water signal in the *zgcppr* spectra was slightly narrower than in the *zgpr* spectra. This is due the composite 90° pulse, which results in a more accurate 90° pulse by compensating for inhomogeneities in the applied B₁ field.

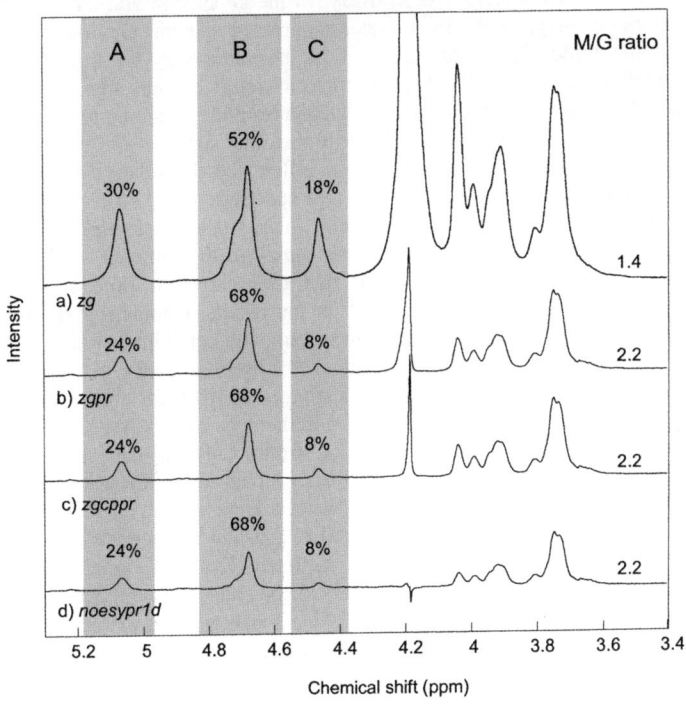

Figure 3 *Stack plot of the ¹H NMR spectra (carbohydrate region) of an alginate sample acquired without (a) and with water suppression (b, c and d). The zg spectrum is vertically scaled by a factor of 4.5 (receiver gain adjustment). The relative areas of the signals in the anomeric region (A, B and C) and the M/G ratios calculated from these are indicated at each spectrum.*

The four spectra presented in Figure 3 were vertically scaled such that the intensities of the TSP-d4 signals in the four spectra were equal. In reality this only required vertical scaling of the *zg* spectrum by a factor of 4.5 to compensate for the lower receiver gain used when compared to the water-suppressed spectra. The intensities of the carbohydrate signals, however, were generally higher in the *zg* spectra than in the water-suppressed spectra. On the other hand, when comparing the intensities of the carbohydrate signals in the water-suppressed spectra, the *zgpr* and *zgcppr* spectra were found to match perfectly when superimposed, whereas the intensities of most of the carbohydrate signals in the *noesypr1d* spectrum were 1.6 times lower than in the *zgpr* and *zgcppr* spectra. This effect is most likely due to relaxation during the mixing time. In the *noesypr1d* spectra the signals close to the water resonance were 2.0-2.3 times lower than the same signals in the *zgpr* and *zgcppr* experiments. This indicates that the *noesypr1d* pulse sequence affects the signals near the water resonance to a higher extend than the *zgpr* and *zgcppr* experiments.

The relative intensities of the signals in the anomeric region (A, B and C) and thereby also the calculated M/G ratios are clearly affected by the water suppression. The M/G ratios calculated from the water-suppressed spectra were all higher (2.2) than the M/G ratios calculated from the non-water-suppressed spectra (1.4). The same trend was observed for all other samples. However, the relative difference was not the same for all samples. The area of signal A, B and C in percent of the total area of the three peaks were 30%, 52% and 18% (1.7:2.9:1.0), respectively, in the *zg* spectra and 24%, 68% and 8% (3.0:8.5:1.0), respectively, in the three water-suppressed spectra. Obviously, the relative intensities of the three signals are affected in the same way by the different water suppression techniques. Generally, the assumption would be that signals closest to the water signal were subjected to the greatest influence from the water suppression, but in this case signal B is significantly enhanced relative to the other signals.

In order to investigate if the receiver gain influenced the relative intensities of the signals in the anomeric region, we recorded a *zgpr* spectrum with a receiver gain optimised for the *zg* experiment. This resulted in overall lower signal intensities, but the relative intensities and thereby the calculated M/G ratio remained the same as with the optimised receiver gain. Moreover, it was observed that the relaxation time and thereby the presaturation time had no significant impact on the results. Comparing spectra acquired with relaxation times of 2 and 20 s, resulted in spectra with equal intensities.

For a better demonstration of the differences observed between spectra with and without water suppression, the *zg* and *zgpr* spectra of two different alginates are shown in Figure 4. The intensities of signal A (H-1 of G) in the *zgpr* spectra have been scaled up to match the intensity of the same signal in the *zg* spectra to reveal the relative differences. Signal A was chosen for scaling since it is furthest away from the water resonance and therefore less likely to be influenced by the water suppression. The spectra in Figure 4 shows that when keeping the intensity of signal A constant an increase in signal B and a decrease in signal C is observed. Moreover, the intensity of the signals just upfield the water resonance decreases, whereas an increase of the signals around 3.7-3.8 ppm, which have been assigned to H-3, H-4 and H-5 of M,[7] is observed. Investigating the relative difference between the *zg* and *zgpr* spectra of all samples showed that water suppression resulted in a 27-44% decrease of the signals near the water resonance (4.38-4.55 ppm and 3.96-4.07 ppm) and a 2-25% increase of the signals originating from M (4.68-4.70 ppm and 3.67-3.83 ppm). All changes are given relative to the H-1 resonance of G. The signal increase for the M protons was unexpected and we are currently investigating this phenomenon in greater detail to understand the underlying mechanisms.

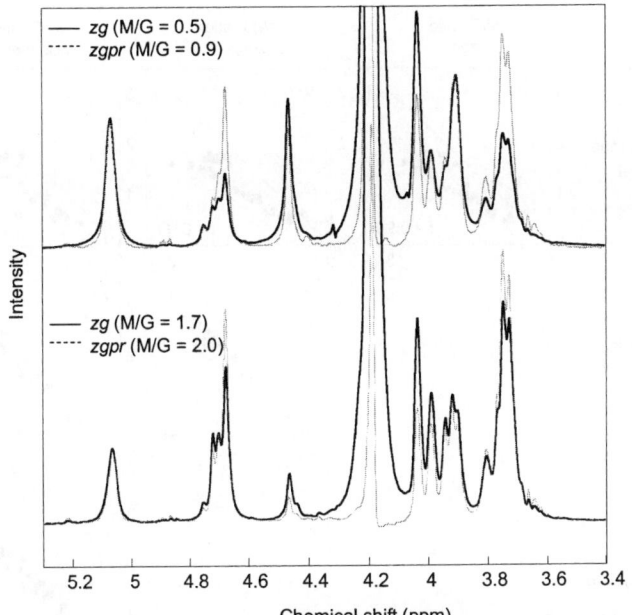

Figure 4 *^{1}H NMR spectra (carbohydrate region) of two alginate samples with high and low M/G ratio recorded without (zg) and with (zgpr) water suppression. Higher M/G ratios are obtained from the zgpr spectra than from the zg spectra.*

Since the increase is observed only for the protons originating from M (and especially H-1 of M sitting next to another M (4.68 ppm)) it is reasonable to assume that the signal increase is related to the molecular geometry and flexibility differences of M- and G-blocks. M and G are distributed along the linear alginate polymer in either blocks of M, blocks of G or as alternating MG units.[15] The geometries of the M-, G- and MG-blocks are related to the particular shapes of the monomers and their mode of linkage in the polymer. The M-blocks have an extended ribbon shape and are flexible,[16] the G-blocks are buckled and stiff[17] and the MG-block regions are of intermediate stiffness. Thus, a plausible explanation for the observed signal increase, may be that the M-blocks interact with water to a higher extend than the G-blocks due to their more linear structure and higher flexibility. Water suppression can potentially disturb this interaction and therefore induce changes for protons in the M-blocks.

Figure 5 shows the M/G ratios of the 40 alginates calculated form the four sets of NMR spectra plotted against each other in order to illustrate how they correlate. The best correlation was found between the M/G ratios calculated from the *zgpr* and *zgcppr* spectra (Figure 5d). This was expected since these spectra were practically identical when comparing within each sample. The M/G ratios calculated from the *noesypr1d* correlate fairly well with the values from the *zgpr* and *zgcppr* experiments with exception of one outlying sample (Figure 5e and f). The correlation coefficients between the M/G ratios calculated from the *zg* spectra and the *zgpr/zgcppr* and *noesypr1d* spectra were 0.90 (Figure 5a and b) and 0.88 (Figure 5c), respectively. It should be noted that samples with high M/G ratio deviate more from the target line than samples with low M/G ratio, indicating that samples with high content of M are influenced more by the water suppression than samples with low content of M.

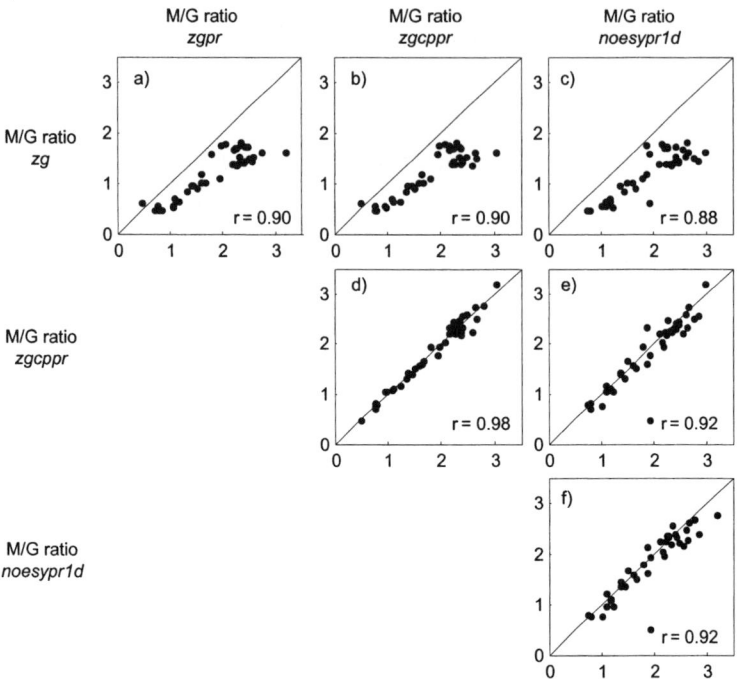

Figure 5 *Scatter plots of M/G ratios calculated from the four different sets of NMR spectra. The correlation coefficients are given in each plot.*

In a previous study[11] we have shown that the M/G ratio of alginates can be determined from Raman spectra of intact alginate powders using PLSR regression and M/G ratios from ^{1}H NMR spectra as response values. In order to elucidate if the M/G ratio values obtained by the non-water-suppressed spectra result in more reliable and consistent M/G ratio values than the values obtained from the water-suppressed spectra, PLSR models between Raman spectra and the M/G ratios calculated from the four sets of data were established. It is assumed that the solid-state Raman spectra, which are measured on a much faster time scale than the NMR spectra, are independent of water interferences and of biopolymer dynamics and therefore will provide a good measure of the validity of the M/G ratios established from the different solution-state NMR experiments. The M/G ratios calculated from the *zg* and *zgpr* spectra versus the M/G ratio predicted from the Raman spectra (using the *zg* and *zgpr* values as reference values, respectively) are shown in Figure 6a and b, respectively. As evident from Figure 6, the model established between the M/G ratios calculated from the *zg* spectra and the Raman spectra significantly improved compared to the model based on the M/G ratios calculated from the *zgpr* spectra. The correlation coefficient was improved from 0.82 to 0.97 (r =0.97) for the non-water-suppressed NMR measurements and the prediction error was reduced from RMSECV= 0.28 to RMSECV=0.08. The same tendency was observed in the models based on the M/G ratios from the *zgcppr* and *noesypr1d* spectra.

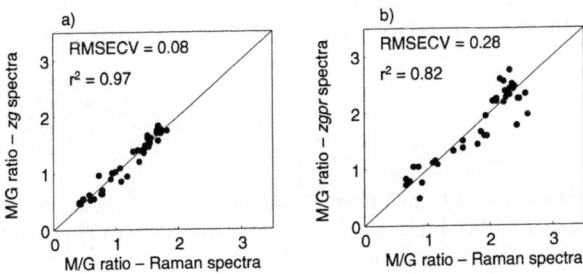

Figure 6 *Measured versus predicted plots from the PLSR analyses. M/G ratio measured by NMR without (a) and with (b) water suppression (zgpr) versus the M/G ratio predicted from the Raman spectra. RMSECV = root mean square error of cross validation. Both models are based on two PLS components.*

4 CONCLUSIONS

From this work, it is clear that the application of water suppression techniques in the NMR analysis of alginates can introduce undesired and uncontrolled variability to the spectra leading to erroneous quantitative results. Our results indicate that it was mainly resonances originating from M-units that were prone to intensity changes. This effect may be due to a closer water association in the M-blocks than the G-blocks.

References

1   H. Grasdalen, B. Larsen and O. Smidsrød, *Carbohydr. Res.*, 1979, **68**, 23.
2   H. Grasdalen, *Carbohydrate Research*, 1983, **118**, 255.
3   A. Haug, B. Larsen and O. Smidsrød, *Carbohydr. Res.*, 1974, **32**, 217.
4   M. Indergaard, G. Skjåk-Bræk and A. Jensen, *Botanica Marina*, 1990, **33**, 277.
5   B. Stockton, L.V. Evans, E.R. Morris, D.A. Powell and D.A. Rees, *Botanica Marina*, 1980, **23**, 563.
6   A. Haug, S. Myklesta, B. Larsen and O. Smidsrød, *Acta Chem. Scand.*, 1967, **21**, 768.
7   C.A. Steginsky, J.M. Beale, H.G. Floss and R.M. Mayer, *Carbohydr. Res.*, 1992, **225**, 11.
8   B.T. Stokke, O. Smidsrød, P. Bruheim and G. Skjåk-Bræk, *Macromolecules*, 1991, **24**, 4637.
9   T.G. Neiss and H.N. Cheng, *ACS Symposium Series*, 2003, **834**, 382.
10 M. Shinohara, H. Kamono, T. Aoyama, H. Bando and M. Nishizawa, *Fisheries Science*, 1999, **65**, 909.
11 T. Salomonsen, H.M. Jensen, D. Stenbæk and S.B. Engelsen, *Carbohydr. Polym.*, 2008, **72**, 730.
12 S. Wold, H. Martens and H. Wold, *Lect. Notes Math.*, 1983, **973**, 286.
13 H. Martens, J.P. Nielsen and S.B. Engelsen, *Anal. Chem.*, 2003, **75**, 394.
14 D.K. Pedersen, H. Martens, J.P. Nielsen and S.B. Engelsen, *Appl. Spectrosc.*, 2002, **56**, 1206.
15 A. Haug, B. Larsen and O. Smidsrød, *Acta Chem. Scand.*, 1966, **20**, 183.
16 E.D.T. Atkins, I.A. Nieduszy and K.D. Parker, *Biopolymers*, 1973, **12**, 1865.
17 E.D.T. Atkins, I.A. Nieduszy, K.D. Parker and E.E. Smolko, *Biopolymers*, 1973, **12**, 1879.

NMR-BASED METABONOMICS APPROACHES FOR THE ASSESSMENT OF THE METABOLIC IMPACT OF DIETARY POLYPHENOLS ON HUMANS

John van Duynhoven[1], Ewoud van Velzen[1,2], Gabriele Gross[1,3], Ferdi van Dorsten[1], Doris Jacobs[1], Max Bingham[1], Richard Draijer[1], Theo Mulder[1], Thea Koning[1], Elaine Vaughan[1], Tom van der Wiele[3], Johan Westerhuis[2], Age Smilde[2]

[1]Unilever Food and Health Research Institute, Unilever R&D, Vlaardingen, The Netherlands
[2]Biosystems Data Analysis, Swammerdam Institute for Life Sciences, University of Amsterdam, The Netherlands
[3]Laboratory for Microbial Ecology and Technology, Ghent University, Belgium

1 INTRODUCTION

1.1 Dietary Polyphenols and Health

In the quest for 'natural' functional ingredients with beneficial health effects, the food industry has developed a strong interest in polyphenols. Polyphenols are abundant in fruits and vegetables, and in food products such as tea, wine and chocolate. Dietary polyphenols show considerable structural diversity, ranging from low molecular weight phenolic acids, to multi-ring aromatic compounds and oligo- and polymers thereof. These compounds have been associated with risk-reduction of degenerative diseases, in particular cardiovascular disorders[1] and cancers[2,3]. Most evidence, however, is based on a limited number of biomarkers and clinical end-points. Flavonoids do have an impact on oxidative stress markers, but there is a growing awareness that the 'antioxidant theory' is a naïve simplification[2,4,5]. Hence there is a need to pursue further research in this area[6,7].

1.2 Bioavailability of Polyphenols

The bioavailability of dietary polyphenols is for a major part determined by their molecular weight and polarity. Whereas low molecular weight and polar phenolic acids are readily absorbed, this is mostly not the case for larger and more apolar species. In order to facilitate excretion, polyphenols are mostly glucuronidated, sulphated and methylated in the gut mucosa and inner organs. Many polyphenols, however, persist to the colon, where they are extensively metabolised by the microbiota into a wide array of low molecular weight phenolic acids, which in their turn also can undergo conjugation[8,9,10]. In order to assess the health effects of dietary intake of polyphenols, it is essential to understand which conjugated derivatives and microbial metabolites are absorbed. Alternatively, some health effects of polyphenols may not require their absorption through the gut barrier, but may act as metal chelators in the gut lumen or modulators of the gut microbiome[11]. Until recently, most research focussed on bioavailability of intact polyphenols and a few identified conjugates. There is a growing awareness that our understanding of the bioavailability of polyphenols will not progress by focussing on a few pre-identified/hypothesized metabolites. Thus more holistic analytical approaches are called for which are able to grasp the full complexity (Figure 1) of the interactions between diet, the gut microbiome and the human host metabolome[12]. In this respect NMR-based metabonomics is a particularly powerful analytical approach, as it combines (chemically) unbiased, global profiling of low MW metabolites in body fluids with advanced multivariate statistical data analysis.

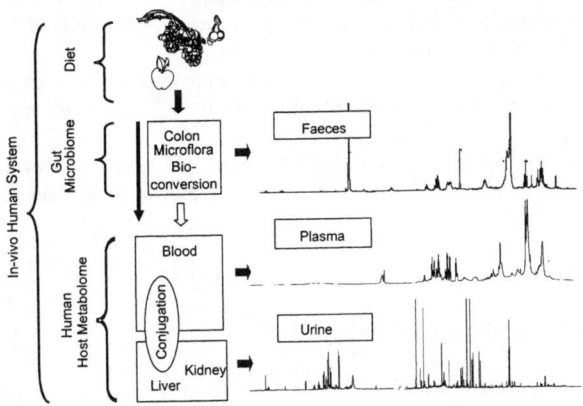

Figure 1 *Schematic overview of the interactions between dietary polyphenols, gut microbiota and the human host metabolome.*

			Approaches			
			Trial design & data analysis	In-vitro/animal models	Profiling	
					NMR	LC/GC
Challenges	Biological variation	• Diet/lifestyle	X			
		• Genetic				
		• Host	X			
		• Microbiome	X	X		
		Co-metabonome interactions		X		
	Analytics	• Sensitivity	X	X		X
		• Variation	X	X	X	
		• Coverage		X	X	
		• Throughput			X	

Scheme 1 *Challenges in assessment of the metabolic impact of polyphenols in humans and various approaches with methodological solutions applied.*

2 BOTTLENECKS AND PITFALLS

Until recently, nutritional research was strongly hypothesis-driven and based on pre-identified markers and benefits. We are now witnessing a movement towards more exploratory, hypothesis generating 'omics' approaches, which assess living organisms in a holistic manner[12]. From all "omics" approaches, metabonomics is the one that is closest to the phenotype. Thus, the nutritional field developed a strong interest in metabonomics[13] as a technology to provide direct feedback on metabolic effects of dietary interventions. Nutritional applications of metabonomics are however challenging due to three main bottlenecks (Scheme 1):

- The magnitude of the dietary interventions is small relative to the genetic and environmental[14] variation in healthy human volunteers. This is certainly the case for the metabolic impact of dietary polyphenols on humans[15].
- Polyphenols undergo extensive bioconversion in the human intestine especially due to the activity of the human colonic microbiota[16] [17] [18]. Unraveling of such co-metabonome interactions is extremely complex due to separation of metabolic events over time and over different (body) compartments.
- Metabolites of polyphenols are known to show wide chemical diversity and occur at a wide range of concentrations within the body[19, 20, 21], thus posing a considerable analytical challenge

In recent years several of these challenges have been addressed, by developing adequate methodologies, which are summarized in Scheme 1 and will be discussed in the next sections.

3 METHODS

3.1 Integration of trial design and data-analysis

Polyphenols are abundant in healthy diets, and their metabolites can be detected in urine and plasma at relatively high concentrations[22] [23] [3]. This is a complicating factor for trials where the effect of dietary polyphenols interventions is studied. Imposing a low-polyphenol dietary regime is unhealthy and is also rather discomforting for human volunteers. Thus medical ethical committees are reluctant to allow low-polyphenol restrictions, in particular for long-term dietary interventions. Typically in trial protocols, only relatively mild restrictions are imposed by asking the volunteers to refrain from abundant sources of polyphenols such as wine, tea and fruit juices. In free living volunteers this inevitable leads to issues with compliancy. Therefore, for short term studies, volunteers are typically kept on-site and provided standard meals. Nevertheless, one is left with the considerable inter-individual variation between the genotypes of humans, as well as their gut microbial ecologies. Thus, nutritional intervention trials are typically carried out in so-called cross-over design, where all volunteers undergo all interventions, including placebo treatment. In such a design all individuals serve as their own control. In conventional studies where only individual biomarkers are considered, the cross-over design is exploited in Analysis Of Variance statistical treatment, where different sources of variation can be separated. Recently, we applied such a hybrid approach[24] [25], coined Multi Level (ML)-PLS-DA, to the multivariate profiles obtained from a polyphenol intervention trial[26]. We demonstrated that both the interpretability of the different variation sources in the data as well as the significance of the observed treatment effect was improved after variation splitting[26].

3.2 In-vitro and animal models

The interactions between dietary polyphenols and the host-gut co-metabonome take place in different metabolic compartments, and take place at different time-scales. Unravelling these interactions in the human super-organisms would require longitudinal sampling in different metabolic compartments (Figure 1). Due to ethical considerations one can mostly only obtain urine, plasma and faeces in human intervention trials. Metabolic profiles from these materials provide a relatively indirect picture of the microbial bioconversions in the gut. Consequently in-vitro models for the human colon, including the Simulator of the Human Intestinal Microbial Ecosystem SHIME[27], the University of Reading 3-stage model[28], and the TIM-2 *in vitro* model of the human intestine[29] have been developed. Although these models simulate the in-vivo situation, their operation is cumbersome and time-consuming, thus also simpler and rapid batch fermentation models are being employed. Besides the advantage of more experimental freedom, metabolites occur at higher levels in in-vitro models than in in-vivo situations, and the range of metabolites is also relatively small due to absence of conjugation. In between the human in-vivo situation and in-vitro models, also gnotobiotic animal models can be deployed[30]. After inoculation with human gut microbiota, these co-metabonome models also allow for more experimental freedom than in intervention trials with human volunteers. Their deployment is bound to more ethical considerations than in-vitro models, however.

3.3 Profiling and identification

In order to enable holistic metabolic assessment of living organisms, one requires methods that can acquire metabolic profiles in a rapid, reproducible and comprehensive manner, without bias towards compound classes. ^{1}H NMR meets these requirements effectively, and can assess metabolites down to the tens of μM level[31]. This may seem high when compared to more targeted methods such as LC(MS) and GC(MS), but NMR has the unique advantage that hundreds of metabolites can be assessed in a reproducible and quantitative manner in a single shot. Metabolite identification is a notorious bottleneck in metabonomics. In anticipation of this challenge, we built a pH dependent AMIX ^{1}H NMR spectral database for gut polyphenols fermentation products for which literature[22] provided clues on their abundance in faeces and urine.

4 APPLICATIONS

4.1 Short term kinetic impact of polyphenols

Ample data exists on the pharmacokinetics of intact polyphenols. Typically, they appear in conjugated form in plasma and urine a few hours after ingestion, but their bioavailability is low (<5%). As was described in the previous section, the bioavailability of gut microbiota fermentation products is much higher, but there is surprisingly little evidence on the kinetics of these metabolites. In a pilot study, a single bolus dose of 800 mg tea extract was taken by 3 volunteers, and excretion of hippuric acid was monitored by NMR[32]. Mass balance calculations indicated that almost half of the ingested flavonoids were excreted as hippuric acid, which started to increase at 11 hours after tea ingestion. The large inter-individual variation in urinary excretion levels and kinetics of these three volunteers prompted us to carry out a more elaborate study (20 volunteers) on the pharmacokinetics of single bolus intakes of black tea and wine extracts.

Subject numbers

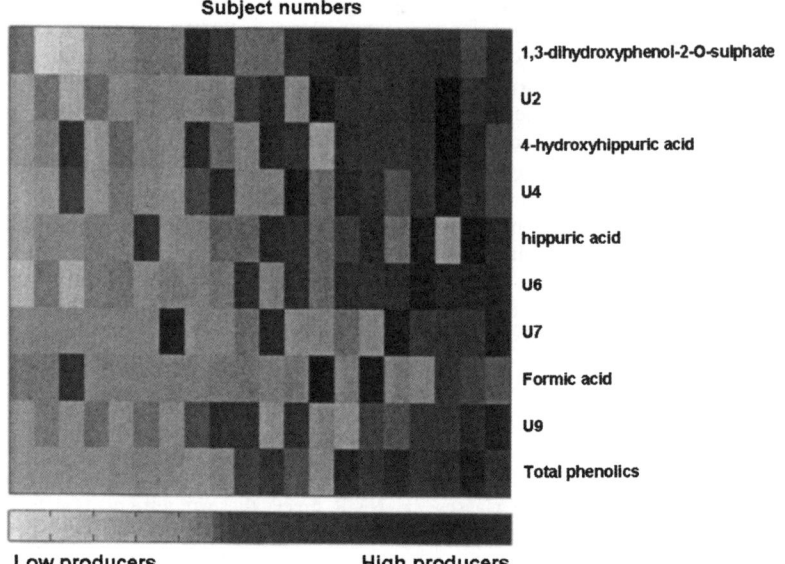

1,3-dihydroxyphenol-2-O-sulphate

U2

4-hydroxyhippuric acid

U4

hippuric acid

U6

U7

Formic acid

U9

Total phenolics

Low producers   **High producers**

Figure 2 *Heat map representing the total urinary output of 10 selected NMR variables/ metabolites measured in 20 subjects after black tea consumption. The dark grey and light grey cells indicate high and low urinary output levels.*

Using the data of the latter study, a new procedure was developed for the untargeted analysis of data from crossover design nutritional intervention studies with a kinetic component. In this approach the kinetic measures of all relevant metabolites are estimated by fitting the appropriate kinetic models to the evolution curves over time. This allows detailed investigations of the inter-individual variation in response to dietary intervention. An example is given in Figure 2 where the total urinary output after consumption of a black tea extract is shown, and a clear distinction can be made between high and low metabolizer phenotypes.

4.1 Impact of mid- and long-term intake of polyphenols

From the aforementioned short-term intervention studies we primarily gained insight in how and when polyphenols metabolites become available. The current consensus is that for inducing significant effects on the endogenous metabolism of the host long term interventions are required. Thus we carried out two of such studies where we investigated both exogenous as well as endogenous metabolic effects. In the first study we compared the impact of polyphenol-rich green and black tea intake for a period of two days in a placebo-controlled crossover study with twenty healthy volunteers[32]. Intake of black and green tea, which differ in polyphenol composition, showed significant but different effects on the urine NMR metabolite profiles. Both the aromatic NMR region, which is dominated by exogenous metabolites, and the aliphatic region, which is dominated by endogenous metabolites, were affected by tea intake. A summary of these effects is given in Scheme 2. After green and black tea ingestion a major excretion of hippurate can be observed as well as that of several other phenolic acids. The pronounced effect of the green tea intervention on the excretion of endogenous metabolites involved in the TCA cycle is remarkable. This striking observation is in line with observations from other intervention studies with green tea that demonstrated effects on energy expenditure and weight loss.

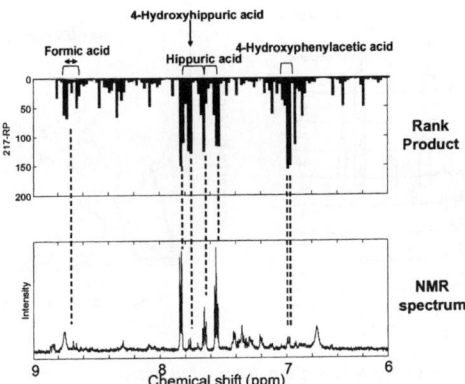

Figure 3 Aromatic *region of a typical urinary 600 MHz 1D ¹H NMR spectrum (bottom) and the corresponding rank product (top) corresponding to a wine/grape extract dietary intervention.*

In another human intervention study we compared the impact of two different polyphenol-rich grape extracts. This was a two-leg placebo-controlled cross-over study with 58 mildly hypertensive volunteers who consumed either one of two polyphenol-rich grape supplements or a placebo for a period of four weeks. In this study, NMR metabolite profiles of 24-h urine samples, plasma as well as faecal extracts were analysed and treatment-related effects were analysed using ML-PLS-DA (Figure 3). After 4-week consumption of one extract, hippuric acid levels in 24-h urine samples were quantified, and showed a significant increase, which was not the case after intake of the other extract. This is remarkable since the total polyphenol content in the two grape extracts was equal. Besides hippuric acid, also changes in low-level phenolic metabolites, including 4-hydroxyhippuric acid 4-hydroxyphenylacetatic acid and 3-hydroxybenzoic acid could only be identified after ML-PLS-DA was applied. In addition, a number of NMR spectral changes in the aliphatic region of the urine profile were observed, which may reflect changes in endogenous metabolism. Endogenous biochemical effects include for example higher urinary excretion of glycine and citrate.

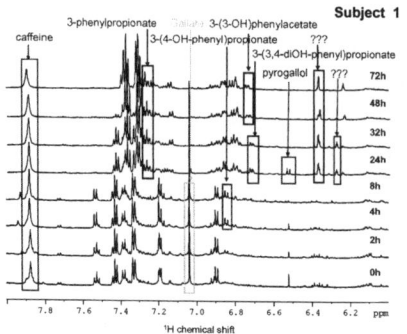 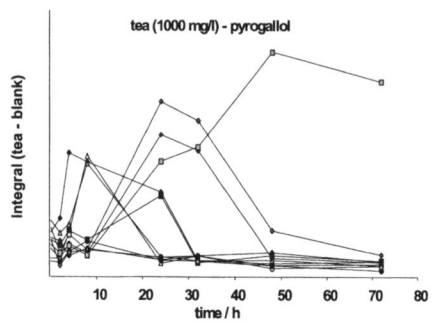

Figure 4 *(A) ^1H NMR time series taken after dosing tea extract to a batch in-vitro fermentation inoculated with faeces of one volunteer. Spectral assignments are indicated. (B) Kinetics of pyrogallol appearance and subsequent degradation during in vitro fermentation of faecal samples from ten volunteers with tea extract .*

4.3 In-vitro models for gut microbial bioconversion of polyphenols

The urinary profiles in the aforementioned human intervention studies allow for only a rather indirect view on the bioconversion of polyphenols by the gut microbiota. Thus we performed *in vitro* faecal batch fermentations of grape and tea extracts under anaerobic conditions by inoculating faecal samples from 10 healthy human volunteers. During the fermentation process samples were taken for NMR profiling (see Figure 4A for aromatic regions). A range of phenolic acids, alcohols, organic acids and amines were identified, and those most directly associated with polyphenol degradation are listed in Scheme 2. In agreement with the human intervention trials (Figure 2), significant inter-individual variation can be observed, which is illustrated by the kinetics of pyrogallol in Figure 4B.

5 CONCLUSIONS AND PERSPECTIVES

- NMR-based metabonomics is able to detect the subtle metabolic effects of dietary interventions of polyphenols in human intervention trials. For this purpose trial design, metabolic profiling and data-analysis need to be well interknitted.
- NMR-profiling is a rapid tool for gaining time-dependent information on bioconversion of polyphenols in in-vitro gut models.
- Whereas NMR-based metabonomics has its merits with respect to measurement speed, straightforward data-preprocessing and availability of spectral databases, eventually limitations will be met with respect to analytical sensitivity. With our currently established methodology we can now detect effects on exogenous urinary metabolites at levels which are as low as 100 μM. Furthermore we recently extended our profiling arsenal with a GCMS method for profiling of phenolic acids in urine, plasma, faeces and in-vitro gut models[33]. In order to cover also higher molecular weight metabolites of polyphenols we will also have to take recourse to untargeted LC-MS methods[34].
- We expect that with this suite of NMR, GCMS and LCMS based methods we will reveal many novel metabolites. Their identification currently exists as a challenge[35] but we expect to benefit from latest developments in hyphenation of separation, enrichment and spectroscopy such as LC-SPE-cryoNMR-MS.

	Exogenous	**GW**	**BT**	**GT**	**Endogenous**	**GW**	**BT**	**GT**
Urine	Hippuric Acid	↑↑↑	↑↑↑	↑↑	Alanine	↓	↑	
	4-OH-hippuric acid	↑	-	-	Glutamin/ate		↑	↑
	4-OH-phenylacetate	↑	-	-	Glycine	↑	↑	↑↑
	Gallic acid	-	↑	-	Valine		↑	↑
	3-OH-benzoate	-	↑	-	α-ketoglutarate		↑	↑
	1,3-diOH-2-O-sulfate	-	↑	-	Citrate	↑		↑↑
					Oxaloacetate			↑
					Pyruvate		↑	↑↑
					Succinate		↑	↑
Plasma					Glucose		↓	↓
					Acetate	↓	↓	
					Lactate			↓
					Lipoprotein		shift	shift
Faeces					Isobutryric	↓	-	-
In-vitro	3-(phenyl)-propionate	↑↑	↑↑	-				
	3-(3-OH-phenyl)propionate	↑	↑	-				
	3-(3,4-diOH-phenyl)propionate	↑↑	↑	-				
	3-OH-phenylacetate	↑	↑					
	Pyrogallol		↑					

Scheme 2 *Overview of endo- and exogenous metabolites associated with the long-term intake of dietary polyphenols in black tea (BT), green tea (GT) and grape/wine(GW) extracts, observed by ^1H NMR in urine, plasma, faeces and in-vitro faecal batch incubations.*

6 ACKNOWLEDGDEMENTS

We acknowledge the financial support of the Community under the Marie Curie Transfer of Knowledge – Industry-Academia Strategic Partnership Scheme, specifically MTKI-CT-2006-042786 (GUTSYSTEM). Sam Possemiers and Rober Kemperman are acknowledged for setting up the gut model and useful discussions.

References

1 M. Dell'Agli, A. Busciala, E. Bosisio, *Cardiovasc. Res.*, 2004, **63**, 593.

2 A. Scalbert, I.T. Johnson, M. Saltmarsh, *Am J Clin Nutr*, 2005, **81**, 215S.

3 G. Williamson and C. Manach, *Am J Clin Nutr*, 2005, **81**, 243S.

4 S.B. Lotito and B. Frei, *Free Radical Biology and Medicine*, 2006, **41**, 1727.

5 B. Halliwell, J. Rafter, A. Jenner, *Am J Clin Nutr*, 2005, **81**, 268S.

6 B. Holst and G. Williamson, *Curr. Opin. Biotechnol.*, 2008, **19**, 73.

7 P.A. Kroon, M.N. Clifford, A. Crozier, A.J. Day, J.L. Donovan, C. Manach, G. Williamson, *Am. J. Clin. Nutr.*, 2004, **80**, 15.

8 P.C. Hollman and M.B. Katan, *Biomed. Pharmacother.*, 1997, **51**, 305.

9 M.R. Olthof, P.C. Hollman, M.N. Buijsman, J.M. van Amelsvoort, M.B. Katan, *J. Nutr.*, 2003, **133**, 1806.

10 K. Keppler and H.U. Humpf, *Bioorganic & Medicinal Chemistry*, 2005, **13**, 5195.

11 A. Scalbert, C. Morand, C. Manach, C. Remesy, *Biomed. Pharmacother.*, 2002, **56**, 276.

12 J.K. Nicholson, E. Holmes, J.C. Lindon, I.D. Wilson, *nature biotechnology*, 2004, **22**, 1268.

13 M.J. Gibney, M. Walsh, L. Brennan, H.M. Roche, B. German, B. van Ommen, *Am J Clin Nutr*, 2005, **82**, 497.

14 M.C. Walsh, L. Brennan, J.P. Malthouse, H.M. Roche, M.J. Gibney, *Am. J. Clin. Nutr.*, 2006, **84**, 531.

15 M.C. Walsh, L. Brennan, E. Pujos-Guillot, J.L. Sebedio, A. Scalbert, A. Fagan, D.G. Higgins, M.J. Gibney, *Am. J. Clin. Nutr.*, 2007, **86**, 1687.

16 M. Blaut, L. Schoefer, A. Braune, *Int. J. Vitam. Nutr. Res.*, 2003, **73**, 79.

17 A.R. Rechner, G. Kuhnle, P. Bremner, G.P. Hubbard, K.P. Moore, C.A. Rice-Evans, *Free Radical Biology and Medicine*, 2002, **33**, 220.

18 A.R. Rechner, M.A. Smith, G. Kuhnle, G.R. Gibson, E.S. Debnam, S.K.S. Srai, K.P. Moore, C.A. Rice-Evans, *Free Radical Biology and Medicine*, 2004, **36**, 212.

19 J.P.E. Spencer, M.M.A. Mohsen, A.M. Minihane, J.C. Mathers, *British Journal of Nutrition*, 2008, **99**, 12.

20 C. Manach, G. Williamson, C. Morand, A. Scalbert, C. Remesy, *Am. J. Clin. Nutr.*, 2005, **81**, 230S.

21 C. Manach, A. Scalbert, C. Morand, C. Remesy, L. Jimenez, *Am. J. Clin. Nutr.*, 2004, **79**, 727.

22 L.I. Mennen, D. Sapinho, H. Ito, S. Bertrais, P. Galan, S. Hercberg, A. Scalbert, *Br. J. Nutr.*, 2006, **96**, 191.

23 G. Williamson and C. Manach, *Am. J. Clin. Nutr.*, 2005, **81**, 243S.

24 J.J. Jansen, H.C.J. Hoefsloot, J. Van der Greef, M.E. Timmerman, J.A. Westerhuis, A.K. Smilde, *Journal of Chemometrics*, 2005, **19**, 469.

25 A.K. Smilde, J.J. Jansen, H.C.J. Hoefsloot, R.J.A.N. Lamers, J. Van der Greef, M.E. Timmerman, *bioinformatics*, 2005, **21**, 3043.

26 E.J.J. van Velzen, J.A. Westerhuis, J.P.M. Van Duynhoven, F.A. Van Dorsten, C.J. Hoefsloot, S. Smit, R. Draijer, C.I. Kroner, A.K. Smilde, *Journal of Proteome Research*, 2008.

27 T. van de Wiele, N. Boon, S. Possemiers, H. Jacobs, W. Verstraete, *FEMS Microbiol. Ecol.*, 2004, **51**, 143.

28 G.T. Macfarlane, S. Macfarlane, G.R. Gibson, *Microb. Ecol.*, 1998, **35**, 180.

29 M. Minekus, M. Smeets-Peeters, A. Bernalier, S. Marol-Bonnin, R. Havenaar, P. Marteau, M. Alric, G. Fonty, i. Huis, V, *Appl. Microbiol. Biotechnol.*, 1999, **53**, 108.

30 S. Possemiers, S. Rabot, J.C. Espin, A. Bruneau, C. Philippe, A. Gonzalez-Sarias, A. Heyerick, F.A. Tomas-Barberan, D. De Keukeleire, J. Verstraete, *J. Nutr.*, 2008, **138**, 1310.

31 J.C. Lindon, J.K. Nicholson, J.R. Everett, *Annual Reports on Nmr Spectroscopy, Vol 38*, 1999, **38**, 1.

32 C.A. Daykin, J.P.M. Van Duynhoven, A. Groenewegen, M. Dachtler, J.M.M. Van Amelsvoort, T.P.J. Mulder, *Journal of Agricultural and Food Chemistry*, 2005, **53**, 1428.

33 C. Grün, F. van Dorsten, D. Jacobs, M. Le Belleguic, E. van Velzen, M. Bingham, H.-G. Janssen, J.P.M. Van Duynhoven, *Journal of Chromatrography*, 2008.

34 R.C.H. De Vos, S. Moco, A. Lommen, J.J.B. Keurentjes, R.J. Bino, R.D. Hall, *Nature Protocols*, 2007, **2**, 778.

35 S. Moco, R.J. Bino, R.C.H. De Vos, J. Vervoort, *Trac-Trends in Analytical Chemistry*, 2007, **26**, 855.

¹H HR MAS NMR: PROFILING METABOLITES IN SINGLE CEREAL KERNELS (OF WHEAT AND BARLEY)

N. Viereck[1], H. Winning[1], H.F. Seefeldt[2] and S.B. Engelsen[1]

[1]Quality & Technology, Department of Food Science, Faculty of Life Sciences, University of Copenhagen, Rolighedsvej 30, 1958 Frederiksberg C, Denmark
[2]Department of Genetics and Biotechnology, Faculty of Agricultural Sciences, Aarhus University, Forsøgsvej 1, 4200 Slagelse, Denmark

1 INTRODUCTION

The functionality and nutritional value of cereals for food and feed is primarily determined by the composition of starch, fibres and proteins. Therefore, it is of utmost importance to obtain detailed knowledge about the composition of the mature grain. Nuclear magnetic resonance (NMR) is a versatile, powerful, non-invasive technique able to quantitatively measure multiple parameters of a sample. Single seeds can be characterised as solid cellular matrices with liquid domains.[1] Indeed, this sample characteristic introduces a challenge for NMR measurements, as the experimental setup for solids and liquids differs significantly. Generally, using high-resolution magic angle spinning (HR MAS) NMR, it is possible to measure NMR spectra from small pieces of solids, including plant tissue with no other pre-treatment than soaking in a locking solvent. The combination of the magic angle and fast rotation of the sample eliminates dipole-dipole couplings and chemical shift anisotropy which induce significant line broadening in the NMR spectra of solids. These effects are averaged out in solutions due to fast isotropic motions. Indeed, HR MAS makes solids appear liquid-like, and the spectral resolution similar to liquid-state NMR. However, one major disadvantage of spinning is that it leads to the presence of spinning sidebands. These are false signals that result from the modulation of the magnetic field at the spinning frequency. The spinning sidebands always appear on either side of any large signal at a separation equal to the spinning rate. The intensity of these sidebands will be proportional to the intensity of the original signal.

Several HR MAS NMR studies of plant, animal and human tissue can be found in the literature,[2-5] however only recently on single cereal seeds.[6,7] In these studies, carbohydrate grain filling in barley and the asynchronous protein metabolism of wheat during grain filling was studied using single-kernel ¹H HR MAS NMR, respectively.

This chapter describes the development of ¹H HR MAS NMR single seed measurements and the analysis will be exemplified in studies of wheat and barley. In order to assign ¹H NMR signals to the various metabolites within the seed, both one-dimensional (1D) and two-dimensional (2D) NMR measurements were performed;[6,7] however, only 1D spectra are presented in this chapter. Considerations about spinning rate, temperature and locking solvent are discussed, as these experimental parameters may influence the resulting spectra. The problems and limitations of the HR MAS method, especially concerning large immobile molecules within the sample, are summarised and exemplified.

2 MATERIALS AND METHODS

2.1 The Samples

Several barley and wheat kernels and flour samples were used in the work presented in this chapter. The details concerning the plant material can be found in the relevant papers.[6,7] The samples were prepared for NMR analysis by placing the kernel part (25-40 mg) or flour (app. 14 mg) in a 50 µl zirconia rotor (4.0 mm o.d.) followed by adding deuterated solvent (D_2O or dimethyl sulfoxide (DMSO)) directly into the rotor. 5.8 mM 3-(trimethylsilyl)propionic acid-d_4 sodium salt (TSP-d4) was added as chemical shift reference.

2.2 NMR Spectroscopy

^{1}H HR MAS NMR spectra of the samples were collected using a Bruker AVANCE 400 (Bruker BioSpin, Rheinstetten, Germany) spectrometer, operating at a frequency of 400.13 MHz for protons equipped with a HR MAS double channel probe. 1D spectra were acquired using a composite-pulse experiment with water suppression. Specific acquisition and processing parameters can be found in the papers,[6,7] but of importance to this work, data were recorded at various temperature (25°, 61° and 75°C) and various spinning rates (2, 3, 5, 7, 10 kHz). Bruker Topspin 1.3 software (Bruker BioSpin, Rheinstetten, Germany) was used for acquisition and processing of NMR data. All spectra were referenced to TSP-d_4 at 0.0 ppm prior to data analysis.

3 RESULTS AND DISCUSSION

NMR is traditionally described as a non-destructive technique, but the HR MAS technique imposes some restricted and possible destructive sample conditions. For this reason, some considerations about the use of HR MAS NMR on biological material and possible damage of the material must be considered.

3.1 Considerations Concerning the Spinning Rate

Two opposite considerations have to be taken into account with regard to the spinning rate. The speed has to be sufficiently high to ensure high-quality signals from all expected compounds and for preventing spinning sidebands in the spectra. On the other hand, since plant material is permeable for solvent, the destructive forces of centrifugation point towards using a low spinning rate. In order to study the artefacts from spinning sidebands, we investigated several spinning rates in the ^{1}H HR MAS NMR spectra of one half-seed mature wheat sample (Figure 1) and found that the spinning rate has to be above 7000 Hz in order to mobilise all possible protons within the sample and to avoid spinning sidebands in the spectral region of interest. The same result was obtained in barley (data not shown). At a spinning rate of 2000 Hz, many of the expected cereal signals can be found, including signals from protons of the aliphatic side-chains of the amino acids and protons from saturated parts of the fatty acids in the aliphatic area of the spectrum (0 - 3 ppm) and carbohydrates (3 - 6 ppm). For a more detailed assignment, see the papers.[6,7] In the aromatic area (6 – 8 ppm), only one signal with intensity similar to signals in the high-field part of the spectrum is found when measured at a spinning rate of 2000 Hz. This signal moves from 6.4 ppm to 8.8 ppm or 1000 Hz when the spinning rate is raised to 3000 Hz, which proves that it is a spinning sideband. In this case, it originates from the CH_2 protons in the fatty acid chains. Spinning sidebands are low-intensity signals spaced at distances which are multiples of the spinning rate to the resonance frequency of the originating signal (indicated with double arrows in Figure 1). Since the intensity of the spinning sideband is related to the intensity of the original signal, only spinning sidebands originating from high-intensity signals are visible in the HR MAS NMR spectrum. However, spinning sidebands deriving from a suppressed signal such as water will also be included in the spectra and will be important to the determination of the proper spinning rate (Figure 1).

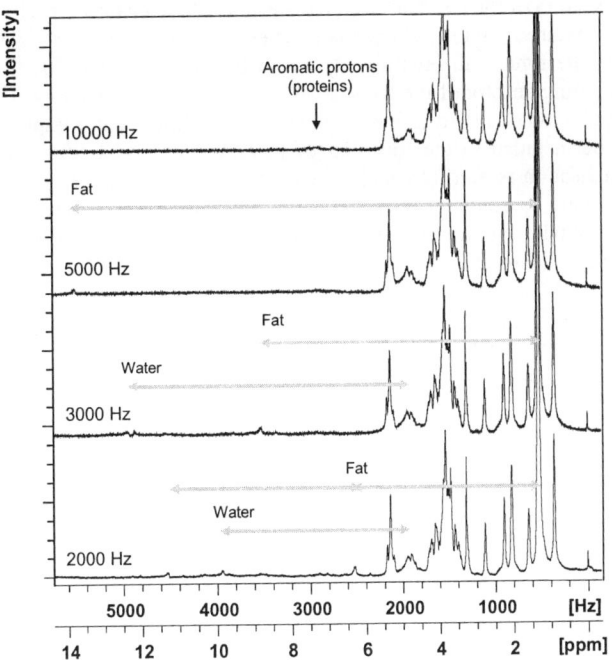

Figure 1   *^{1}H HR MAS NMR spectra of one mature half-seed wheat sample (Shamrock) at various spinning rates (2000, 3000, 5000 and 10000 Hz). Spectra were measured at t = 25°C and the sample was soaked in D_2O. Spectra were normalised with respect to the TSP signal at 0 ppm. Double arrows indicate distance between spinning sidebands and the original signal equal to the spinning rate.*

Already at a spinning rate of 5000 Hz, all possible spinning sidebands are moved outside the spectral window where the relevant signals are expected. On a 400 MHz spectrometer a resonance frequency of 5000 Hz corresponds to 12.5 ppm. Thus, a signal with a chemical shift of 0 ppm (0 Hz) will have the first spinning sideband at 12.5 ppm (5000 Hz) at a spinning rate of 5000 Hz. When the spinning rate is raised to 10000 Hz, the signal intensity increases in the aromatic region of the HR MAS NMR spectrum. This is due to mobilising of aromatic protons presumably from aromatic amino acids in proteins. All other signals are unchanged in shape and position; hence, the seed matrix appears intact also at high spinning rate. However, all signals are also visible in ¹H HR MAS NMR spectra of mature wheat measured at a spinning rate of 7000 Hz (spectrum not shown), so 7000 Hz was therefore chosen as the proper spinning rate for further HR MAS NMR measurements of whole grain.

3.2 Considerations Concerning Sample Presentation

The natural size of the kernel studied and the relatively small dimension of the rotor used as sample holder in HR MAS NMR measurements are not always compatible. Except for immature, green seeds, the seeds are normally too large to fit in the rotor. Therefore, ¹H HR MAS NMR spectra of the two half-seed parts of the whole mature kernel halved along the crease were measured (Figure 2). The main difference is the size of the two halves, but as apparent from the figure, all bulk signals are included in both spectra. Care must be taken when comparing spectra of the two halves, as normalising with respect to TSP will influence the results. Since the rotor has to be completely filled with sample and solvent, more TSP will be included in the spectrum of the small half-seed part and *vice versa*. Therefore, normalisation is a critical and difficult issue in quantitative evaluation of HR MAS NMR. In Figure 2, the two spectra were normalised to a similar noise level. Another approach is to include another semi-solid internal standard which is not lost in the procedure of packing the rotor.

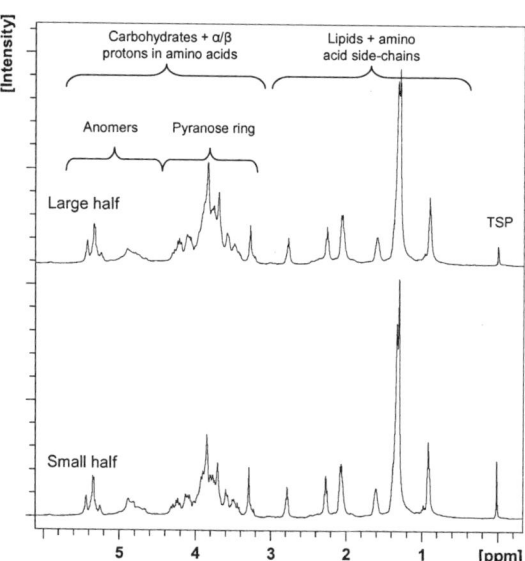

Figure 2   *¹H HR MAS NMR spectra of the two halves of one mature wheat kernel (Shamrock). Spectra were measured at t = 25°C and samples were soaked in D₂O. Spectra were normalised to similar noise level.*

Another way of overcoming the problem of cutting the seed in halves is to use flour instead of whole seeds. By using flour, variation between halves and also between seeds is averaged out. HR MAS NMR measurements of flour should have the additional advantage of increasing the mobility of especially the crystalline compounds (starch). Consequently, spectra of flour should have more intense signals in the carbohydrate area, due to increased water accessibility of the molecules in the flour. The two ^{1}H HR MAS NMR spectra presented in Figure 3 clearly show that the NMR spectra of flour are better resolved in the carbohydrate area (3 - 6 ppm) compared to the spectra of single kernels. In addition, the relative distribution between the lipid and carbohydrate signals within each spectrum indicates that the highly mobile protons from lipids always are 100% represented in the ^{1}H HR MAS NMR spectra of single kernels in contrast to the more immobile starch signals. Wheat and barley seeds only contain fat and lipids at approximately 3%. Moreover, the intensity of the signals in the aromatic region containing the signals from the aromatic amino acids are slightly higher in the single kernel spectrum; hence, a decision about using single kernels or flour must be based on the desired information.

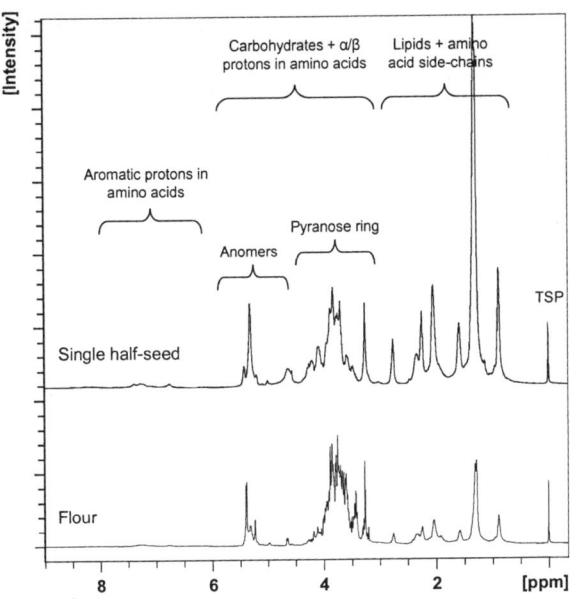

Figure 3 *^{1}H HR MAS NMR spectra of mature wheat flour and a single half-seed (Shamrock) at spinning rate 7000 Hz. Spectra were measured at t = 25°C and samples were soaked in D₂O. Spectra were normalised with respect to TSP.*

One important remaining question to be discussed in relation to sample preparation of HR MAS NMR is the choice of solvent and exposure time to the solvent. We found that especially mature dry seeds tend to absorb the added deuterated water (D₂O) and exchange it with internal water (H₂O). The result of this process is a poorer shimming and in turn increasingly lower quality spectra over time. In order to evaluate if another solvent with advantage could be used to replace D₂O, wheat flour samples were measured using deuterated DMSO with ^{1}H HR MAS NMR (Figure 4). DMSO was chosen, since most published NMR work on starch is measured in DMSO. The main differences in the NMR spectra of wheat flour using either DMSO or D_2O are observed in the carbohydrate region of the spectra. The D_2O spectrum is better resolved in this region, whereas the DMSO spectrum is very similar to a fully gelatinised starch spectrum due to the better solubilised starch polymers (see Figure 5, 75°C). Furthermore, the use of DMSO left the sample glassy and in the case of whole kernels, very fragile seeds remained. In contrast, D_2O resulted in a minimum of seed or flour damage. A steady state of water exchange as measured by stabilised shimming and identical spectra was found after 2 hours of soaking (data not shown). Samples should thus be left in the packed rotor for a minimum of 2 hours before HR MAS NMR measurements.

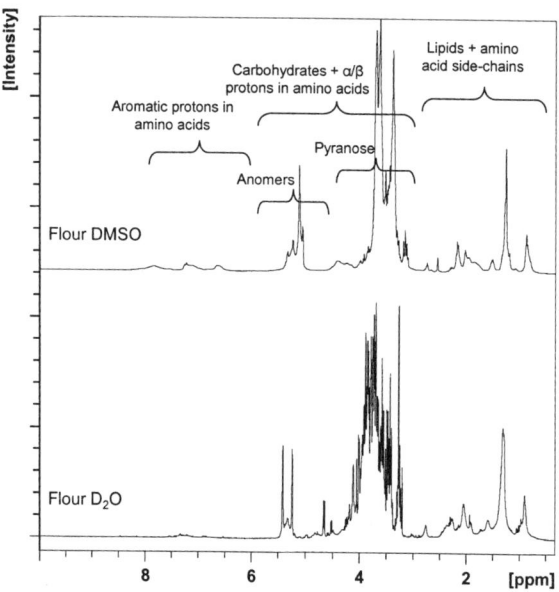

Figure 4   *^1H HR MAS NMR spectra of mature wheat flour samples (Shamrock) soaked in either DMSO or D_2O at spinning rate 7000 Hz. Spectra were measured at t = 25°C and normalised to similar noise level.*

3.3 Considerations Concerning the Temperature

Especially at maturity, the seeds can be regarded as a solid with a very rigid cellular and granular matrix leading to fewer and broader signals compared to immature seeds.[6,7] Mobilisation of particularly the carbohydrates (starch and fibres) can be enhanced by heating of the samples. Heating to 75°C induces gelatinisation of the starch granules where rigid or semi-crystalline protons become mobile and therefore observable in the HR MAS NMR spectrum. This process is shown in Figure 5 where the same seed sample was heated from room temperature to 75°C and spectra continuously collected (for clarity, only spectra at 25°, 61° and 75°C are shown in Figure 5). At 25°C and 61°C, only mobile protons from lipids and smaller carbohydrates are visible. However, at 75°C signals from mobilised protons in the starch polymers are dominant in the carbohydrate region, whereas the signals from lipids remain unchanged in intensity. The signals from the protons in smaller carbohydrates are still present, but overlap with signals from starch. This is most pronounced in the pyronose area, since the chemical shift of the anomeric protons are baseline separated and are characteristic for each sugar compound.

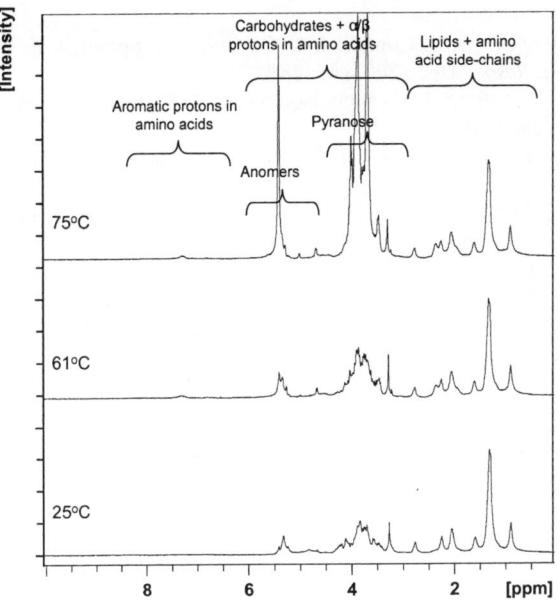

Figure 5 *¹H HR MAS NMR spectra of one mature half-seed wheat sample (Shamrock) measured at various temperature (25°, 61° and 75°C) at spinning rate 7000 Hz. The sample was soaked in D₂O and normalised with respect to TSP.*

A weak effect of raising the temperature is also observed in the aromatic region of the spectrum due to denaturation and unfolding of the proteins and hence mobilisation of the various amino acids.

4 CONCLUSION

From the work presented in this paper, it is obvious that the use of HR MAS NMR on single cereal kernels introduces unique possibilities in studies of cereals. However, sample presentation (single-kernel or flour, and solvent) and critical parameters (spinning rate and temperature) has to be optimised for the specific problem to be studied, especially in order to ensure reproducible spectra including all mobilisable compounds. This study has underlined that the HR MAS NMR technique does not provide signals from all protons in the solid sample. Only signals from protons that are mobile by themselves (protons from fat), protons that are naturally in contact with cellular water or protons that are accessible by the soaking (D_2O) liquid are visible in the ¹H HR MAS NMR spectra. The latter can be manipulated by solvent type, soaking time and temperature. For quantitative purposes it is thus imperative to keep these factors constant.

References

1. M. Bardet, M. F. Foray, J. Bourguignon and P. Krajewski, *Magn. Reson. Chem.*, 2001, **39**, 733.
2. S. Garrod, E. Humpher, S. C. Connor, J. C. Connelly, M. Spraul, J. K. Nicholson and E. Holmes, *Magn. Reson. Med.*, 2001, **45**, 781.
3. J. L. Griffin, J. C. M. Pole, J. K. Nicholson and P. L. Carmichael, *Biochim. Biophys. Acta, Gen. Subj.*, 2003, **1619**, 151.
4. J. C. Lindon, E. Holmes and J. K. Nicholson, *Prog. Nucl. Magn. Reson. Spectrosc.*, 2001, **39**, 1.
5. M. A. Brescia, G. D. Martino, C. Fares, N. D. Fonzo, C. Platani, S. Ghelli, F. Reniero and A. Sacco, *Cereal Chem.*, 2002, **79**, 238.
6. H. F. Seefeldt, F. H. Larsen, N. Viereck, B. Wollenweber and S. B. Engelsen, *Cereal Chem.*, 2008, **85**, 571.
7. H. Winning, N. Viereck, B. Wollenweber, F. H. Larsen, S. Jacobsen, I. Søndergaard and S. B. Engelsen, *J. Exp. Bot.*, 2008, accepted.

MICROSTRUCTURE INVESTIGATION OF CONCENTRATED DAIRY GELS BY REAL-TIME NMR DIFFUSION EXPERIMENTS

F. Mariette[1,2], S. Le Feunteun[1,2]

(1) Cemagref, TERE, 17, avenue de Cucillé, F-35044 Rennes, France
(2) Université européenne de Bretagne, France

1 INTRODUCTION

By studying the diffusion of probe molecules of various sizes, information can be obtained on the microstructure of a sample at different length scales [1, 2]. Such experiments have therefore been applied to a great variety of matrices, including polymers and proteins in both liquid and gel states [3-12]. The diffusion of poly(ethylene glycol)s (PEGs) measured by Pulsed Field Gradient (PFG)-NMR is probably the most widely used method to perform these investigations.

Previous pulsed field NMR studies have shown that PEG diffusion in casein suspensions and gels is greatly dependent on both the volume fraction occupied by casein particles and the probe size [5, 8]. For a given volume fraction of casein particles, the reduction in the diffusion coefficient induced by the obstacles is smaller for smaller probes. Recently, it has been demonstrated that the relative change of the self-diffusion coefficient depends on the final gel microstructure [8]. This could be explained by considering that the diffusion of larger probes are more sensitive to the extent of the casein aggregate compaction and thus to the extent of certain rearrangement processes. However, as in all the studies quoted above, the diffusion of a molecule was measured at equilibrium, before and after the perturbation of the system. This method cannot provide any dynamic information on the evolution of the sample microstructure although the modifications of the matrix in reaction to the perturbation can be very progressive. This is precisely the case when casein suspensions are coagulated by addition of chymosin and/or acidifying agent.

In the present study, we designed an original method to investigate how and when probe diffusion rates vary during the coagulation induced by chymosin action and during acid coagulation. Self-diffusion measurements were repeated throughout the coagulation processes by means of pulsed field gradient (PFG)-NMR techniques. This method enabled us to reveal dynamic information on the sample microstructure modifications during sol-gel transition. This study therefore constitutes a new illustration of the potential of probe diffusion measurements to reveal structural changes in complex and evolving matrices.

2   MATERIALS AND METHODS

2.1 Materials

Native phosphocaseinate powder (INRA, Rennes, France) was used (powder composition was described in ref [8]. The PEG polymer was obtained from Polymer Laboratories (Marseilles, France), with an average molecular mass of 96750 g/mol. Sodium azide (NaN3) (Merck, Darmstadt, Germany), sodium chloride (NaCl) and Glucono-Delta-Lactone (GDL) with a purity above 99% (Sigma-Aldrich, Steinheim, Germany) were used without further purification. The chymosin solution used was Chymax-Plus purchased from Chr-Hansen (Arpajon, France).

2.2 Preparation of Native Phosphocaseinate Suspension (NPCS).

Native phosphocaseinate powder was added to a 0.1 M NaCl water solution containing 0.2% (w/w) NaN$_3$ to prevent bacterial growth. The solution was vigorously stirred at room temperature for 36 hours to ensure total rehydration of the powder. After this period, 0.1% (w/w) of PEG was added. The casein suspension concentrations in all samples were 16.75 $\pm$ 0.35 g of casein for 100 g of H$_2$O for acid gel and 16.18 $\pm$ 0.35 g for chymosin gel, and their initial pH was 6.80 (Schott, pH combination electrode type No N6280, Germany).

2.3 Coagulation Processes.

For the acid coagulation, GDL was added to the casein suspension at the proportion of 4.10 g to 100 g. This procedure provided a pH of around 4.6 at t = 24h. For enzymatic coagulation, a chymosin dilution (1 mL in 99.0 g of distilled water) was prepared and stored at 4°C approximately 20 min before each inoculation to allow more accurate dosage. To start the coagulation process, the chymosin dilution was added to the casein suspension at the proportion of 700 μL for 100g of NPCS. For both coagulation processes, a chronometer was started at the same time (t = 0 s) and the solution was vigorously stirred for 3 min. Samples were then rapidly prepared for NMR and dynamic rheological measurements, whereas for SEM samples were stored at 20°C before analysis.

2.4 pH Measurements.

In order to follow the acidification kinetics, the pH-electrode was placed in an extra amount of sample maintained at 20°C in a water bath. The pH-meter was connected to a data logger to record the pH every five minutes. An example of the curves obtained is presented in Figure 1. The equation relating the time and pH was obtained by fitting all the data by Table Curve. This procedure enabled us to plot our diffusion and viscoelastic results against pH.

2.5 Gel Time Determination.

The coagulation processes were characterized by dynamic rheological measurements with a controlled stress rheometer (Rheostress RS150, Haake, Germany) using a double gap cylinder sensor (DG41). The temperature was maintained at 20°C and the surface of the sample was covered with silicone oil to prevent evaporation. The storage modulus (G') and the loss modulus (G") were recorded at a frequency of 1 Hz and the rheometer was programmed to adjust the stress automatically to provide a strain of 0.5%, which was found to be within the linear viscoelastic region of the samples. The gel time was defined as the time at which G' equal G''.

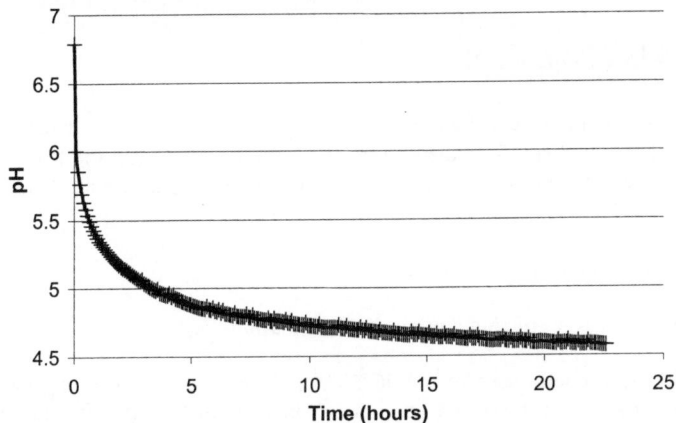

Figure 1. *Example of curve representing the recorded values of pH as a function of time after the addition of GDL to a casein suspension maintained at 20°C. The solid line represents fitting of the data obtained by Table Curve.*

2.6 Scanning electron microscopy (SEM).

Images were obtained with a scanning electron microscope (Jeol JSM 6301F) operating at an acceleration voltage of 9 kV. The images were produced by CMEBA (France, Rennes). The detailed protocol concerning sample preparation was previously given in Ref [8].

2.7 Self-Diffusion Measurements.

PEG self-diffusion measurements were performed on a 500 MHz Bruker spectrometer equipped with a field gradient probe with 5 mm NMR tubes. Diffusion spectra were acquired at $20 \pm 0.1°C$ with a stimulated echo sequence using bipolar gradients (STE-BP) and a 3-9-19 WATERGATE pulse scheme to suppress the water signal. Experiments were carried out with 16 different values of g, ranging from ~ 0.25 to 5.00 T/m and with $\delta = 1.0$ ms. Sixteen scans were undertaken and the recycle delay was set at five T_1. The diffusion interval Δ was adjusted to 173 ms to obtain a diffusion distance z ~ 1.5 μm in the casein suspensions, in accordance with the Einstein equation $z = (2\ D_{PEG}\ \Delta)^{1/2}$. This procedure enabled molecular probes to cover a much greater distance than the casein micelle diameter (diameter around 150 nm).

2.8 NMR processing methods.

All the data processing was performed with Matlab software. Monte-Carlo simulations were used for error calculations with 200 iterations. All self-diffusion coefficients were calculated from the following equation:

$$I/I_0 = \exp(-k\ D) \qquad (1)$$

where D is the molecular self-diffusion coefficient and where I and I_0 are the signal intensities in the presence and the absence of gradients respectively. The values taken by k in a STE-BP NMR sequence with a WATERGATE scheme are given by the following equation:

$$k = \gamma^2 g^2 \delta^2 \left(\Delta - \delta/3 - \tau/2\right) \tag{2}$$

Here, γ is the gyromagnetic ratio (for protons, $\gamma = 26.7520 \times 10^7$ rad.T^{-1}.s^{-1}), g the amplitude of the gradient, δ gradient duration, Δ the time between the leading edges of gradients and τ the time between the end of each gradient and the next radiofrequency pulse.

3   RESULTS AND DISCUSSION

3.1   Self-diffusion during Chymosin Coagulation

The self-diffusion coefficients of the 96750 g/mol PEG measured during the chymosin coagulation process are presented in Figure 2 in relation to time after the addition of enzyme. Two different phases clearly appeared. During the first period (from t = 0 to about 5 h) the diffusion coefficient of the polymer remained stable while the transition from solution to gel occurred between approximately t = 2 and 3h as revealed by rheological measurements (data not shown). The gel time was found to be t_{gel}= 2 h 35 ± 3 min. During the second period the diffusion coefficient then regularly increased with time. After 24h, the diffusion coefficient was enhanced by approximately 20%, which is consistent with results previously reported in casein gels made with heavy water [5, 8].

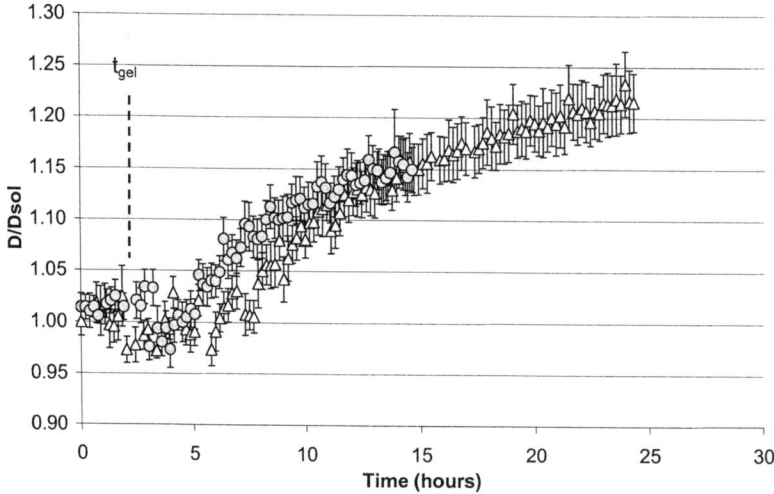

Figure 2. *Evolution of the 96750 g/mol PEG self-diffusion coefficient as a function of time after chymosin addition. Circles and triangles are repetitions. The Gel time is indicated. Error bars represent the uncertainties given by the Monte-Carlo simulations.*

SEM images of the gel are shown at different times after enzyme addition in Figure 3. Figure 3-a show that the network soon after gel formation was constituted of casein aggregates composed of small spherical particles with a mean diameter around 100-150 nm. At this point the gel was coarse, with small pores. At t = 8 h 45, the network organization was very similar, but larger empty spaces were observed at the same time, indicating that the gel structure had evolved (Figure 3-b). Finally, 24 h after the addition of chymosin, the microstructure appeared clearly more branched (Figure 3-c). The casein particles had partially fused together, thin strands were formed and pore sizes were increased. A clear evolution of the network towards a more "open" microstructure was thus observed during the hours following the gel formation.

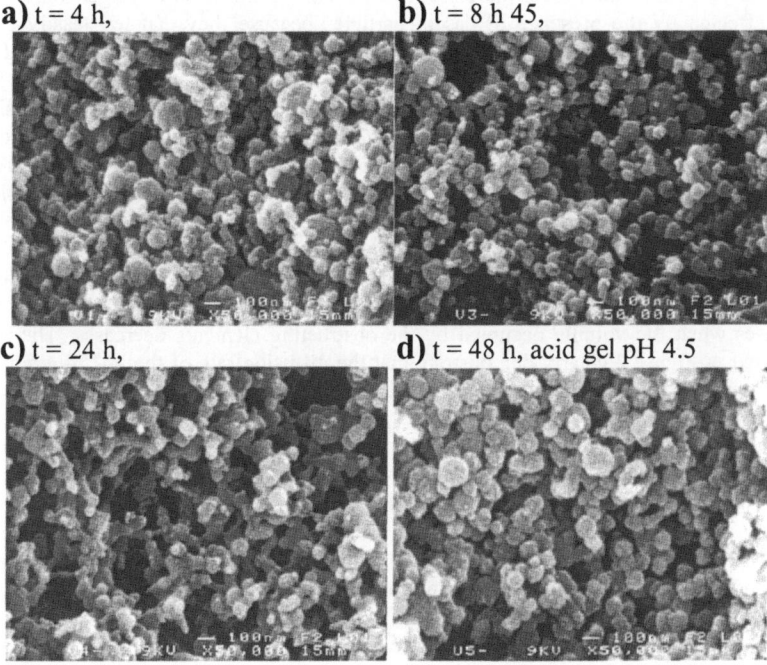

a) t = 4 h, **b)** t = 8 h 45, **c)** t = 24 h, **d)** t = 48 h, acid gel pH 4.5

Figure 3. *SEM images (× 50 000) of enzymatic and acid gels at different times. a) t = 4 h, b) t = 8 h 45 and c) t = 24 h after addition of chymosin. d) at pH 4.5, 48 h after addition of GDL.*

The coagulation induced by chymosin action is generally broken down into three phases: enzyme action, aggregation and gel ageing [13, 14]. First, the enzyme specifically splits off the κ-casein which is at the surface of the supramolecular edifice. This reduces the steric and electrostatic repulsions between the casein particles and the suspension becomes unstable. In the second phase, the resulting para-casein micelles spontaneously aggregate and form a macroscopic network. Finally, the third phase corresponds to the ageing of the gel and is characterized by the occurrence of structural changes in the casein network. To describe this particular phase, Mellema et al. [15] proposed a general model based on several types of rearrangement occurring at different length scales. According to them, particle fusion and inter-particle rearrangements result in the formation of straight and progressively thinner strands, with more bonds per casein aggregate and hence stronger junctions. This consequently leads to local matrix fusion and compaction, resulting in a gel with larger pores and a higher permeability [15-18].

Previous studies have shown that probe diffusion in casein suspensions and gels is greatly dependent on both the volume fraction occupied by casein particles and the probe size [5, 8]. The reduction in diffusion coefficient for a given volume fraction of casein particles is smaller for smaller probes. This phenomenon was previously explained by assuming a model with two diffusion pathways, one around and one through the casein micelles [5]. According to the "two site" model, variations in the diffusion rate of a molecule only depend on its ability to diffuse through the casein particles and on the volume fraction occupied by them. This model therefore implies that the diffusion of larger molecules is more affected by the presence of casein particles because they can less easily diffuse through them. According to this model, the compaction of the casein strands should have two opposite effects on probe diffusion. On the one hand, more molecules should freely diffuse because more empty spaces are created inside the gel microstructure. This should result in an enhanced diffusion coefficient. On the other hand, the "through the casein aggregates" diffusion component would be expected to decrease as they become denser, and this should lead to a reduction in the diffusion rate. By following this interpretation, it becomes possible to explain the increase in diffusion coefficient of the PEG. Because of its large size, it would be dangerous to make assumptions regarding the ability of this large PEG to diffuse through casein particles. The self-diffusion coefficient of this polymer can therefore only increases because the volume fraction that is accessible to this probe increases when the volume occupied by the obstructing elements decreases. The results presented in Figure 2 clearly demonstrate that the diffusion rate of the 96750 g/mol PEG was very sensitive to the progressive increase in gel porosity as shown by the SEM images (Figure 3).

3.2 Self-diffusion during acid-induced coagulation.

Above pH = 4.7, although the rheological properties (data not shown) of the sample remained stable since the sol-gel transition occurred at pH = 4.60, wide variations in the self-diffusion coefficients of PEG occurred, as illustrated in Figure 4. The curve could be broken down into three different phases. Between the initial pH and pH ~ 5.9, the diffusion coefficient of the PEG was enhanced. Below pH ~ 5.9, the diffusion rate then decreased until pH = 4.9 was reached. The PEG diffusion coefficient increased thereafter until the end of the experiments (at pH ~ 4.55).

In the hours following the onset of gelation, the gel was not solid enough to produce SEM images. This experiment was therefore performed only at around t = 48 h when the pH of the sample was stabilized just above 4.5. At this pH, the gel structure consisted of an assembly of small and spherical particles (Figure 3-d). Although these particles were necessarily connected since a gel was formed, no fusion between them was observed.

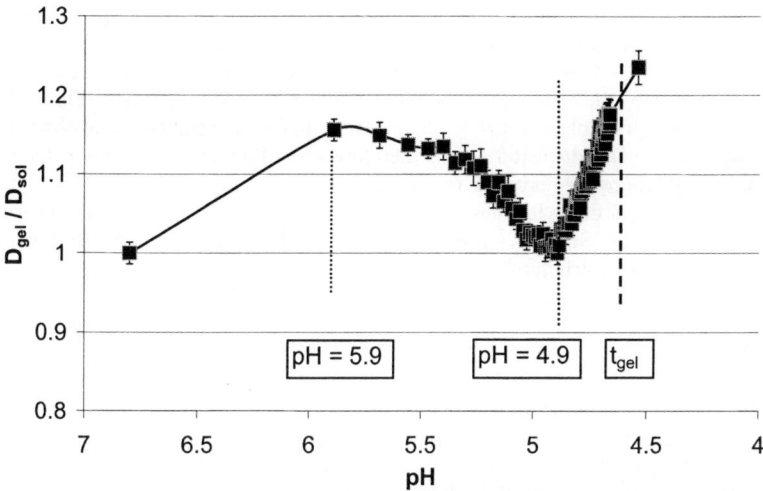

Figure 4. *Evolution of the 96750 g/mol PEG self-diffusion coefficient as a function of pH during acid coagulation. Error bars represent the uncertainties given by the Monte-Carlo simulations. The solid line is a guide for the eyes.*

In the case of enzyme-induced coagulation, the diffusion rate of PEG did not vary before the network formation was completed. This could be explained simply since no great changes are thought to occur in the sample microstructure before gelation. However, during acidification the diffusion of the probe was greatly modified before the sol-gel transition (Figure 4). This is particularly interesting since several important physicochemical changes are known to take place during this period in the sample.

From the initial pH to 5.9, the most remarkable event taking place is a progressive decrease in the casein particle hydration, which reaches a minimum at pH ~ 5.9. At this pH value, about 15% [19-23] of the water contained by the colloidal particles has been released in the aqueous phase and this causes a reduction in the mean particle size as observed by diffusing wave spectroscopy experiments [24-26]. Since the 96750 g/mol PEG cannot or can hardly diffuse through casein particles, the reduction in the casein particle size can be viewed as a decrease in the volume occupied by obstructing elements. More space was thus accessible to this large polymer, and its diffusion coefficient consequently increased (Figure 4).

From pH = 5.9 to 4.9, three main phenomena occur. First of all, solubilization of casein molecules takes place during this phase and reaches a maximum at pH ~ 5.5 at 20°C [20, 21]. Although this should directly affect the PEG diffusion, since there are fewer soluble proteins at pH ~ 4.9 than at pH ~ 5.9, such a phenomenon cannot explain the overall decrease observed in this pH range (Figure 4). Secondly, particle hydration is known to return progressively to about its initial value. The details of the mechanism explaining this swelling is generally attributed to the third important event occurring during this phase : the solubilization of the colloidal calcium phosphate (CCP). In fact, the CCP, which contributes considerably to casein micelle integrity [20, 21, 27], is progressively solubilized as soon as acidification is in progress, but process of this dissolution becomes faster below pH ~ 6.0, and at pH = 4.9 all the calcium phosphate has been transferred into the serum phase [20, 21, 28, 29]. As during the first phase, the increase in particle hydration seemed to be the key factor explaining the results we obtained.

Below pH = 4.9, the amount of water contained by the casein particles again sharply decreases because of the neutralization of the general charge until the isoelectric pH of caseins is reached (pHi ~ 4.64 [19-22]). As explained for above pH = 5.9, the reduction in particle hydration caused the diffusion of the molecule investigated to increase (Figure 4). Gelation of the system took place at pH = 4.6 but no particular changes could be observed in the diffusion curves. Our diffusion results therefore show that acid coagulation should be viewed as a very progressive process, in accordance with the description of a gradually increasing tendency of particles to interact as the pH is reduced [26]. This description was moreover reinforced by the SEM image which indicates that the network structure has not been subjected to extensive structural rearrangements (Figure 3-d).

4   CONCLUSION

Throughout both acid and chymosin coagulation processes, the diffusion of a 96750 g/mol PEG was very sensitive to variations in the size of the casein particle and in the appearance of the network. Many of the structural modifications occurring in the sample at different levels could thus be revealed from PFG-NMR, and interesting conclusions on several key points of the milk coagulation processes could be drawn. In addition, it is of note that the duration of diffusion measurements can be reduced and that the use of lactic bacteria instead of GDL seems to be feasible. Both of these modifications would allow the acquisition of more diffusion data at high pH values.

Our results also proved that probe diffusion and rheological experiments are highly complementary. As for rheometry, Pulsed Field NMR experiments can be performed continuously on the same sample. This is a great advantage, especially when studying systems that do not react instantaneously to "disturbances" (e.g. acidification of a milk system). In addition, whereas rheology permits characterization of the macroscopic behavior of the sample, the diffusion of probes can be used to provide structural information on smaller scales with equal efficiency in both the liquid and gel states. Finally, it should be noted that, in contrast to most of the techniques commonly employed to characterize modification of a microstructure, probe diffusion is even more sensitive to structural changes when samples are concentrated. All these features demonstrate that kinetic probe diffusion experiments are very valuable to study the coagulation of milk systems, and might also provide advantages to investigate the coagulation of other systems.

REFERENCES

1   P. Stilbs in *Mesoscale Phenomena in Fluid Systems,* ed.   Amer. Chemical soc., Washington, 2003, p 27.
2   R. Trampel, J. Schiller, L. Naji, F. Stallmach, J. Kärger and K. Arnold, *Biophys. Chem.*, 2002, **97**, 251.
3   S. G. Baldursdottir, A.-L. Kjoniksen and B. Nystrom, *Euro. Polym. J.*, 2006, **42**, 3050.
4   R. Colsenet, M. Cambert and F. Mariette, *J. Agric. Food Chem.*, 2005, **53**, 6784.
5   R. Colsenet, O. Soderman and F. Mariette, *Macromolecules*, 2005, **38**, 9171.
6   P. Croguennoc, T. Nicolai, M. E. Kuil and J. G. Hollander, *J. Phys. Chem. B*, 2001, **105**, 5782.
7   L. Johansson, C. Elvingson and J. E. Lofroth, *Macromolecules*, 1991, **24**, 6024.
8   S. Le Feunteun and F. Mariette, 2007, **55**, 10764.

9   L. Masaro, M. Ousalem, W. E. Baille, D. Lessard and X. X. Zhu, *Macromolecules*, 1999, **32**, 4375.
10 S. Matsukawa and I. Ando, *Macromolecules*, 1996, **29**, 7136.
11 B. Newling, *J. Phys. Chem. B*, 2003, **107**, 12391.
12 M. Nyden, O. Soderman and G. Karlstrom, *Macromolecules*, 1999, **32**, 127.
13 A. O. Karlsson, R. Ipsen and Y. Ardo, 2007, **17**, 674.
14 P. Zoon, T. Van Vliet and P. Walstra, *Neth. Milk. Dairy. J.*, 1988, **42**, 249.
15 M. Mellema, P. Walstra, J. H. J. Van Opheusden and T. Van Vliet, *Adv. Colloid Interface Sci.*, 2002, **98**, 25.
16 J. A. Lucey, M. Tamehana, H. Singh and P. A. Munro, *J. Dairy Res.*, 2000, **67**, 415.
17 J. A. Lucey, M. Tamehana, H. Singh and P. A. Munro, *Int. Dairy J.*, 2001, **11**, 559.
18 M. Mellema, J. W. M. Heesakkers, J. H. J. van Opheusden and T. van Vliet, *Langmuir*, 2000, **16**, 6847.
19 S. Banon and J. Hardy, *J. Dairy Res.*, 1991, **58**, 75.
20 M. H. Famelart, F. Lepesant, F. Gaucheron, Y. Legraet and P. Schuck, *Lait*, 1996, **76**, 445.
21 E. Gastaldi, A. Lagaude, S. Marchesseau and B. T. DelaFuente, 1997, **62**, 671.
22 N. Vetier, S. Banon, J. P. Ramet and J. Hardy, *Lait.*, 2000, **80**, 237.
23 S. Banon and J. Hardy, *J. Dairy Sci.*, 1992, **75**, 935.
24 M. Alexander, M. Corredig and D. G. Dalgleish, *Food Hydrocolloids*, 2006, **20**, 325.
25 M. Alexander and D. G. Dalgleish, *Colloid Surf. B-Biointerfaces*, 2004, **38**, 83.
26 D. Dalgleish, M. Alexander and M. Corredig, *Food Hydrocolloids*, 2004, **18**, 747.
27 A. C. M. Van Hooydonk, H. G. Hagedoorn and I. J. Boerrigter, *Neth. Milk Dairy J.*, 1986, **40**, 281.
28 D. G. Dalgleish and J. R. Law, *J. Dairy Res.*, 1989, **56**, 727.
29 W. J. Lee and J. A. Lucey, *J. Dairy Sci.*, 2004, **87**, 3153.

DEVELOPMENTS IN TIME DOMAIN AND SINGLE SIDED NMR

G. Guthausen[1], A. Kamlowski[2]

[1]SRG10-2, IMVM, Universität Karlsruhe, D-76128 Karlsruhe, Germany
[2]Bruker BioSpin GmbH, Silberstreifen, D-76287 Rheinstetten, Germany

1 INTRODUCTION

Recent developments of time domain NMR (TD-NMR) and single sided NMR can be classified with respect to their impact on application, hardware, and data processing. Well known applications, developed in the 1970 and 1980's, experience extensions due to improved hardware and data processing capabilities. One prominent example is the determination of water and fats in objects with high water content.

In the following, the term "low field NMR system" is used to describe the whole variety of NMR systems, from simple (two-pulse) to complex (R&D) systems with either huge or vanishing (static) magnetic field gradients. Specifically, these low field NMR systems are characterized by permanent magnets providing a static magnetic field corresponding to a ^{1}H Larmor-frequency of 2-60 MHz. In the present contribution, the focus is not primarily on the variety of applications of TD-NMR and single sided NMR in food science and industry[1], but on selected methodological and data processing developments as well as equipment innovations.

2 COMMON APPLICATIONS: DEVELOPMENTS

2.1 Solid Fat Content – Characterization of Fat Blends

One of the most important and well known applications of TD-NMR is the determination of solid fat content (SFC) in fat blends[2,3]. Melting curves and crystallization rates[4] of fat compositions (even chocolate) are assessed by measuring the free induction decay (FID) of samples as function of temperature and time, respectively. Crudely, the FID is assumed to consist of a linear superposition of two components, a solid (fast relaxing) and a liquid (slow relaxing) one. In the direct SFC method, a ratio of the total signal amplitude measured directly after the probe's dead time and the liquid like signal at about 70 µs is obtained. Magnetic field inhomogeneity and dead time effects are corrected for by phenomenological k and F factors, to provide the SFC value at a given temperature.Since the development of the SFC application in the 1970's, capabilities of low field NMR systems have improved, allowing a more throughout analysis of fats and their blends.

2.1.1 Crystallinity

SFC characterization of a fat blend is only one parameter; the polymorphism is of great importance as well. Several forms with specific morphologies can be distinguished by their NMR properties. Approximate description of the FID or the FID-CPMG signals by an analytical function allows the quantification of the crystalline forms, especially α, β, and β'[5]. The signal decay is described as a sum of three functions: an exponential decay (assigned to the liquid part), an 'Abragam sinc' function (assigned to the β' forms for spins interacting via modulated dipolar couplings), and a Gaussian function (assigned to the α form). The relative amounts of the different crystalline phases are revealed. Good agreement is observed when comparing the results of this deconvolution approach to classic SFC determination. Furthermore, the crystallization process can be monitored for each phase by time dependent FID measurements and subsequent analysis[6].

2.1.2 Monitoring of crystallization process

Currently, cost saving is aimed for by tailoring the processes, requiring a detailed understanding of process parameters and their consequences onto the product. Thus, the attempt was made to monitor the SFC of fats under shear[7], which is usually the case in production. A Couette rheometer was built into the probe of a standard 20 MHz TD-NMR system. Shear rates up to 450 s^{-1} have been reported. Under shear, the SFC values were measured via both the direct and indirect SFC methods. Interestingly, both methods did not provide equivalent results under shear: A shear rate dependence was found which was interpreted as being due to the different influence of sample temperature on direct and indirect SFC measurements: In case of direct SFC, the Boltzmann factor is essentially eliminated, only the materials changes are measured, whereas the result of the indirect method is influenced by both effects. The conclusion is that in the isothermal minispec environment, the heat dissipation in the viscous-elastic medium due to stirring leads to an effective sample temperature increase of up to 12°C[7].

Online monitoring of the crystallization process was followed by single sided NMR in another study[8]. To this end, a CPMG multi-echo train was acquired to obtain the SFC values. Good agreement was achieved between offline TD-NMR, online single sided NMR as well as ultrasonic measurements.

2.2 Moisture and Oil Determination – The classic approach and the case of high water content

Based on the fact that the relaxation properties of ^{1}H nuclei depend strongly on their environment, determination of water and fat/oil content of products becomes feasible. Two approaches are discussed, the direct which is applicable to water contents up to about 10% and a more comprehensive generalized approach for any water content.

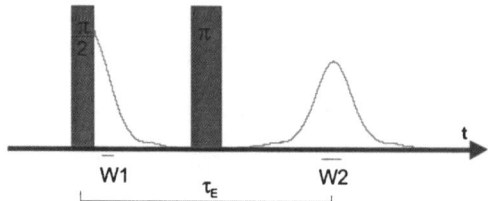

Figure 1: *Pulse sequence and a typical oil seed signal measured at 20 MHz. The echo time τ_E is 7 ms. The FID is attributed to "hard" materials as well as moisture and oil. The echo signal amplitude at the time window W2 can be correlated with the oil content. The moisture content is calculated as difference of the signals at W1 and W2.*

2.2.1 Oil or Fat Determination at low water content by TD-NMR

In this case water can be regarded as partially bounded (moisture). A simple Hahn-echo sequence with two acquisition windows can simultaneously quantify the moisture and the oil or fat content (Figure 1). After some tens of µs, the FID signal is made up essentially of water and oil, whereas the echo with an echo time of about 7 ms correlates to ^{1}H spins of oil. This is because of the comparably fast transverse relaxation time of the bound water protons. The measuring protocol is successfully applied to a whole variety of products as chocolate, dried dairies, oil seeds[9].

In case of 20 MHz systems, which are used for dried diaries, chocolate etc., the system related reproducibility is usually equal or better than 0.1% relative standard deviation (RSD), the RSD of accuracy is in the range < 0.3 % with a measurement time of about 10 s and a sample volume of about 11.5 cm^3. Spatial inhomogeneity within the sample does not compromise the quality of measurements as the rf-field amplitude varies less than 1 % over the specified sample volume. Temperature effects have to be considered as they influence both relaxation times and Boltzmann factor.

In case of oil determination in oil seeds, special care has to be taken that the sample volume is representative, leading to the use of 10 MHz or even lower frequency TD-NMR systems. Several studies showed that the accuracy of measurements can be drastically improved by measuring larger quantities, especially in case of large seeds such as sunflower seeds[10].

2.2.2 Samples with high water content

Meat products, fruits, dairies are examples for food products with large amount of water. The simple transverse relaxation filter will fail as free water will exhibit similarly long transverse relaxation times as oils. An obvious approach would be drying and subsequent fat determination by TD-NMR, a time-consuming two-step approach, however. For a one-step quantification of water and oil, several attempts have been made[11-14]: The sample's magnetization, simultaneously influenced by T_1 and T_2 relaxation processes, is recorded, since the combination of both processes drastically increases the contrast at low magnetic fields. This can either be done by applying inversion recovery, each point followed by a CPMG train[12,15], or by saturation and subsequent observation of recovery by CPMG[13,14]. Most often, the data sets are processed via partial least squares regression (PLS), the accuracy and repeatability depends on the type of samples as well as on the calibration.

Typically, a satisfying reproducibility of the order of 1-3 % is obtained. An interesting spin-off of this approach is the whole body composition analysis (BCA) of small animals[15].

3 ONLINE APPLICATIONS OF TD-NMR

3.1 Online Measurement of Single Seeds

Seeds as an agricultural product are inhomogeneous per se. A definition of a representative volume is therefore hardly to achieve. Therefore, one attempt is a 100 % control, i.e., oil quantification of each and every seed. Since the amount of seeds is enormous, an online fast control is to be considered[9]. On basis of steady state free precession (SSFP), the oil content of single seeds was determined. It was shown, that the amplitude of the SSFP signal can be correlated to the oil mass, as in case of the classic application described above. Up to 20,000 single seeds could be measured in an hour. However, the online NMR was done on a 2.1 T magnet which obviously leads to a better signal-to-noise ratio as compared to the usual TD-NMR fields for seeds analysis of 10 MHz or less. Nonetheless, it demonstrates the unique opportunities of NMR as online tool.

3.2 Online weighing

A true online low field NMR example is the NCCW machine ("non-contact check weighing") used in pharmaceutical production[15]. The task is to weigh each substance filled vials in a pharmaceutical filling line, 100% check-weighing. The classical weighing by balances requires the tare weight of each vial; additionally it is sensitive to air flow, which, however, has to be present in the filling line due to hygiene and safety regulations. Optical methods are sensitive to surface-texture and colour and require additionally a uniform density. NMR as a volumetric and contrast-rich method overcomes these drawbacks. Inline NMR check weighing has been realized for liquid and powdery drugs. However, the task is challenging: In the case of a powdery drug described here, 150 moving filled vials have to be measured per minute with one scan. A vial is closed with a rubber stopper exhibiting an even larger NMR signal and longer transverse relaxation than the drug itself. Since floor space in pharmaceutical production is expensive, an extended pre-polarisation zone was not an option. Due to the limited built-up of Boltzmann polarization, the signal is about a factor of two weaker as in equilibrium state. The main challenges were: (a) enhancing of signal-to-noise ratio in the given measurement time while keeping motional artefacts at a minimum, (b) suppressing unwanted signals from neighbouring vials and rubber stoppers and (c) guaranteeing reproducibility and tolerance to instabilities of the filling line.

Task (a) is addressed by multiple acquisition and co-addition of solid echoes according to the sequence of Ostroff and Waugh[17]. Fast repetition of pulses in the OW4 sequence leads to an effective slowing down of the apparent transverse relaxation (spin locking in a quasi steady state) such that the gain is over-proportional. Combined with an active Q-switch for dead time reduction, a number of echoes (up to 20) could be detected for averaging.

Task (b) is solved by a proprietary open probe design[18]. As the number of spins is counted in the drug's total volume, preferably the rf field B_1 has to be constant over that volume and has to fall off at the position of the stopper. An about 25 dB attenuation of the rubber signal has been achieved. In the direction of the conveyor belt, similar construction elements lead to an almost vanishing nearest neighbour effect. The open top of the probe fulfils the criteria of air flow and cleaning in place.

The result of efforts in (a-c) is shown in Figure 2. Care was taken with respect to temperature effects, electromagnetic noise, and stability of the electronic components. Additionally, the vials speed and consequently the magnetic history are essential parameters to be considered in the measuring sequence.

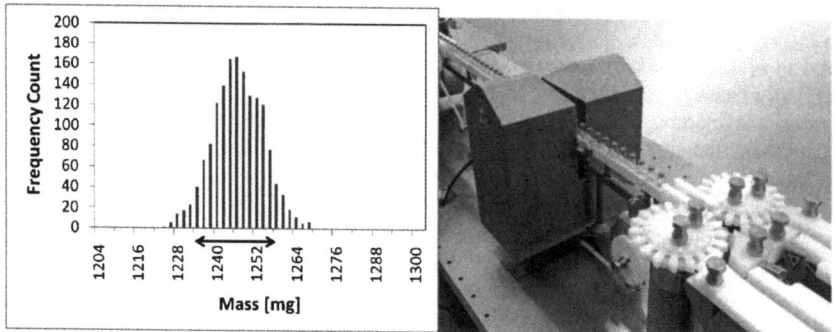

Figure 2: *Right: Frequency distribution of mass measured by NMR. The distribution function is approximately Gaussian, proving stability towards environmental influences. The distribution's half-width is better than the required specification (red arrow). Left: NCCW for a powdery drug.*

4 LOW FIELD MR IMAGING

4.1 Imaging and Quality Control

With the availability of hardware, questions in quality control of foods are addressed also with LF-NMR *imaging*. One example is non-destructive fruit inspection, where a volumetric, but spatially resolving technique is needed. Extensive research has been done in seeds detection[19,20], with the main focus on the online capability[19] and the methodical development[20] to fulfil the constraints. In[19], the focus is the examination of motional effects in the online application: Mandarins are conveyed through a 4.7 T NMR system with 54 mm/s, applying fast imaging sequences (FLASH) to minimize motional artefacts.

Turbo FLASH, fast spin echo, and gradient recalled echo images[19] were conducted at 1 T. By exploiting the biological specialities of seeds grow, imaging parameters can be optimally chosen. The images were masked and filtered to minimize artefacts. A contrast optimization was achieved by multivariate image analysis which allows for a remarkably good prediction. In summary, seeds detection via online low field NMR becomes realistic.

4.2 Single Sided NMR: Spatial Resolution and Flow

Single sided NMR is characterized by large, intrinsic static magnetic field gradients. B_0 and B_1 gradients were utilized for gaining spatial resolution[21]. However, conventional MRI methods like slice selection and read gradients, cannot be applied here because the static gradients are just too large. The sensitive volumes are usually small, leading to issues in sensitivity and reproducibility. Pulse sequences were developed to improve the overall sensitivity in even very small volumes, leading to the "Profile MOUSE"[21]. For imaging, the principle of phase encoding was exploited, resulting in three dimensional images within reasonable measurement times. One key example is the measurement of flow velocity distributions in a given volume which may be quite promising[21].

5 SPECTROSCOPY AT LOW MAGNETIC FIELDS

Attempts to realize spectroscopy in low field NMR date back to the early days of NMR. Reducing the sample volume increases resolution at the expense of sensitivity. Other approaches address magnetic field shimming either by mechanical (geometrical adjustment of soft or hard magnetic permanent magnets) or by electrical means (via current-adjustable shim coils). Recently, even in single sided NMR, spectral resolution was achieved. The key question, however, is: Does the increased complexity and technology improve the applicability of low field NMR in such a way as to justify the investment and running costs. An example for the use of spectral resolution is given in this section.

5.1 Single Sided NMR: Field Match, Shim and dedicated sequences

Single-sided NMR devices are convenient tools for completely non-destructive testing, because of their open design. On the other hand, a plenty of chemical information is available in chemical shift and also *J*-coupling. Chemical shift interactions are obscured by B_0 inhomogeneity. To overcome these apparent contradictions, several attempts mostly on de-shimmed conventional magnets have been made for gaining spectral resolution by dedicated pulse sequences and field design[21,22,23,24]. The approaches require three steps: the spatial dependence of B_0 and B_1 has to be designed and in the second step optimized by electric shims. Third, the rf excitation and gradients have to be tuned for obtaining the desired spin manipulation. Often, the pulse sequences are based on mixed echoes.

The direct approach is to shim B_0 in a defined volume outside the magnet[24]. This technique in a similar MRI approach is known as 'localized spectroscopy' or 'chemical shift imaging'. As in commercially available low field magnets, mechanical shims are realized in form of smaller permanent magnets. Fine tuning of field homogeneity is achieved by electrical linear shims (*x, y, z*). The region of interest is finally selected by slice selection, realized by a soft preparation pulse in combination with a gradient pulse. Experimentally, a spectral resolution of about 0.25 ppm was achieved over a sample volume of 5x5 mm^2 and a thickness of 0.5 mm, located about 2 mm above the surface of the single-sided NMR device. The low ^1H frequency of 8.33 MHz and the small sensitive volume rendered averaging necessary. A toluene NMR spectrum is given as example[24] with a measurement time of one minute and an S/N of about 10-14. The chemical shift differentiation of oil and water was shown as well[24] together with an intriguing way to compensate for the temperature dependent magnetic field drifts.

5.2 Bench-top Instruments: Component Quantification during a Chemical Reaction

In the context of improving the efficiency of use of natural resources and of profit optimization, the question arises whether a chemical process can explicitly be monitored in real time by low field NMR spectroscopy. Typical requirements are: A resolution of about 0.1 ppm which is often sufficient to detect and quantify concentrations of reactants, a measurement time short enough for temporal resolution of the reaction (several 10 s). Questions like pre-magnetization and accuracy have to be addressed. The most obvious way to follow the reaction by NMR is via a bypass line. Representative sampling is here a crucial point, which was addressed in a recent publication[26]. The esterification reaction of crotonic acid and 2-butanol in toluene was monitored. The concentrations were determined by the peak integrals, being normalized to the methyl signal area of toluene. The average statistical errors (RSD) were found to be in the range of a few percent in the online, flow-through study[26].

6 DATA PROCESSING: NEW APPROACHES IN LOW FIELD NMR

Apart from discrete modelling of relaxation processes, 1D and 2D Inverse Laplace Transformation (ILT) is gaining more and more interest. Moreover, soft and hard modelling data processing tools like PLS or multivariate curve resolution (MCR) are applied to low field NMR data. Special algorithms were developed for the needs in relaxation modelling, for example DOUBLESLICING[27].

6.1 Inverse Laplace Transform: Classification and Correlation of Relaxation and Diffusion Parameters

This approach of data processing considers distributions of relaxation rates or other physical parameters rather than a sum of limited and discrete characteristic quantities. Prerequisite is that the decaying function is multi-exponential. ILT may not be applied to Gaussian decays, which are observed for instance in dipolar coupled systems (hard polymers). Prominent examples for the ILT are relaxation studies of liquid-like samples and investigations of (restricted) diffusion by varying the gradient amplitude g. Even in case of restricted diffusion, like in droplets size studies, the relation between signal attenuation and g^2 is often exponential.

Two dimensional 2D ILT was applied in a study addressed to the composition and the physical behaviour of water and fat mixtures[28]. Several dairy products (like milks, cheeses) were measured in the stray field of a superconducting magnet (^{1}H Larmor frequency 5.034 MHz). At $g = 0.545$ T/m, the samples diffusion properties can be observed, characterized by sample-specific diffusion coefficient (D) distributions.

The D–T_2 distributions were measured by a stimulated echo sequence followed by a CPMG echo train. The T_1-T_2 correlation measurements were performed by inversion recovery preparation and a subsequent CPMG. As relaxation and diffusion parameters are different for water, fat, proteins and other macromolecules, separate data treatments can be made for fat and water behavior in the different dairy products:

(a) Behavior of fat: The T_2 distributions get broader with increasing firmness of the sample, whereas the diffusion distribution is relatively sharp. This can be interpreted in terms of a wide molecular weight distribution of triglycerides in milk fats and by a geometrical confinement of fat in globules with a mean diameter of about 3 μm.

(b) Behavior of water: The diffusion depends strongly on sample, i. e. firmness. In case of restricted diffusion on the long time scale (tortuosity limit), the diffusion coefficient depends on the square root of water concentration, which is related to the protein concentration. Also T_2 is observed to be a function of the protein content.

The experiments give evidence for the influence of the micro-scalar structure on the NMR parameters and allow a characterization of the different dairy samples. This study is also an example for the measurement of the influence of a sample component (proteins) which is otherwise hardly to be observed.

6.2 Soft Modelling

In recent years, chemometrical data processing was increasingly applied to low field NMR data. As matter of fact, transverse and combined relaxation data were analysed by partial least squares regression (PLS), both for single sided NMR as well as for conventional TD-NMR data (see also 2.2.2). On the other hand, a variety of statistical soft modelling approaches is available nowadays, which can in principle be explored for their use in low field NMR relaxometry and spectroscopy. Attempts have been made applying tools like DECRA[29] or DOUBLESLICING[27] instead of the approximative approaches of multi-exponential fitting[30] or ILT in relaxation studies.

Soft modelling tools[31,32] have been applied to low field NMR spectra obtained from emulsions with varying water to oil ratio. The data were acquired on a 20 MHz TD-NMR system (Bruker minispec mq20), equipped with electrical shims and a prototype proton probe, containing an external reference sample for lock purposes. Eight scans were accumulated in an experiment time of 16 s. Each scan was phase and frequency corrected to account for minute field and phase fluctuations. The spectral resolution was better than 0.2 ppm with an S/N of about 400 (Figure 3, left). This also demonstrates the general S/N advantage of NMR in homogeneous fields vs. single-sided NMR (cf. section 5.1).

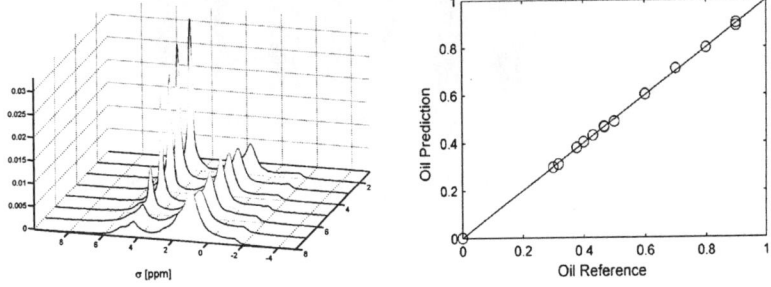

Figure 3: *Left: spectra of water and oil emulsions, right: correlation of reference values and "predicted" oil values (CPR), measured by NMR.*

Correlation of the spectra to the oil content of the emulsions was established (Figure 3, right). The best correlation was obtained by applying continuum power regression (CPR). The correlation coefficient R2 was 0.999. In addition, PLS as a soft modelling procedure was applied to the data (Figure 4).

6.3 Indirect Hard Modelling

Another approach for data processing involves simulation of pure spectra. These model spectra are then taken for a quantitative description of the mixture spectra. This procedure is referred to as indirect hard modelling (IHM)[32]. Obviously, changes in line shape, line width, and chemical shift may occur as function of concentration and due to system imperfections which are taken into account by IHM. The peaks are modelled by Voigt-functions with variable Gaussian to exponential ratio. The main advantage of IHM is that it allows a limited physical interpretation of the models. Further, unlike PLS based methods, IHM only requires reference spectra of the pure compounds, reducing the calibration effort drastically.

6.4 Comparison of methods for Spectral Data Analysis

In Figure 4 the root mean square error (RMS) and the root mean square error of cross validation (RMSECV) of different data processing methods and parameters are shown. As expected, the RMSECV is larger than the RMS for each method. The larger errors of the IHM are due to the non-perfect description of the pure spectra. Interestingly, CPR shows for a set of ranks (number of components used for description of the spectra) and power coefficients the lowest errors. In this example of the mixture of water and oil, this is attributed to the fact, that CPR not only considers the correlation, but also the variance with a power coefficient.

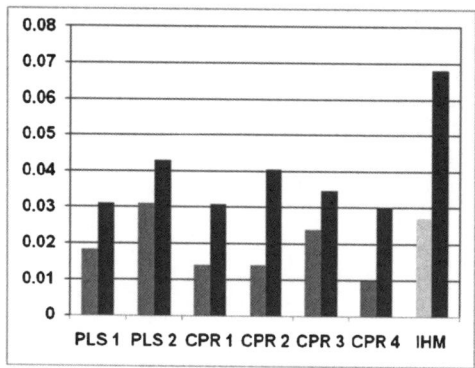

Figure 4: *Comparison of the different methods and processing parameters, using the root mean square error (left bars) and the root mean square error of prediction (right bars) as quality characteristic.*

7 CONCLUSION

Recent developments in low field NMR can be grouped with respect to data processing, hardware development, and new applications.

In well known application fields, different signal interpretation and hardware modifications allow deeper insight into processes and sample characteristics. Hardware improvements implicate an improved accuracy and repeatability.

Efforts are made to improve the online capability of low field NMR. Applications in agricultural and pharmaceutical areas show the feasibility even in demanding environments and specifications.

A significant widening occurred in single sided as well as in TD-NMR due to the improved field homogeneity. Limited spectroscopic resolution was shown allowing the monitoring of processes, quantification of sample's components. This sets a new and promising stage to low field NMR which previously has not been possible with just relaxometric or diffusion weighted data.

The potential of data processing is increasingly recognized as important as hardware or NMR pulse sequence inventions. Statistical as well as approximative approaches provide deeper insight and understanding also for well known applications and classic TD-NMR challenges.

It is the authors understanding that the progress made in low field NMR will lead to a wealth of new and exciting as well as an improvement of existing applications in food science and industry.

Acknowledgements

The Shared Research Group SRG10-2 is supported financially by the "Concept for the Future" of the German Excellence Initiative (DFG). Prof. Dr. Dr. B. Blümich, Prof. Dr. M. J. McCarthy, Dr. J. Duynhoven, D. Engel, Dr. H.-P. Juretschke, and Dr. N. Nestle are thanked for sharing results during the preparation of this manuscript. Prof. Dr. H. Buggisch, Dr. E. H. Hardy and D. Mertens are acknowledged for discussions and reading of the manuscript.

References

Please consider also the references in the cited publications.

[1] for example: A. M. Haiduc, E. Trezza, D. van Dusschoten, A. A. Reszka, J. P.M. van Duynhoven, LWT, 2007, 40, 737–743 ; R. A. Prestes, L. A. Colnago, L. A. Forato, L. Vizzotto, E. H. Novotny, E. Carrilho, Anal. Chim. Acta, 2007, 596, 325–329; N. Margheto, L. Venturi, B. Hills, Postharv. Bio. Technol., 2008, 48, 331; M. J. McCarthy, J. H. Walton, J. S. de Ropp, S. D. Collins, M. V. Shutov, A. G.Goloshevsky, *Food Sci. Biotechnol.*, 2004, 6, 848.

[2] K. van Putte, J. van den Enden, *J. Am. Oil Chem. Soc.*, 1973, **51**, 316

[3] Official Methods: AOCS Cd 16b-93, ISO 8292, IUPAC 2.150

[4] P. Wassell, N. W. G. Young, *I. J. food Sci. and Techn.*, 2007, **42**, 503

[5] a) E. Trezza, A. M. Haiduc, G. J. W. Goudappel, J. P. M. van Duynhoven, *Magn. Reson. Chem.* 2006, 44, 1023

b) F. P. Duval, J. P.M. van Duynhoven, A. Bot, *J. Am. Oil Chem. Soc.*, 2006, 83, 905

[6] E. Trezza, A. M. Haiduc, J. P. M. van Duynhoven, *Magnetic Resonance in Food Science*, 2005, Eds. RSC Books, London.

[7] G. Mazzanti, E. M. Mudge, E. Y. Anom, *J. Am. Oil Chem. Soc.*, 2008, 85, 405

[8] S. Martini, M. L. Herrera, A. Marangoni, *J. Am. Oil Chem. Soc.*, 2005, 82, 313

[9] Official Methods: AOCS Ak 4 – 95, ISO 10565, ISO/CD 10632

[10] L. A. Colnago, M. Engelsberg,‡ A. A. Souza, L. L. Barbosa, *Anal. Chem.*, 2007, 79, 1271

[11] H. T. Pedersen, S. Ablett, D. R. Martin, M. J. D. Mallett, S.B. Engelsen, *J. Magn. Res.*, 2003, 165, 49

[12] D. N. Rutledge, A. S. Barros, *Analyst*, 1998, 123, 551

[13] A. Guthausen, G. Guthausen, H. Todt, W. Burk, A. Kamlowski, D. Schmalbein, *J. Am. Oil Chem. Soc.*, 2004, 81, 727

[14] G. Guthausen, J. König, A. Kamlowski, *Bruker Spin Report*, 2004, 154/155, 41

[15] F. C. Tinsley, G. Z. Taicher, M. L. Heiman, *Obesity Research*, 2004, 12, 150

[16] J. Corver, G. Guthausen, A. Kamlowski, *Pharmaceutical Engineering*, 2005, 25, 18

[17] E. D. Ostroff, J. S. Waugh, *Phys. Rev. Lett.*, 1966, 16, 1097

[18] US patent 7,397,247 B2

[19] N. Hernandez-Sanchez, P. Barreiro, J. Ruiz-Cabello, *Biosystem Engineering*, 2006, 95, 529

[20] R. R. Milczarek, M. J. McCarthy, to be published (2008);
S. M. Kim, R. R. Milczarek, M. J. McCarthy, *Modern Physics Letters B*, 2008, 22, 941

[21] B. Blümich, J. Perlo, F. Casanova, *Progress in NMR*, 2008, 52, 197

[22] E. L. Hahn, D. E. Maxwell, *Phys. Rev.*, 1952, 88, 1070

[23] C. A. Meriles, D. Sakellariou, H. Haise, A. J. Moule, A. Pines, *Science*, 2001, 293, 82

[24] B. Shapira, L. Frydman, *J. Magn. Reson.*, 2006, 182, 12

[25] J. Perlo, F. Casanova, B. Blümich, *Science*, 2007, 315, 1110

[26] A. Nordon, A. Diez-Lazaro, C.-W. L. Wong, C. A. McGill, D. Littlejohn, M. Weerasinghe, D. A. Mamman, M. L. Hitchman, J. Wilkie, *Analyst*, 2008, 133, 339–347

[27] L. Andrade, E. Micklanfer, I. Farhat, R. Bro, S. B. Engelsen, *J. Magn. Reson.*, 2007, 189, 286

[28] M. D. Hürlimann, L. Burcaw, Y.-Q. Song, *J. Coll. Interf. Sci.*, 2006, 297, 303

[29] W. Windig, B. Antalek, *Chemom. Intell. Lab. Syst.*, 1997, 37, 241

[30] H.T. Pedersen, R. Bro, S.B. Engelsen, *J. Magn. Reson.*, 2002, 157, 141; http://www.models.kvl.dk/source

[31] M. Daszykowski, S. Serneels, K. Kaczmarek, P. van Espen, C. Croux, B. Walczak, *Chemom. Intell. Lab. Syst.*, 2007, 85, 269

[32] E. Kriesten, F. Alsmeyer, A. Bardow, W. Marquardt, *Chem. Intell. Lab. Syst.*, 2008, 91, 181

[33] J. Jaumot, R. Gargallo, A. de Juan, R. Tauler, *Chem. Intell. Lab. Syst.*, 2005, 76, 101

INVESTIGATION OF SODIUM IONS IN CHEESES BY ^{23}Na NMR SPECTROSCOPY

M. Gobet[1], L. Foucat[2], C. Moreau[1]

[1] INRA-ENESAD-University of Burgundy, UMR1129 FLAVIC, F-21000 Dijon, France
[2] INRA, UR370 QuaPA F-63122 Saint Genès Champanelle, France

1 INTRODUCTION

The reduction in the salt content of foods without affecting their production methods and sensorial properties is a current challenge for the food industry.[1] In foods, sodium ions can adopt various motional states through interactions with macromolecules such as hydrocolloids or exchange between two phases in heterogeneous systems. A correlation between sodium ion mobility and the saltiness perception has been suggested, the role of sodium ions with restricted motion or "bound" sodium being of particular consideration.[2] There is therefore a need to investigate the mobility states of the different pools of sodium ions and to quantify these fractions to develop a better understanding of their respective roles on salt perception. In this context, as it is one of the most salted foodstuffs and as it is consumed worldwide, cheese is a target of choice for investigations on reducing sodium content.

NMR spectroscopy of the sodium nucleus is a useful tool to investigate the structural and dynamics properties of sodium ions in a non-invasive way.[3,4] Several NMR approaches have been used to evidence sodium ions experiencing slow or restricted motion in biological or food samples. Dynamic states of sodium ions have been investigated through [23]Na T_1 and T_2 relaxation time measurements on ionic hydrocolloid systems.[2,5-8] In their studies, Rosett et al.[2,5] showed that the addition of cation-binding hydrocolloids has a reductive effect on the perception of saltiness of soups. Their results suggest a link between salt perception and more than one motional state of sodium, revealed by their [23]Na T_2 transverse relaxation time behaviour. The Double-Quantum Filtered (DQF) experiment allows the discrimination of those sodium ions in a "bound" state (or in an "ordered" state) amongst all the sodium ions present.[9,10] In a recent work, Mouaddab et al.[11] (2007) demonstrated the possibility to access the absolute quantification of bound sodium ions in a cationic exchange resin. The investigation of sodium ions in carrageenan gels has also been successfully applied to correlate the gelation mechanism with the binding states of sodium ions (Gobet et al., submitted).

Herein, we report preliminary results on the dynamics of total and 'bound' Na$^+$ ions using ^{23}Na NMR single-quantum (SQ) and double-quantum-filtered (DQF) experiments applied to 3 semi-hard cheeses. The impact of sodium motional state is discussed in regard to their various compositional parameters, such as salt or moisture content.

2 METHOD AND RESULT

2.1 Cheese samples

An unsalted semi-hard cheese with 60% w/w moisture in non-fat substance (MNFS), 2.8% calcium in non-fat dry matter (Ca/NFDM) and 40% fat in the dry matter (FDM) was made from pasteurised cow's milk. This unsalted cheese was divided into three equal portions. To obtain cheeses of varying salt (NaCl) content, each portion was soaked in a brine solution for a different duration. Cheeses of differing salt content were produced: 0.93 % w/w (Cheese1), 1.63% w/w (Cheese2) and 2.30% w/w (Cheese3). Mass compositions of the cheeses are presented in table 1. It should be noted that the NaCl content presented in this table was deduced from the quantification of chloride ions (Corning 926 chloride analyzer[12]).

Table 1. *Mass compositions of the semi-hard cheeses*

Composition (% w/w)	cheese1	cheese2	cheese3
Salt	0.93	1.63	2.30
Water	49.19	47.92	47.42
Dry Mater	50.81	52.08	52.58
Fat	19.94	19.94	19.95

The salting process induces a loss of water in the cheese, and consequently the water content of the three cheese samples was found to differ.

2.2. ^{23}Na NMR experiments on cheeses

A cylindrical section ($\varnothing$ = 9-10 mm and length = 5 cm) was taken from the core of the cheese and inserted in a 10 mm NMR tube. A coaxial 5 mm NMR tube containing a solution of $Na_7Dy(PPP)_2$ in D_2O ($[Na^+]$ = 0.40 M) was inserted as an external reference. ^{23}Na SQ and DQ NMR experiments were recorded at 132.29 MHz on a Bruker Avance-500 spectrometer equipped with a 10 mm broad-band probe. A $\pi/2$ pulse length of 12.5-14 µs was used for ^{23}Na NMR experiments. The ^{23}Na resonance peak of $Na_7Dy(PPP)_2$ appears at a low frequency ($\Delta\delta \approx$ 30 ppm) from the peak of the sample (shown in figure 1). The Na^+ content of the reference solution was checked by using an aqueous solution of NaCl (0.203 mol.L^{-1}). All NMR experiments were carried out at 25°C.

2.2.1. Total sodium from ^{23}Na SQ experiments. Figure 1 shows the typical SQ spectrum obtained for the three semi-hard cheeses.

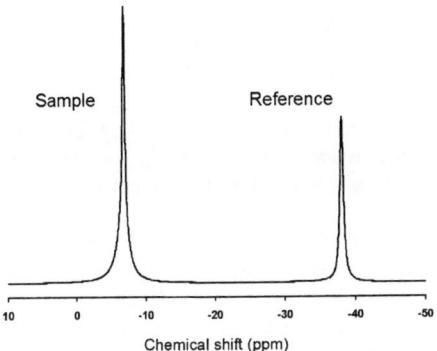

Figure 1 *^{23}Na SQ spectrum of semi-hard cheese n°2 (d1 = 300 ms, ns = 2 k)*

Validation of total sodium quantification. The ratio of the integrals of the sample and reference SQ signals were determined. From the known Na$^+$ concentration of the reference solution and the ratio between the analysed volumes (3.26) in both tubes, the Na$^+$ content (in mol.L^{-1}) of each cheese was calculated. Figure 2 shows the NaCl content (in g.L^{-1}) determined by NMR as a function the NaCl mass content (in g.kg^{-1}) obtained by the quantification of Cl$^-$ ions. The plot is closely approximated by the linear regression: [NaCl]$_{NMR}$ = 1.088.[NaCl]$_{Cl}$ - 0.105 (R^2 = 0.999). The slope (1.088) should be a good evaluation of the cheeses density. The non-null offset (-0.105) may be due to a Cl/Na molar ratio greater than 1 in unsalted cheese, since this ratio is close to 1.3 in cow's milk.[13] A difference in penetration of Na$^+$ and Cl$^-$ ions in the cheese during the brining process could also be considered. Nevertheless, these results validate this process of total sodium detection by ^{23}Na SQ NMR in these cheese samples.

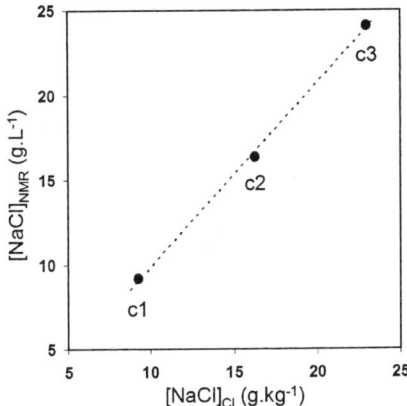

Figure 2. *NaCl concentration in cheeses determined by NMR ([NaCl]$_{NMR}$, in g.L^{-1}) as a function of the concentration determined by chloride quantification ([NaCl]$_{Cl}$, in g.kg^{-1}). The dashed line represents the linear regression [NaCl]$_{NMR}$ = 1.088.[NaCl]$_{Cl}$ - 0.105 (R^2 = 0.999).*

SQ line shape. For spin 3/2 nuclei, the interaction between the nuclear electric quadrupole moment and fluctuating electric field gradients can produce a bi-exponential relaxation. Outside the extreme-narrowing limit, the outer SQ transitions (-3/2 → 1/2 and 1/2 → 3/2) relax faster than the inner transition (-1/2 → 1/2). The resulting SQ signal consists of a superimposition of a narrow and a broad peak accounting for 40% and 60% respectively of the signal integral. In some heterogeneous systems, for example biological systems, the peak owing to outer components is broader than the detection range, thus leading to a loss of 60% of the SQ signal.[3,14]

As shown in figure 3, the deconvolution of the cheeses SQ signals using PeakFit software (v4.12, Seasolve) resulted in a 38:62 (±5) ratio between the integrals of the sharp ($\Delta v_{1/2} = 55 \pm 3$ Hz) and the broad ($\Delta v_{1/2} = 224 \pm 16$ Hz) components.

This feature shows clear evidence of the presence of at least two pools of sodium ions, with one such pool experiencing restricted motion in these samples.

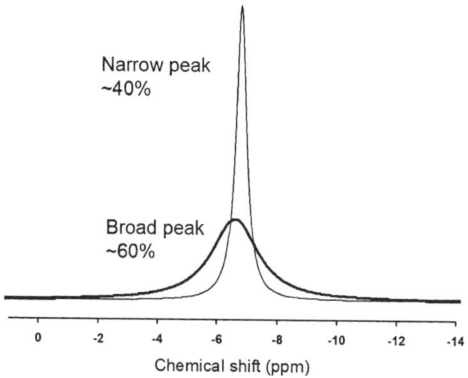

Figure 3. *deconvoluted SQ signal of cheese n°2. Note that the slight gap between the two lines chemical shifts is probably due to the magnetic field inhomogeneity.*

[23]Na SQ Relaxation times. [23]Na SQ transverse relaxation times were measured from Carr-Purcel-Meiboom-Gil (CPMG) experiments (128 echo times from 1 to 128 ms). The 'fast' (T_{2f}^{SQ}) and 'slow' (T_{2s}^{SQ}) components were determined by bi-exponential fitting of the decaying curve. The 'fast' T_{2f}^{SQ} relaxation characterises the central transition (-1/2→1/2) while the 'slow' T_{2s}^{SQ} relaxation characterises the outer transitions (-3/2→-1/2 and 1/2→3/2) of the sodium nucleus. The $T_{2f}^{SQ} : T_{2s}^{SQ}$ populations for the three cheeses (62:38 ± 2) calculated by this fitting are in accordance with the proportions of the two components found by the deconvolution of SQ signals.

[23]Na longitudinal relaxation times (T_1^{SQ}) of the cheese samples were determined by the mono-exponential fitting of the recovery curves obtained from inversion-recovery (IR) experiments (8 inter-pulse delays from 5 to 300 ms). The mono-exponential T_1^{SQ} relaxation behaviour could be attributed to total sodium ions evolving in an isotropic environment, as found in aqueous solutions. However, the two Lorentzian lines of the SQ spectra and the detection of DQ signals (see 2.2.3) for the three cheeses indicates that this

is a weight-averaged T_1^{SQ} relaxation of sodium ions in exchange between at least two environments.

Figure 4 shows the dependence of T_1^{SQ} and T_2^{SQ} relaxation times on water content. Total sodium ions were less mobile (T_1^{SQ}) in saltier cheeses (12.2 ms in cheese 1 *vs.* 11.2 ms in cheese 3). As expected, T_1^{SQ} of total sodium ions increases with increasing water content. The two components T_{2f}^{SQ} and T_{2s}^{SQ} display the same behaviour as T_1^{SQ} relaxation times with regard to water content.

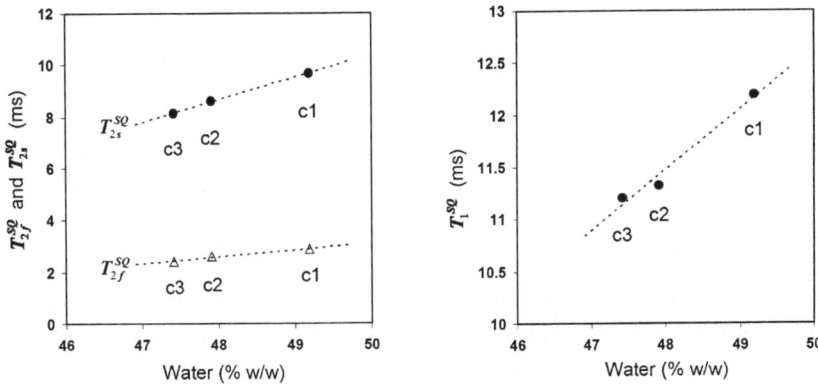

Figure 4. T_{2f}^{SQ}, T_{2s}^{SQ} (lef)and T_1^{SQ} (right) values as functions of water content (% w/w)

The addition of salt content during brining stage induces the expulsion of water molecules from the cheese. Thus, cheese with higher salt concentration exhibits a lower water content.

2.2.3 "Bound" sodium ions from ^{23}Na DQF experiments. By discarding the signal obtained from ions that exhibit isotropic motion, the DQF sequence permits the selective study of ions in a restricted motional state. DQ coherences were selected using the phase-cycled pulse sequence:

$$(\pi/2)_\varphi - \tau/2 - (\pi)_{\varphi+90} - \tau/2 - (\pi/2)_\varphi - \delta - (\pi/2)_0 - Acq(t)_{\varphi'}$$

where τ is the DQ creation time and δ the DQ evolution time. The basic four-step phase cycle of the sequence to eliminate SQ coherences is: $\varphi = 0°$, $90°$, $180°$, $270°$; $\varphi' = 2 (0°, 180°)$. ^{23}Na DQF NMR spectra were acquired with $\delta = 10$ μs for 32 τ values, ranging from 100 μs to 30 ms.

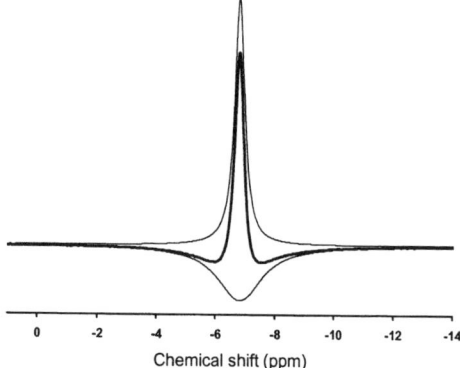

Figure 5. ^{23}Na *DQF spectrum of cheese 2. The thick line represents the experimental spectrum. The thin lines represent the two components obtained from deconvolution*

As shown in figure 5, the DQF sequence places 2 Lorentzian lines (a narrow and a broad line) in antiphase, with equal areas but different widths (56 ± 5 Hz and 258 ± 39 Hz). These two DQ components are thus characterized by two different relaxation times: the 'fast' (T_{2f}^{DQ}) and the 'slow' (T_{2s}^{DQ}) DQ relaxation times corresponding to the broad and the narrow lines respectively. T_{2f}^{DQ} and T_{2s}^{DQ} components were determined by curve fitting of the DQ signal intensity (I_{DQ}) as a function of τ according to equation (1) (Figure 6):

$$I_{DQ} = k * \left[\exp\left(\tau / T_{2s}^{DQ}\right) - \exp\left(\tau / T_{2f}^{DQ}\right) \right]$$

(1)

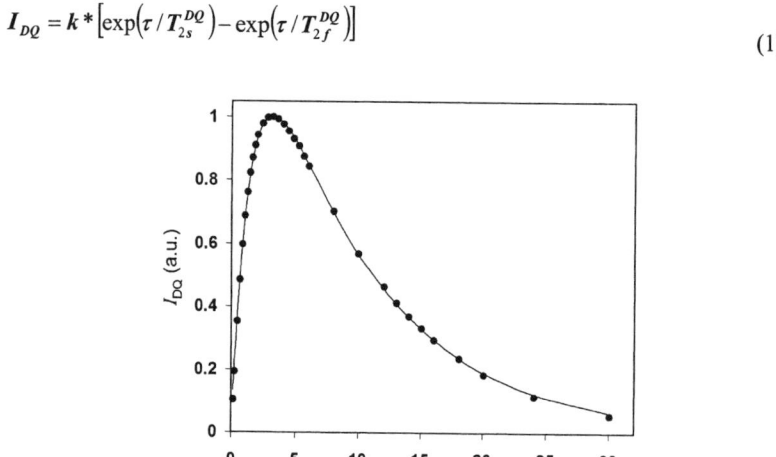

Figure 6. *Evolution of the DQ signal intensity (I_{DQ}) of cheese 2 as a function of the creation time τ (ms). The curve was fitted according to equation (1)*

The optimum duration of τ (τ^{opt}) that maximizes the amplitude of the DQ signal was calculated from the equation (2):

$$\tau^{opt} = \frac{\ln(R_{2f}/R_{2s})}{R_{2f} - R_{2s}} \qquad (2)$$

where $R_{2S} = 1/T_{2s}^{DQ}$ and $R_{2F} = 1/T_{2f}^{DQ}$

τ^{opt} values of 3.70, 3.19 and 2.91 ms were determined for the cheese 1, 2 and 3, respectively (Figure 7). τ^{opt} decreases as the water content decreases, indicative of more "ordered" sodium ions in cheese 3. Figure 7 shows that there is a substantial dependence of T_{2f}^{DQ} and T_{2s}^{DQ} relaxation times on water content of the samples. Such a relationship is non-trivial and warrants further investigations.

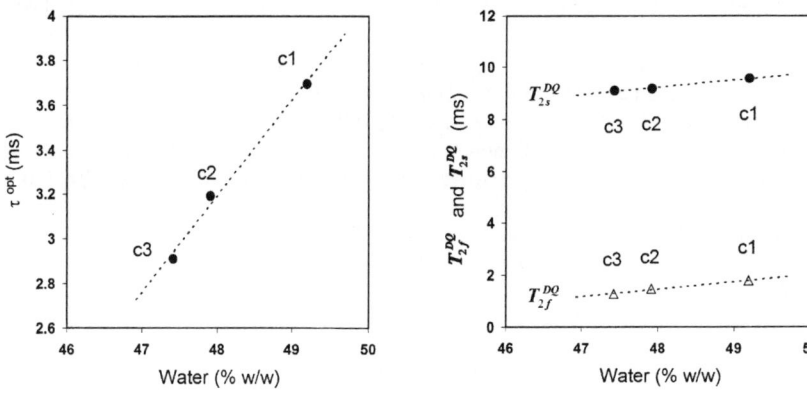

Figure 7. τ^{opt} *parameter(left) and* ^{23}*Na DQ* T_2 *relaxation parameters (right) with* T_{2f}^{DQ} T_{2s}^{DQ} *values as functions of water content*

3 CONCLUSION

^{23}Na NMR spectroscopy permits a direct and non-invasive quantification of sodium ions in cheeses. A linear correlation was found between total sodium concentration as determined by SQ signals and chloride ion concentrations. As expected, relaxation times from SQ experiments were found to correlate with water content. The DQ experiments highlighted the presence of 'bound' sodium ions, or sodium ions with restricted motion, in such food samples. The term 'bound' for sodium ions in these samples could be considered an overstatement, as it is difficult to discriminate between restricted motion induced by the viscosity of the medium, and those induced by sodium binding to proteins.

We now intend to apply this procedure to a larger range of cheeses, presenting more varied parameters (moisture, fat, calcium and salt content) that are assumed to have an impact on the dynamics of total and bound sodium ions. We will also finalise the calculations for absolute quantification of the bound fraction.

Acknowledgments

The authors are indebted to the Bruker BioSpin Company, the European Social Fund and the Regional Council of Burgundy for financial support (PhD grant to MG).

References

1. D. Kilcast and F. Angus, *Reducing salt in foods*, Woodhead Publishing Limited, Cambrigde, England (GBR), 2007.
2. T. R. Rosett, L. Shirley, S. Schmidt and B. Klein, *J. Food Sci.*, 1994, **59**, 206.
3. R. Kemp-Harper, S. P. Brown, C. E. Hughes, P. Styles and S. Wimperis, *Prog. Nucl. Magn. Reson. Spectrosc.*, 1997, **30**, 157.
4. D. E. Woessner, *Conc. Magn. Reson.*, 2001, **13**, 294.
5. T. R. Rosett, S. L. Kendregan, Y. Gao, S. J. Schmidt and B. P. Klein, *J. Food Sci.*, 1996, **61**, 1099.
6. S. J. Schmidt, N. Ayya and L. Shirley, *Lebens.-Wiss. u.-Technol.*, 1992, **25**, 540.
7. L. Piculell, S. Nilsson and P. Ström, *Carbohyd. Res.*, 1989, **188**, 121.
8. L. L. Shirley and S. J. Schmidt, *Food Hydrocol.*, 1993, **7**, 147.
9. L. Foucat, J. P. Donnat and J. P. Renou in *Magnetic Resonance in Food Science : Latest Developments*, ed. P. S. Belton, Royal Society of Chemistry, Cambridge, 2003, p 180.
10. G. Navon, H. Shinar, U. Eliav and Y. Seo, *NMR Biomed.*, 2001, **14**, 112.
11. M. Mouaddab, L. Foucat, J. P. Donnat, J. P. Renou and J. M. Bonny, *J. Magn. Reson.*, 2007, **189**, 151.
12. C. Bugaud, S. Buchin, Y. Noël, L. Tessier, S. Pochet, B. Martin and J. F. Chamba, *Le Lait*, 2001, **81**, 593.
13. F. Gaucheron, *Reprod., Nutr., Dev.*, 2005, **45**, 473.
14. M. Shporer and M. M. Civan, *Biophys. J.*, 1972, **12**, 114.

APPLICATIONS OF FIELD CYCLING RELAXOMETRY TO FOOD CHARACTERIZATION

S. Baroni[1], S. Bubici [1], G. Ferrante [2], and S. Aime[3]

[1] Invento S.r.l., spin off of University of Torino, Via Nizza 52, 10126, Torino, Italy
[2] Stelar S.r.l., Via Fermi 4, 27035 Mede (PV), Italy
[3] Dipartimento di Chimica I.F.M., Università di Torino, Via Pietro Giuria 7, 10125 Torino, Italy

1 INTRODUCTION

The last decades have witnessed a growing interest in the search of new tools for food characterization. Any progress in analytical instrumentation is exploited to obtain more advanced description of foodstuff. Often the diffusion of a new methodology is hampered only by the complex sample preparation and the associated costs. Therefore methods that do not require any (or minimal) sample treatment are highly desired. Analysis of food products should require the use of non-invasive techniques because sensorial and safety properties are related to the structural and compositional complexity and heterogeneity of the matrix, which must be preserved as much as possible in its original state.

Along the years, High Resolution NMR spectroscopy has gained an important position among the spectroscopic methods for food characterization thanks to its unique ability to identify (and quantify) all the major low-medium molecular weight components by a single spectrum acquisition without the need for any separative or destructive procedure.

Besides the High Resolution approach, NMR has provided important contributions to the field by the use of its Low Resolution (or Wide Line) version. Low resolution methods lack of the chemical shift information and the acquired ^{1}H signal consists of a single absorption containing the information arising from all protons present in the specimen. This approach has been shown to be particularly useful to assess the occurrence of different phases (solid/liquid, water/fat, etc.)[1,2]. However the relaxation characteristics of the single absorption bring about a number of information on the solute molecules that are highly relevant in the characterization of a given (liquid) specimen. Moreover the possibility of recording the proton relaxation rate over an extended range of magnetic fields further improves the potential of this approach as it allows identifying the occurrence of different contributions assignable to various components of the foodstuff.

This chapter aims at showing the potential of the Field Cycling Relaxation approach in the food analysis.

2 FAST FIELD CYCLING NMR

2.1   The FFC Experiment

The mobility of proton containing molecules in foods can be investigated by the acquisition of Nuclear Magnetic Resonance Dispersion (NMRD) profiles that report about the changes in the ^{1}H-spin-lattice or longitudinal relaxation rate ($R_1 = 1/T_1$) as function of the applied magnetic field strength.

A dedicated NMR spectrometer (Stelar Spinmaster-FFC relaxometer, Stelar S.n.c., Mede (PV), Italy) allows us to acquire relaxation data over an extended range of proton Larmor Frequencies (from 0.01 to 20 MHz). The benefit of exploring the range of low Larmor frequencies is to detect typical relaxation features associated with molecular process characterized by very long correlation time, such as molecular surface dynamics and collective effects. Additional relaxation data can be obtained on commercially available instruments operating at higher field strength.

In the basic NMR Field Cycling experiment the magnetic field B_0, which is applied to the sample, is cycling through three different values (figure 1). First the sample is polarized in a high magnetic field B_p (polarization field) for a time t_p until the nuclear magnetization achieves saturation, i.e. t_p should be longer than 4 times T_1 at B_p. Second, the magnetic field is switched to a value B_r (relaxation field, fixable to any desired value) for a time τ during which the magnetization reaches a new equilibrium value (Bloch equation). Third, the magnetic field is switched to the detection field B_d and the equilibrium magnetization is measured with a 90° pulse followed by acquisition. For every given B_r the signal is measured as a function of τ, the time dependence (T_1) is due exclusively to the relaxation at the B_r field. This sequence, named pre-polarized (PP), is useful at low relaxation field strengths; otherwise, a non-polarized (NP) sequence is sufficient at higher relaxation field strengths, where the signal intensity is higher.

2.2   Theoretical Aspects

Several reviews and books have appeared in the literature on the relaxation mechanism and theory[3-7]. The focus of the current paragraph is to recall some basic concepts useful in the analysis of NMRD profiles.

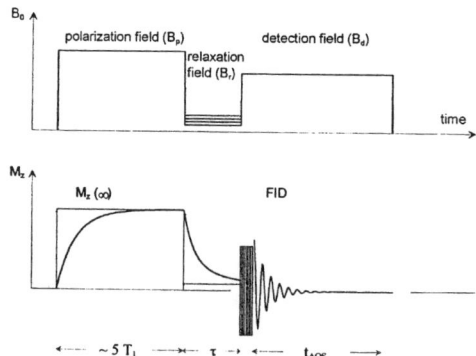

Figure 1   *The basic NMR Field Cycling experiment*

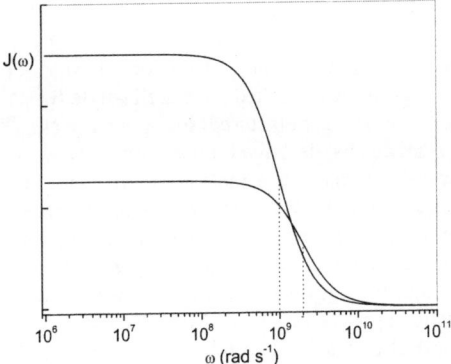

$J(\omega)$

ω (rad s^{-1})

Figure 2 *The spectral density function J(ω) as a function of frequency. The J(ω) profiles for τ$_C$ =1·10^{-9} s and 5·10^{-10} s are reported. Note that the area under the curves is the same.*

The relaxation parameters are of great interest as they report on the various magnetic interactions through which proton spins are involved, the most common being the classical dipolar interaction (between the magnetic dipoles associated with nuclear spins). In order to be efficient for relaxation, these interactions must be time dependent, *via* molecular motions. Therefore relaxation parameters not only encompass structural information (through the interactions themselves), but also dynamical information (through this time modulation). This latter information can than report about (*i*) overall motions, translational or rotational, governed essentially by the molecular volume and the viscosity of the medium in which they are embedded the molecules under investigation, (*ii*) local motions and (*iii*) chemical exchange.

The fundamental requirement for longitudinal relaxation of a proton nucleus is a time dependent magnetic field fluctuating at the Larmor frequency of the nuclear spin. In the fast motion limit, the frequency distribution of the fluctuating magnetic fields associated with the molecular motion, i.e. the spectral density function J(ω), has, in the simplest case, a Lorentzian shape, as described by Eq. (1), and is characterized by the correlation time τ$_C$:

$$J(\omega) \propto \frac{\tau_C}{1+\omega^2\tau_C^2} \qquad (1)$$

where $\omega=\gamma B_0$ is the resonance angular frequency depending on the gyromagnetic ratio γ of the nuclei and the external magnetic field B_0. The functional form of Eq. (1) provides a field dependence of the NMRD profile called *dispersion*; its inflection point (i.e., at half the maximum of the dispersion curve) occurs at ωτ$_C$ =1 (figure 2).

In general, τ$_C$$^{-1}$ is the sum of the inverse correlation times of each process involved in the relaxation (proton exchange, rotation, electron spin relaxation, and diffusion) and is dominated by the fastest process.

NMRD profiles are particularly valuable to assess the interactions of water molecules with paramagnetic and large-size macromolecular systems. The large and fluctuating local magnetic field in the proximity of a paramagnetic center provides additional relaxation pathways for solvent nuclei.

2.3   Applications

Godfroy *et al.* [8] studied the dynamics and interaction of water bound to milk proteins (caseins) as a function of aging. A dry salted mozzarella-style (pasta filata or hot-streched) and a gouda-style cheese were manufactured for this project. The proton spin-lattice relaxation rate of the samples is described by a Power Low function in the Larmor Frequency. All the curves terminate in a plateau at low field and its relaxation rate is directly related to the degree of hydration of the protein that increases with ripening.

Laghi *et al.*[9] carried out a quantitative analysis of the proton relaxation rate of raw hen eggs. The albumen water content decreases during storage (partly by loss of water by diffusion of vapour through the eggshell and partly by water transport from the albumen into the yolk) and liquefies upon aging. The NMRD data were analyzed according to the model free approach by Halle *et al.*[10]. The relaxation profiles of fresh albumen samples display lower R_1 values than aged ones.

The field of studying alcoholic and soft drinks is largely unexplored. Figure 3 reports the profiles of three wines to show the differences one may expect. The major determinants of the observed relaxation rates are associated to the presence of paramagnetic ions and large macromolecules and sugars that affect their viscosity that, in turn, determines the molecular reorientational time of water molecules. Moreover on the some wine one may follow, through the changes of the NMRD pattern, oxidation processes that take place over time. In the following we will discuss in more details the results obtained from the NMRD relaxometric study of vinegars.

2.3.1   A case study: Balsamic Vinegars. ^{1}H-NMR-Field Cycling Relaxometry has been applied to characterize Traditional Balsamic Vinegars (TBVM) and Balsamic Vinegars of Modena (BVM) of different aging process. Balsamic Vinegar is essentially made of wine vinegar added with caramel with possible aging in casks for less or more than three years (on this basis they are labeled with red or white stamp, respectively).

TBVM is a completely different product, made of cooked must and aged in series of decreasing volume casks made of different wood for at least 12 years and up to 25 years or longer. TBVM is a high quality product, whose production is ruled by law (Italian D.M., February 9, 1987) and its cost on the market is rather high dependent upon its ageing process.

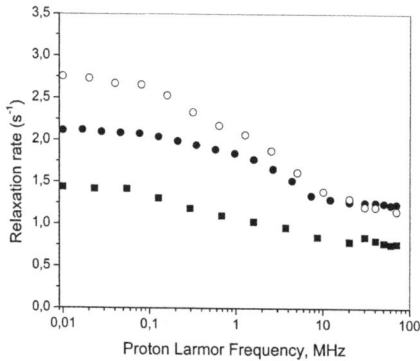

Figure 3   *NMRD profiles of wines: Tokaji wine (Hungary)(filled circle), white table-wine (filled square) and white Port wine (open circle).*

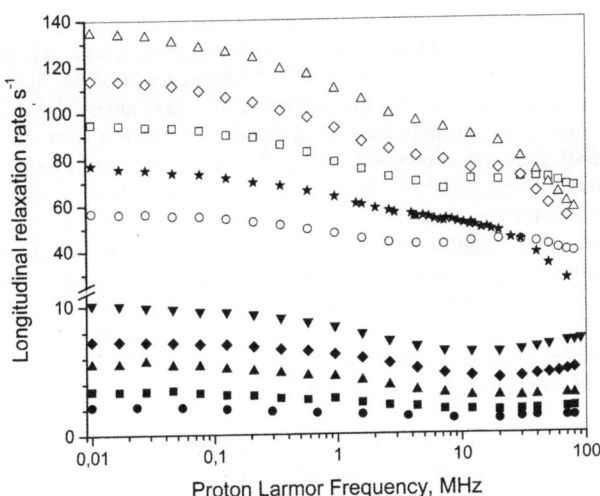

Figure 4 *NMRD profile of BVM, TBVM and suspected counterfeit TBVM samples. BVM samples: 1 (filled circle) and 2 (filled square) with an ageing process shorter than 3 years; 3 (filled upward triangle) and 4 (filled diamond) with an ageing process longer than 3 years; 5 (filled downward triangle) with an ageing process longer than 8 years. TBVM samples: 6 (open circle) with an ageing process longer than 12 years; 7 (open square) with an ageing process of 20 years; 8 (open upward triangle) with an ageing process longer than 25 years and sample 9 (open diamond) with an ageing process of 34 years. Suspected counterfeit TBVM sample: 10 (filled star).*

In this respect frauds may occur by altering young vinegars that will be commercialized as old or extra old products upon modification of their physico-chemical properties, like density, color, etc.

In this work, 8 BVM and 7 TBVM samples were analyzed; 4 suspected counterfeit TBVM samples (labelled as old) were also investigated. They have undergone an in-depth characterization by NMR spectroscopy[11-13]. Some selected NMRD profiles (25°C) are reported in Figure 4.

In general, the observed NMRD profiles of the BVM samples resemble those ones obtained from aqueous solutions of paramagnetic metal complexes that show dispersion in the 1-10 MHz range. The dissolved paramagnetic metal ions are represented mainly by Fe(III), Mn(II) and Cu(II)[14]. BVM is rich in species able to coordinate paramagnetic metal ions, in particular organic acids and sugars, thus yielding to an increase of the relaxation enanchement of water protons. Upon increasing the age of BVM specimen, there is a progressive increase of the observed relaxation rates and, in the high field region (>20 MHz), the relaxation hump become more pronounced because of the elongation of the molecular reorientational time (τ_R) of water protons upon interacting with a paramagnetic metal ions. Such elongation may result from either the interaction with macromolecular substrates or the overall increase of the viscosity of the solution.

The NMRD profiles for TBVM samples showed much higher longitudinal relaxation rate values with respect to the BVM samples over the whole proton Larmor frequency range explored. Furthermore, in respect to BVM samples, there is a shift to lower frequency of the high field relaxation hump. The higher longitudinal relaxation rates reflect the differences in the production process of TBVM respect to BVM. In fact, must cooking gives rise to a concentrated solution of many constituents (sugars, organic acids, nitrogen compounds, metal ions and polyphenols), together with the formation of neo-compounds, such as melanoidines and condensed tannins[15]. Then, during the ageing process, besides the slow degradation processes, TBVM undergoes a further slow concentration process by water evaporation through the staves that contributes to the matrix modification. About the hump position, the elongation of τ_R may be so dramatic to cause a larger low-frequency shift of the hump that now appears as a shoulder of the main dispersion (see sample 8).

Figure 4 reports the NMRD profile for one of the four suspected counterfeit TBVM samples analyzed (sample 10). In particular the counterfeit TBVM sample profiles showed the same order of relaxation rates (ca. 25-80 s^{-1}) as the TBVM (aged for 12-20 years) samples. Their behavior is characterized by a steep decrease at high magnetic field strengths.

In order to get more insight into the relationship between ageing and paramagnetic relaxation we went to acquire the transverse relaxation rates of both BVM and TBVM samples (data not shown). In general, an increase in the relaxation rates of both vinegars with ageing is observed. The comparison between longitudinal and transverse relaxation times measured at the same magnetic field (70 MHz in this case, selected for the best discrimination between samples) revealed a correlation with ageing process. Moreover, three of the four suspected counterfeit samples are well separated from the main group of the TBVM samples.

In conclusion, the present studies clearly showed that the use of Field Cycling NMR relaxometry may be a powerful tool to characterize BVM and TBVM samples. The ageing process experienced by the samples, in particular by TBVM samples, is mostly characterized by water loss and progressively concentration that affected the longitudinal and transverse proton relaxation times of this product. A relationship may be extracted that relates the age of the sample and the observed T_1 and T_2. In general, very useful insights can be gained from the frequency dependent hump that occurs in the high field region of the NMRD profile. We assign this hump to the occurrence of slowly moving paramagnetic adducts because of both the higher viscosity with ageing and the interaction with macromolecular substrates. Clearly this hump may also be generated by adding macromolecular species, most likely sugar syrup, Arabic gum or caramel, to young vinegar but the overall shape of the resulting profile does not fit with those of the genuine samples.

3 DATA ANALYSIS

Very often (and this is particular true for foodstuff) one has to deal with heterogeneous systems consisting of multiple compartments. Each micro-environment has characteristic T_1 values for their ^{1}H-decaying signals. This results in a T_1 multi-components distribution. Hence to relate the major relaxometric characteristics of a given environment to its physic-chemical properties, a clear understanding of a sample through NMR relaxometry it is crucial to pursue a reliable deconvolution of all the T_1 components. The NMR decay associated with a continuous distribution of relaxation time is often described in literature in terms of a Laplace transform:

$$s(t) = \int_0^\infty P(\alpha)e^{-\alpha t}d\alpha \qquad (2)$$

Given the NMR s(t) data, the aim in solving the Eq. (2) is the determination of the distribution function P(α). Several algorithms and software packages can be found in literature for the numerical inversion of Laplace equation. One of the most successful regularization process applied to NMR relaxation data is UPEN[16,17].

We have applied UPEN to the analysis of NMRD data obtained from several foodstuff and found significant advantages in respect to the crude simulation of the experimental results.

First UPEN software was used to analyze the relaxation data of synthetic samples to assess its potential on data acquired by FFC spectrometer. Figure 5 shows the results obtained from two GdCl$_3$ solutions, with T$_1$ difference of 39 and 330 ms, respectively. The samples were put in concentric NMR tubes (5 mm and 10 mm diameter tubes, respectively). The UPEN analysis showed to be able to reproduce both T$_1$ values and volumes of the two samples.

Then we applied the same methodology to other data regarding different foodstuff. As an example, we report here the analysis of "grana padano" cheese (core and under-crust parts) (Figure 6).

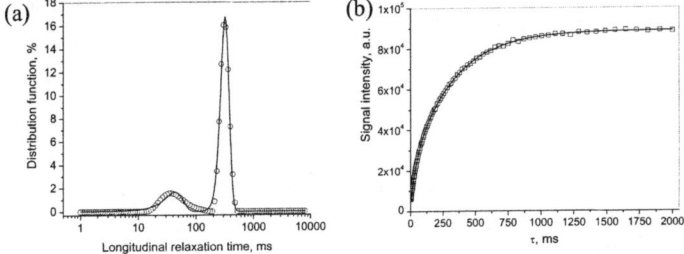

Figure 5   *Gadolinium solutions. (a) The distributions of relaxation times T_1 obtained by the inversion of the relaxation data using the UPEN algorithm. The fitted line represents the deconvolution of the distribution function, according with a Gaussian model. (b) The signal intensity recovery as a function of time. The fitted line is obtained by a bi-exponential function.*

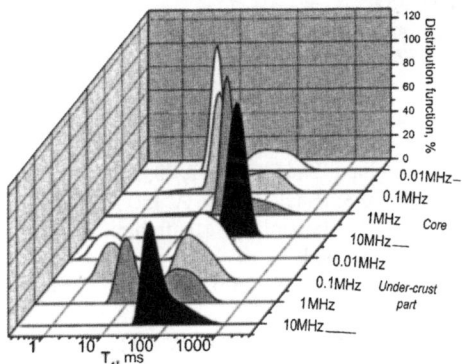

Figure 6   *"Grana Padano" cheese. The distributions of relaxation times T_1 (obtained by the inversion of the relaxation data using the UPEN algorithm) acquired at 0.01, 0.1, 1 and 10 MHz for the core and under-crust parts, respectively.*

Magnetic Resonance in Food Science

4 CONCLUSION

FC Relaxometry provides novel and reliable information to the characterization of foodstuff and it has the potential to enter the armoury of methodologies currently applied in food analysis. The analysis of NMRD profiles allows the acquisition of unique information about the interaction between water molecules and macromolecules and paramagnetic systems. The NMRD profiles may represent excellent fingerprints for genuine samples and may contribute to identify frauds and counterfeit products. Last but not least the sample does not require any preparation/transformation for the acquisition of the NMRD profile and the volume necessary for this analysis is just less than 1 ml. The acquisition of a NMRD profile takes from few tens of minutes to few hours according to the number of points and the relaxation times of the given specimen. The reproducibility can be quite high (±0.5%).

Acknowledgements

We gratefully thank Dr. Roberto Consonni (CNR, Milano, Italy) for the supply of vinegars specimens and for helpful discussions.

References

1   H. Todt, G. Guthausen, W. Burk, D. Schmalbein and A. Kamlowski, *Food Chem.*, 2006, **96**, 436.
2   M.L. Johns and K.G. Hollingsworth, *Progress NMR Spectroscopy*, 2007, **50**, 51.
3   R. van Eldik, I. Bertini (Eds.), *Relaxometry of Water-Metal Ion Interactions*, Advances in Inorganic Chemistry, vol. 57, Elsevier, San Diego, 2005
4   L. Banci, I. Bertini, C. Luchinat, *Electron and nuclear relaxation*, VCH, Weinheim, 1991.
5   L. Helm, *Progress NMR Spectroscopy*, 2006, 49, 45.
6   R.B. Lauffer, *Chem. Rev.* 1987, **87**, 901.
7   J.P. Korb and R.G. Bryant, *J. Chem Phys.*, 2001, **115**, 10964.
8   S. Godefroy, J.P.Korb, L.K. Creamer, P.J. Watkinson, and P.T. Callaghan, *J. Colloid Interf. Sci.*, 2003, **267**, 337.
9   L. Laghi, M.A. Cremonini, G. Placucci, S. Sykora, K. Wright, and B. Hills, *Magn. Reson. Imaging*, 2005, **23**, 501.
10   B. Halle, H. Jóhannesson, and K. Venu, *J. Magn. Reson.*, 1998, **135**, 1.
11   R. Consonni, L.R. Cagliani, F. Benevelli, M. Spraul, E. Humpfer, and M.Stocchero, *Anal. Chim. Acta,* 2008, **611**, 31.
12   R. Consonni and L.R. Cagliani, *Talanta,* 2007, **73**, 332.
13   R. Consonni, L.R. Cagliani, S. Rinaldini, and A. Incerti, *Talanta* 2008, **75**, 765.
14   M. Cocchi, G. Franchini, D: Manzini, M. Manfredini, A. Marchetti, and A. Ulrici, *J. Agric. Food Chem,.* 2004, **52**, 4047.
15   A. Piva, C. Di Mattia, L. Neri, G. Dimitri, M. Chiarini, G. Sacchetti, *Food Chem.*, 2008, **106**, 1057.
16   G.C. Borgia, R.J.S. Brown, and P. Fantazzini, *J. Magn. Reson.*, 1998, **132**, 65.
17   G.C. Borgia, R.J.S. Brown, and P. Fantazzini, *J. Magn. Reson.*, 2000, **147**, 273.

A LOW RESOLUTION ^{1}H NMR STUDY TO INVESTIGATE THE PROTECTIVE MECHANISM OF SORBITOL DURING VACUUM DRYING OF A PROBIOTIC MICRO-ORGANISM

P. Foerst and U. Kulozik

Chair of Food Process Engineering and Dairy Technology, Technische Universität München, Weihenstephaner Berg 1, D-85354 Freising, Germany

1   INTRODUCTION

Lactic acid bacteria with defined functionalities play a major role as starter cultures for the production of fermented foods such as yoghurt, cheese, bread or fermented drinks. Additionally, specific strains of lactic acid bacteria are also widely used as health promoting additives. These so-called probiotics are added to the food in high cell numbers to enhance their functional properties or are marketed in isolated form as special dietary supplements. Probiotics are frequently distributed as dry powder.

Starter cultures and probiotics are produced and distributed world wide by specialised companies. Therefore, stabilization of the bacteria is essential. Drying is a common stabilization method as it leads to an increased storage stability and facilitated transport compared to freezing. However, drying leads to an inactivation of the cells. The dehydration damage is especially pronounced at low water content where structural water is removed from the biomolecules such as proteins and hydrophilic surfaces of biomembranes.

The survival of bacteria after drying can be improved by employing appropriate drying technology and optimal process conditions. Low temperature vacuum drying has been shown as potential alternative to freeze drying.[1] Another means to improve survival after drying is the addition of protective agents such as disaccharides and polyols to the drying medium.[2-5] Several mechanisms have been suggested for the stabilizing action of sugars during drying[6]. For air drying and vacuum drying, especially sorbitol has shown to be an effective protectant.[2;3;7]

One mechanism for the protective mechanism of sugars that has been experimentally proven for model biosystems such as liposomes or isolated proteins is the water replacement hypothesis. According to this hypothesis the hydroxyl groups of the sugar specifically interact with the phospholipid headgroups at the surface of the phospholipid bilayer[8] and replace the water towards the end of drying. The sugar effects that the molecular packing density of the acyl chains does not change with the removal of water and therefore prevents a drying-induced increase in gel to liquid-crystalline phase transition temperature T_m which normally becomes visible at water contents lower than 0.2 g H_2O g^{-1} dry weight [9]. It is suggested that a phase transition would involve a leakage of the cell membrane [10;11] leading to an inactivation of the cell.

For dried Artemia cysts as a model system for desiccation tolerant species it has been found out by pulsed NMR measurements that the mobility of water protons unexpectedly increased below a certain residual water content [12]. The increased mobility was explained by the existence of intrinsic protective substances such as trehalose and glycerol in the cyst that replace water at the surface of biomolecules and therefore increase molecular mobility.

The aim of the study was to use NMR to investigate the protective mechanism of sorbitol during vacuum drying of a probiotic test microorganism (*L. paracasei* ssp. *paracasei*). Therefore, ^1H mobility and fractions of different mobility were determined for both the samples dried with and without sorbitol. Furthermore, the influence of process conditions on both survival and relaxation parameters was studied as it is known that the protective effect of sorbitol depends on process conditions.

2   MATERIAL AND METHODS

2.1 Microorganism and fermentation conditions

Commercial *Lactobacillus paracasei* ssp. *paracasei* (F19) from Medipharm AB (Kaageröd, Sweden), known for its probiotic properties, was used as a model microorganism. The cells were provided as a frozen concentrate with a titre of $\sim10^{11}$ cfu/g. Concentrated cells were stored at -40 °C. For the experiments the cells were cultured at 37 °C in a one litre batch fermentation process in MRS broth (Merck, Darmstadt, Germany). The fermentation process was carried out without neutralising, the initial pH was 5.8. The inoculum was prepared by suspending concentrated frozen cells in MRS broth. An initial cell concentration of $\sim10^8$ cfu/ml was adjusted by adding an appropriate amount of inoculum to the fermentation medium. The amount of inoculum was determined by measuring the optical density of the bacterial suspension photometrically (Lambda 2, Perkin Elmer, Wellesley, USA) at a wavelength of $\lambda = 600$ nm with sterilised MRS broth as reference. The corresponding initial optical density of the fermentation broth was 0.3. The fermentation was stopped after 10h at the end of the exponential growth phase by cooling the cell suspension to 4 °C. The final microorganism concentration in the fermentation broth was 10^{10} cfu/ml.

2.2 Sample preparation and protective agents

After fermentation, the cells were separated from the fermentation broth by centrifugation (4000 g, 10 min, 4 °C) and concentrated to $\sim10^{11}$ cfu/ml. After harvesting, the cells were washed twice with phosphate buffer saline (0.1 M K_2HPO_4/KH_2PO_4 with 0.15 M NaCl, pH 7) in order to remove residual fermentation broth. As protectant, D(-)Sorbitol (Merck, Darmstadt, Germany) was added to the cell suspension. The concentration was 25 % related to dry biomass.

2.3 Vacuum Drying

Vacuum drying was carried out in a batch vacuum dryer (VO 400, Memmert, Schwabach, Germany) for condition 1 and in a vacuum freeze dryer for condition 2 and 3 (Gamma 1-20, Christ, Germany) that was operated above the triple point (without freezing of sample). The conditions applied for the vacuum drying are listed in Table 1. The conditions were chosen such that the product temperature during secondary drying covers a range between 5 and 40 °C. Drying was stopped at different drying times in order to achieve different water contents.

Table 1: Process Conditions applied for the vacuum drying experiments

Condition	1	2	3
Pressure /Pa	1500	2500	2500 (for 3h) then 1000
Shelf Temperature/ °C	15	30	25 (for 3h) then 40

After drying, one glass vial was weighed on a precision balance, transferred to a drying chamber and further dried for 48 h at 105 °C in order to determine the residual water content after drying. The residual moisture content X was calculated on wet weight basis as following

$$X = \frac{m_1 - m_2}{m_1 - m_e} \tag{1}$$

Here, m_1 is the weight after vacuum drying, m_2 the weight after drying for 48h at 105 ° and m_e the weight of the empty (pre-dried) vial. All data have been determined from three independent experiments.

2.4 Determination of survival rate

Colony forming units (cfu) of the cell suspensions were determined before and after vacuum drying. For the analyses of the dried samples the content of one vial was rehydrated with sterile double-distilled water to its original weight before drying at 25 °C. Dilutions of the samples were prepared with Ringer's solution (1/4 strength), plated on MRS agar (Merck, Darmstadt, Germany) and incubated at 37 °C for 48 h under anaerobic atmosphere. The survival rate after vacuum drying was determined as the colony forming units of the rehydrated sample after drying, relative to the initial viable colony forming units before drying. All survival rate data are means from 3 independent experiments.

2.5 Low Resolution ^{1}H NMR measurements

Low resolution ^{1}H NMR analyses were carried out with the dry cell powder in order to understand the role of the protectant with regard to the proton mobility during the drying process. To assess this effect, 12 ml of cell suspension with/without added protectant (25 % sorbitol, referred to dry biomass) were dried for different times in the vacuum dryer at the conditions described above. The dried cell suspension was ground to a powder and ca. 0.5 to 0.7 g of each sample were sealed in NMR tubes with a diameter of 10 mm. For the measurements, a low resolution NMR spectrometer system, Minispec mq20 (Bruker Analytik GmbH, Rheinstetten, Germany) was used. The radio-frequency (rf) of the used NMR spectrometer is 20 MHz. All measurements were carried out at a temperature of 20 °C. A number of 4 scans were accumulated.

The spin-spin relaxation parameters were obtained from a combination of FID (Free Induction Decay) and CPMG (Carr-Purcell-Meiboom-Gill) Sequence. In total, 5000 pulses were applied with a pulse spacing of $\tau = 0.05$ ms. The dead time between the rf pulse excitation and the data acquisition was determined to be 9 μs.

If the protons exist in different binding states and if the exchange of protons between different states is slow (compared to the NMR timescale) than it is possible to distinguish fractions of different proton mobilities and relaxation times. A non-linear fitting routine was applied to the resulting decay curve U(t) which was best described using a sum of a Gaussian and an exponential component: The Gaussian component refers to a 'solid' component with decay times of ca. 10 μs and the exponential component refers to 'liquid' protons with T2 times in the order of magnitude of 100 μs. The proton fractions f_i and the relaxation times $T_{2,i}$ are calculated according to equation (1):

$$U(t) = a + f_S\, e^{-\left(\frac{t}{2T_{2,1}}\right)^2} + f_L\, e^{-\left(\frac{t}{T_{2,2}}\right)}$$

with

$$f_S + f_L = 1$$

(2)

In eq. (2), U(t) is the measured decay signal, f_S refers to the proton fraction of the solid component, f_L to the proton fraction of the liquid component and $T_{2,1}$ and $T_{2,2}$ are the corresponding relaxation times. The term 'solid' refers to molecules of limited mobility such as the biomass C-H protons while the liquid component refers to the more mobile protons from water and mobile sugar molecules.

3   RESULTS AND DISCUSSION

3.1 Influence of process conditions on survival

In order to study the impact of the protectant at different process conditions, drying processes were carried with and without sorbitol for different process conditions. The results for the survival are shown in Fig. 1. Fig. 1 A shows the survival for the lowest drying temperature (15 °C shelf temperature corresponding to 6 °C product temperature during secondary drying). As can be seen, the survival rate remains on a high level even at low residual water contents and no difference in survival between the samples with and without sorbitol seem to be detectable. It should be noted that the survival rates for the sample with sorbitol are slightly higher. However, the residual water content for these samples is also higher. It is assumed that the higher survival is due to the higher water content as the water content is a very critical parameter in region.[1] Fig. 2 A shows the survival for condition 2 that corresponds to a shelf temperature of 30 °C (product temperature 17 °C during secondary drying). Here, a small difference in survival between the different preparations is detectable, especially for low residual water contents. The largest difference between the samples is detectable for condition 3 where the shelf temperature is highest (40 °C during secondary drying). The difference only occurs for low water contents (< 20 %) and becomes larger with decreasing water content. Without protectant, a sharp decrease in survival is observed.

The reason for the sharp decrease lies in the removal of structural water at this water content[1] that is essential for the functionality of the biological structure[13].

The protective effect is dependent on the process conditions. Especially the product temperature during secondary drying seems to be a critical parameter for survival. As the protective effect only occurs for low residual water contents when structural water is already removed, the water replacement mechanism may play a role in protecting cells from desiccation damage. In order to further investigate the mechanism of protection, pulsed NMR measurements were carried out to measure proton mobility of samples with and without added protectant.

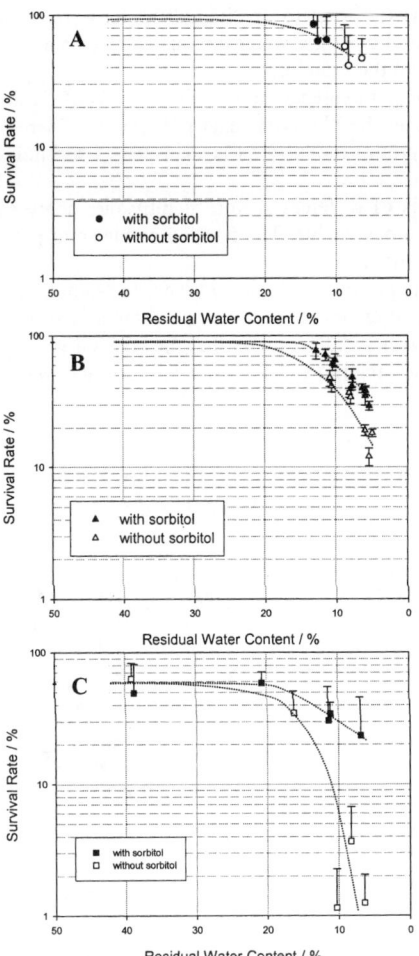

Figure 1: Survival Rate of *L. paracasei* ssp. *paracasei* after vacuum drying with and without sorbitol for different process conditions. A: Cond. 1 (15 °C); B: Cond. 2 (30 °C); C: Cond. 3 (40 °C)

3.2 Determination of T2 relaxation times and fractions of different proton mobility in dried cell preparations

The T2 decay curves revealed two-component behaviour of the dried samples with distinct relaxation times of the single components for both the samples dried with and without sorbitol. One component relates to a solid proton fraction with a very low T2 relaxation time in the range of 10 µs that is independent on residual water content (data not shown). The second component relates to a more mobile proton fraction with a T2 relaxation time in the order of 100 µs that is dependent on water content and becomes lower with decreasing water content. It is assumed that the solid protons mainly relate to the protons of the biomass and the more mobile protons (referred to as "liquid" protons) relate to the water and soluble sorbitol protons. The mobility of the water protons is decreasing as drying proceeds and the free water is removed from the sample.

No clear difference between the preparations with and without sorbitol as well as no influence of process conditions could be observed (data not shown). Therefore it seems that sorbitol does not influence the proton mobility as is expected if sorbitol would replace the water protons.

However, if the fractions of different proton mobility are analysed according to eq. 2, an influence of sorbitol becomes visible. The fraction of the solid protons is plotted against the residual water content in Fig. 2.

Fig. 2 shows that the solid proton fraction is lower for the samples dried with sorbitol with the difference becoming smaller for lower water contents. No clear influence of process conditions is observed. This result suggests that the sorbitol protons partly contribute to the liquid protons and seem to undergo a phase transition for lower water contents.

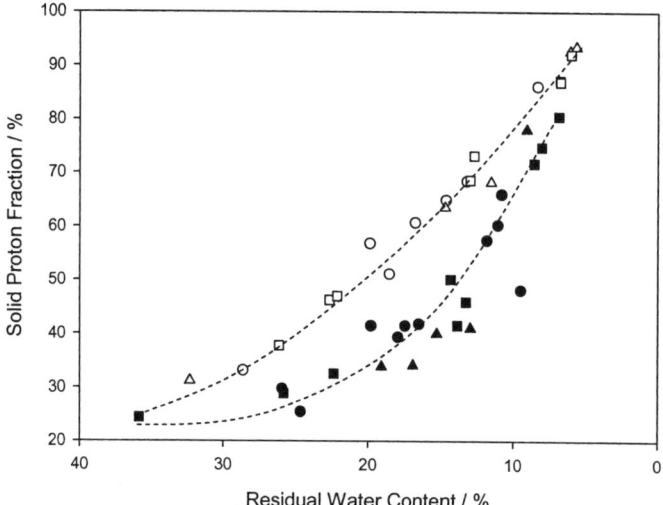

Figure 2: Solid proton fraction with dependence on residual water content for different process conditions and for samples dried with and without (w/o) sorbitol. Cond. 1(15 °C): with sorbitol (●); (w/o) sorbitol (○); Cond. 2(30 °C): with sorbitol (▲), (w/o) sorbitol (△) Cond. 3 (40 °C): with sorbitol (■), (w/o) sorbitol (□).

In order to further investigate this hypothesis, solid proton fractions were calculated according to the composition of the sample taking into account the proton densities of the different domains. As the samples were identified to be composed of two distinct fractions of different proton mobility and the fractions were mainly related to the biomass and the water, the sorbitol protons must be either part of the liquid or of the solid proton fraction. Therefore proton densities were calculated for both cases and compared to the experimental results. The proton density of the biomass could not be derived theoretically and was therefore estimated by comparing the calculated and measured results for the samples without sorbitol. The exact procedure is published elsewhere.[14] In case that the sugar protons are not part of the solid protons, the solid fraction $f_{s,1}$ is calculated as follows:

$$f_{S,1} = \frac{w_B \cdot d_B}{w_B \cdot d_B + w_S \cdot d_S + w_W \cdot d_W} \quad . \tag{3}$$

In case that the sugar protons are part of the solid protons, the solid fractions $f_{s,2}$ is calculated according to

$$f_{S,2} = \frac{w_B \cdot d_B + w_S \cdot d_S}{w_B \cdot d_B + w_S \cdot d_S + w_W \cdot d_W} \quad . \tag{4}$$

Here, w_i is the weight fraction of the component and d_i the proton density. The index B refers to the biomass, S to sorbitol and W to water. The comparison between the calculated and the measured fractions is shown in Fig. 3.

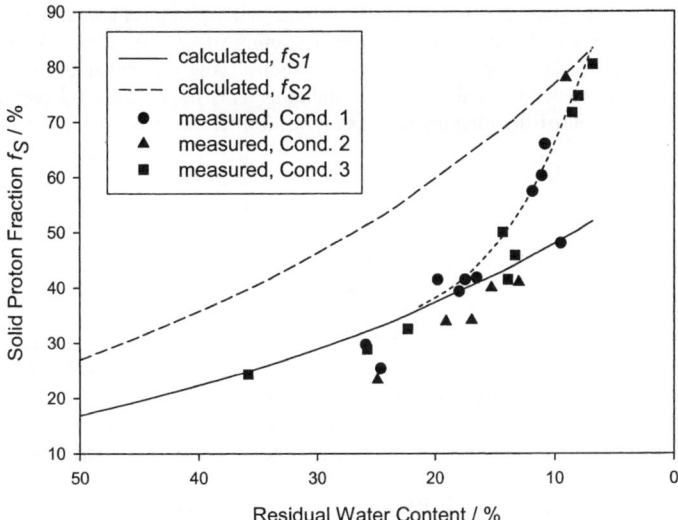

Figure 3: Calculated proton fractions for the cases that sorbitol is part of the liquid protons (f_{s1}, solid line) or solid protons (f_{s2}, dashed line) in comparison to the measured proton fraction for different drying process parameters.

Fig. 3 shows that for high residual water contents the measured values for the solid proton fraction are in accordance with f_{s1}, the calculated fraction for the case that sorbitol is part of the liquid protons. Below water contents of approx. 20 % a transition occurs and the measured values are approaching the curve f_{s2} that is calculated for the case that sorbitol is part of the solid phase. The transition is independent on the process parameters during vacuum drying and occurs for all process conditions at approximately the same water content.

The observed transition suggests that above the water content of 20 % all sorbitol protons are dissolved in the aqueous phase. Below this water content they undergo a gradual phase transition until they are completely incorporated into the solid phase at water contents below 7 %. The onset of the transition coincides with the onset of a protective effect on survival (see

Fig. 1c), therefore suggesting a relationship between survival rate and NMR relaxation parameters exists.

3.3 Comparison of NMR and survival data

In Fig. 3 a transition of sorbitol protons from dissolved into solid state is observed in the water content range between 20 and 7 % suggesting that sorbitol becomes incorporated into the dry biomass and therefore confirming the water replacement hypothesis.

However, this transition is independent on process conditions during drying (Fig. 3) whereas the protective effect of sorbitol depends on processing conditions (Fig. 1).

This seems to be contradictory at first sight as a potential interaction between sorbitol and bio-molecules should lead to a higher survival. The discrepancy between survival and NMR data could possibly be explained by the physical state of the cell membrane during drying. At physiological temperature the membrane exists in the liquid crystalline state when excess water is still present. Upon removal of excess water the membrane phase transition temperature increases when no protectant is added.[10] With protectant, the membrane remains in the physiological liquid crystalline state during drying. However, if drying is carried out far below physiological temperature as is the case for condition 1 (see tab.1), then it is likely that the membrane is already in the gel state at the beginning of drying when excess water is present. Therefore, no phase transition does occur upon drying and sugar has no effect on survival even when interacting with the cell membrane. This hypothesis still remains to be proven by measurement of the membrane phase transition temperature in the dry and liquid state.

REFERENCES

1. B. Higl, C. Santivarangkna, and P. Foerst, *CIT*, 2008, **80**, 1157.
2. L. J. M. Linders, W. F. Wolkers, F. A. Hoekstra, and K. Vantriet, *Cryobiol.*, 1997, **35**, 31.
3. C. Santivarangkna, U. Kulozik, and P. Foerst, *Lett.Appl.Microbiol.*, 2006, **42**, 271.
4. E. Selmer-Olsen, S. E. Birkeland, and T. Sorhaug, *J.Appl.Microbiol.*, 1999, **87**, 429.
5. G. Zhao and G. Zhang, *J.Appl.Microbiol.*, 2005, **99**, 333.
6. C. Santivarangkna, B. Higl, and P. Foerst, *Food Microbiol.*, 2008, **25**, 429.
7. L. J. M. Linders, G. I. W. deJong, G. Meerdink, and K. Vantriet, *J.Food Eng.*, 1997, **31**, 237.
8. A. K. Sum, R. Faller, and J. J. de Pablo, *Biophys.J.*, 2003, **85**, 2830.
9. G. Bryant, K. L. Koster, and J. Wolfe, *Seed Science Res.*, 2001, **11**, 17.
10. J. H. Crowe, J. F. Carpenter, and L. M. Crowe, *Annual Review of Physiology*, 1998, **60**, 73.
11. S. B. Leslie, E. Israeli, B. Lighthart, J. H. Crowe, and L. M. Crowe, *Appl.Environ.Microbiol.*, 1995, **61**, 3592.
12. J. S. Clegg, P. Seitz, W. Seitz, and C. F. Hazlewood, *Cryobiol.*, 1982, **19**, 306.
13. F. A. Hoekstra, E. A. Golovina, and J. Buitink, *Trends Plant Sci.*, 2001, **6**, 431.
14. P. Foerst, J. Reitmaier, and U. Kulozik, *Biotechnology Progress*, 2009, **submitted**.

A LOW-FIELD-NMR CAPILLARY RHEOMETER

Dirk Mertens, Edme H. Hardy, Bernhard Hochstein and Gisela Guthausen

IMVM, Universität Karlsruhe (TH), Adenauerring 20b, 76131 Karlsruhe

1 INTRODUCTION

Due to the high information content of nuclear magnetic resonance (NMR) experiments the application of NMR-based process control systems becomes increasingly attractive.[1] Especially low field NMR systems enhance the applicability of NMR based methods in laboratory and industry, being small, mobile, maintenance friendly, inexpensive and easier adaptable to their surroundings.

The flow behaviour of many fluids depends drastically on to the state of the respective goods. Therefore rheological measurements will improve the capability to understand and control manufacturing processes. The NMR capillary rheometer described is based on a NMR time domain analyzer equipped with an one axis gradient system, allowing for measurement of velocity probability density functions (VPDFs) and subsequent calculation of the corresponding flow curves (i.e. shear stress vs. shear rate) of flow through a cylindrical tube.[2-4]

Advantages of NMR-based rheometry are the direct detection of phenomena like wall slip or yield stress and the model free evaluation of the measured data. While conventional capillary rheometry is usually used to obtain data at very high shear rates the NMR-capillary-rheometer covers a shear rate range up to approximately 500 s^{-1}, a range suggesting applications in food or pharmaceutical industry where shear sensitive goods are handled.

2 NMR-METHODS

A pulsed gradient spin echo (PGSE)-sequence was to be used to measure the VPDF. Due to strong interactions between the pulsed field of the unshielded gradient coils, the magnet's pole shoe and surrounding conductors, the application of pulsed gradient fields results in time shifted echo maxima as well as varying width of the echo signal.[5-7] We were able to separate and quantify two distinct effects of pulsed gradients of amplitude G_y.

First a permanent gradient ($G_{perm} = f(G_y)$) becomes noticeable, second, effects of transient eddy currents ($G_{mism} = f(G_y,t)$) interfere with the desired measurement of spin echo magnitude and phase. Both effects were quantified using the pulse sequence shown in Fig. 1.

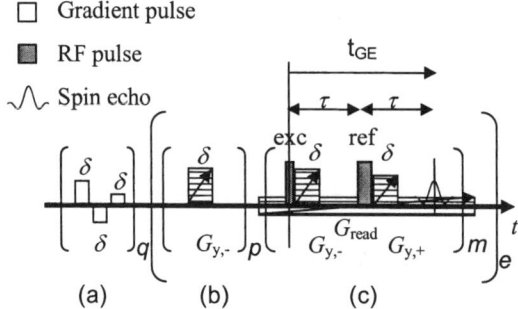

□ Gradient pulse

■ RF pulse

∿ Spin echo

(a)   (b)   (c)

Fig. 1 *Pulse sequence used to quantify permanent and transient effects of gradient pulses on permanent magnet systems. t_{GE}: time of the echo maximum to occur, τ: half echo time, δ: duration of gradient pulses of amplitude G_y before and after (-, +) refocusing, q: number of demagnetisation sequences (a), p: number of dummy gradient pulses G_y(b), exc, ref: excitation and refocusing rf-pulse, G_{read}: amplitude of compensation gradient, m: number of read gradient steps, e: number of flow encoding gradient steps, t: time.*

The sequence consists of three parts: First (cf. (a) in Fig. 1) a defined state of preferably low initial permanent magnetization G_{perm} is established using gradient pulses of alternating polarity and decreasing amplitude. Second, a series of dummy gradient pulses of the same amplitude as the flow encoding gradient $G_{y,-}$ is applied (cf. (b) in Fig. 1), creating a G_{perm} associated with $G_{y,-}$. This step is repeated p times for every encoding gradient step e. In the last part of the sequence a PGSE sequence (cf. (c) in Fig. 1) with echo time $t_{SE} = 2*\tau$ is run m times using a varied gradient G_{read} with both flow encoding gradients set to the same amplitude, i.e. $G_{y,-} = G_{y,+}$.

To obtain the calibration data for the gradient G_{perm} as well as the gradient mismatch G_{mism} several read gradients have to be applied ($m>1$, normally 64 steps between two times the lowest and two times the highest expected G_{perm}) with $G_{y,-} = G_{y,+}$. Considering a constant background gradient G_{perm} an echo signal occurs with $\phi(y, t) = 0$ for all y at $t = t_{SE} = 2*\tau$. This holds for all amplitudes of G_y and $G_{perm} = f(G_y)$. The phase acquired between excitation/refocusing equals the phase acquired in the second interval τ in magnitude, but possesses opposite sign summing to $\phi(y, t) = 0$ at $t=2\tau$.

Assuming a total background gradient $G_{back} = G_{perm} + G_{read}$ and effective rectangular gradient pulses of amplitude $G_{y,-}$ and $G_{y,+}$ (i.e. not disturbed by eddy currents), respectively, the overall phase at the time t during acquisition may be written:

$$\phi(y,t) = \gamma \{[G_{y,-} y \, \delta + (G_{perm} + G_{read}) y \, \tau] - [G_{y,+} y \, \delta + (G_{perm} + G_{read}) y \, (t - \tau)]\} \qquad (1)$$

The effect of eddy currents induced by preceding gradient pulses $G_{y,-}$ is included in eq. 1 replacing $G_{y,+}$ with

$$G_{y,+} = G_{y,-} - G_{mism}. \qquad (2)$$

Case 1: For $G_{mism} \neq 0$ and $G_{read} + G_{perm} = 0$ the phase along y will never be unwound and no echo is observable.

Case 2: If $G_{mism} \neq 0$ and $G_{read} + G_{perm} \neq 0$ the phase acquired due to the mismatched gradient pulses $G_{y,-}$ and $G_{y,+}$ generates a time shift of the echo signal maximum whereas the amplitude of the background gradient G_{back} determines the width of the echo as well as it affects the magnitude of the time shift within the acquisition time window.[8] If eddy currents extend into the acquisition period due to the time dependent background gradient a deformation of the echo shape occurs.

The time shift of the echo magnitude maximum at t_{GE} may be calculated by:

$$t_{GE} - t_{SE} = \delta\, G_{mism}/\, (G_{perm} + G_{read}) \qquad . \tag{3}$$

By fitting equation (3) to the echo maximum positions for a series of G_{read}, G_{perm} and G_{mism} can be determined for the corresponding G_y. The obtained data on G_{perm} and G_{mism} may then be used to measure flow or diffusion using the described pulse sequence while applying a shim gradient compensating the permanent gradient. Further on an intentionally switched gradient pulse mismatch is used to overcome the eddy current induced gradient pulse amplitude mismatch:

$$m=1,\ G_{read} = -G_{perm}\ \text{(i.e. } G_{back} = 0) \text{ and } G_{y,+} = G_{y,-} + G_{mism}. \tag{4}$$

3 EXPERIMENTAL

The experimental setup used is based on a Bruker (Rheinstetten, Germany) mq10 NMR-Analyzer (the minispec software, Version 2.58 Rev. 36/NT/XP) operating at a proton resonance frequency of 9.96 MHz. A probe accepting samples of up to 10 mm diameter was used. It is equipped with a planar gradient system driven by a Bruker Great 60 gradient amplifier (Alouette software version V 060725.B.3). The gradient hardware shows a gradient per unit current of 0.05 T A^{-1} m^{-1} resulting in a maximum magnetic field gradient 3 T m^{-1} along the flow direction y (cf. Fig. 2).

A DC50 / K10 thermostat (Thermo Electron GmbH, Karlsruhe, Germany) was used for temperature control. Further on an extruder pump (3RD8, ViscoTec GmbH, Töging am Inn, Germany) was connected to the flow line. This pump allows to pump a multitude of even highly viscous samples (max. pressure at the pump: 60 bar) with flow rates of up to 0.8 ml s^{-1} with low pulsation.

The sample flows through an inner capillary consisting of a 4.7 mm inner diameter PMMA tube (cf. Fig. 2). Through the annulus between the inner capillary and an outer PVC tube of 9 mm inner diameter temperature control liquid was circulated by a peristaltic pump (Watson Marlow, Wilmington, MA, USA). The temperature control liquid consisted of Gadovist (Bayer, Leverkusen, Germany) doped water (26 mmol/dm^3) showing a T_2 relaxation time of 7 ms thus enabling the use of a water coolant as long as the echo time of the PGSE experiments does not go below 28 ms (i.e. $t_{SE} = 2*\tau \geq 28$ ms).

Above and below the magnet, pressure (Motorola MPX5100DP) and temperature sensors (class A PT100 resistance, ES Electronic Sensor GmbH, Heilbronn, Germany) were mounted. Pressure drop measurements were performed at 525 mm distance. Data recording was done by LabView 8 (National Instruments). Data processing was performed using MatLab R2007b (The MathWorks Inc.). The above mentioned programs were running on the same PC used to control the NMR analyzer an IBM PC using Microsoft Windows XP. A schematic diagram of the setup used is shown in Fig. 2.

Reference measurements were conducted on a MARS rotational rheometer (Thermo Scientific), using a Z20DIN (DIN53019, Searle type) measuring system.

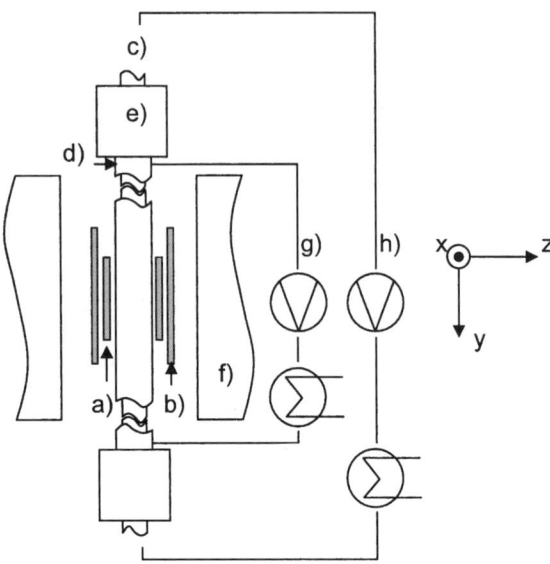

Fig. 2 *Schematic diagram of the setup used: a) rf-Coil, b) gradient coil, c) inner capillary, d) outer tube, e) temperature and pressure sensors, f) mq10 permanent magnet, g) peristaltic pump and heat exchanger for temperature control liquid and h) extruder pump and heat exchanger for sample liquid. The coordinate system designates the coordinates given by the magnet and gradient system.*

The samples consisted of commercially available foodstuffs. The starch sample was prepared from maize starch and tap water, being heated up to 80 °C for 15 minutes while stirring.

Tab. 1 *Relaxation data and measurement times for the different samples used in this study. The repetition time of the experiments was adjusted to exceed 5*T_1.*

Sample	Signal averages	T_1 / ms	T_2 / ms	Exp. time / min
Olive oil	8	150	140	30
Cake batter	8	90	62	22
Mayonnaise	8	180	127	30
Starch	4	1500	366	70

The echo time t_{SE} of all experiments was adjusted to 28 ms with an encoding time (time between the rising edges of the gradient pulses) of 14 ms, a gradient pulse duration $\delta = 3$ ms and 128 encoding steps of G_y. The repetition time in all experiments was larger than 5*T_1. NMR as well as rotational rheometer measurements were performed at a temperature of 303 K. The quite long experimental times indicated in Tab. 1 can be considerably reduced without corrupting the information by reducing the number of signal averages or encoding steps. The VPDF is obtained by magnitude calculation of the Fourier transformation of the series (here: 128) of the complex echo maximum data.

Measurements of G_{perm} and G_{mism} (cf. Fig. 3) were performed using a cylindrical sample of sugar solution the length of which was large in comparison to the length of the RF-coil of the probe.

4 RESULTS

4.1 Quantification of the Permanent and Transient Effects

Application of the pulse sequence (Fig. 1) results in a series of echo maximum positions with respect to the applied gradient G_{read} for every step of flow encoding gradient G_y. Results showing the variation of the echo top position with respect to the applied read gradient (for the example of an encoding gradient of $G_y = 0.42$ Tm^{-1}, $\delta = 5$ ms and $\tau = 8$ ms, encoding time 8 ms) are shown in the left part of Fig. 3.

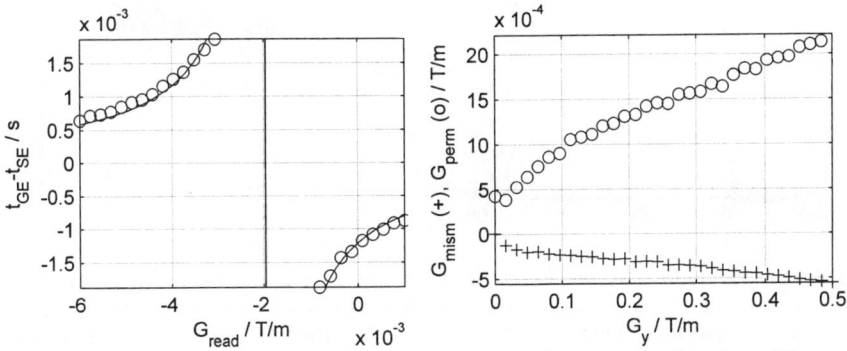

Fig. 3 *Left: Deviation of the echo top occurrence at t_{GE} from t_{SE} as a function of G_{read}. The solid line shows a plot of equation (3) using the parameters resulting from the fit (for the example of an encoding gradient of $G_y = 0.42$ Tm^{-1}, $\delta = 5$ ms and $\tau = 8$ ms). Right: Values for G_{perm} (o) and G_{mism} (+) as a function of the flow encoding gradient.*

Performing the fit of equation (3) to data resulting from several values of the flow encoding gradient G_y leads to G_{perm} and G_{mism} data as a function of G_y, as shown in the right half of figure 3. As long as the sequence timing of the PGSE sequence is not changed, the calibration (cf. Fig. 3, right) curves remain applicable. A change in sequence timing may give rise to the need of a new calibration as the eddy currents of the second gradient pulse may extend into the acquisition time. In this case it will be misinterpreted as being part of G_{perm}.

While the described method (cf. Fig. 1) gives the opportunity to control echo width as well as position, the simple application of a series of dummy gradient pulses drives eddy currents into saturation. This results in a steady state during the PGSE-sequence which does not change the echo position if the timing is chosen correctly. As we were not interested in echo width in this work part a) of the pulse sequence was dropped, ten dummy gradient pulses were used to saturate eddy currents and no shim gradient was used (i.e. $G_{read} = 0$ T/m). A more detailed description of the sequence displayed in Fig.1 as well as the saturation scheme used here will be given elsewhere.[9]

4.2 Results of Rheological Measurements

4.2.1 Olive Oil. Measurement of the olive oil using a rotational rheometer showed a constant viscosity of 0.048 Pa s for shear rates higher than 100 s^{-1} which is in good agreement with the NMR-measurement (cf. Fig. 4, right). Both methods reveal Newtonian flow behaviour of the olive oil which can easily be concluded from the velocity probability density function (cf. Fig 4) showing a constant probability for all velocities between zero and the maximum velocity. A distortion of the VPDF at velocity zero is attributed to Gibbs artefacts. The mean velocity was measured to be 25.9 mm/s volumetrically in good agreement with the velocity calculated from NMR-data: 26.1 mm/s.

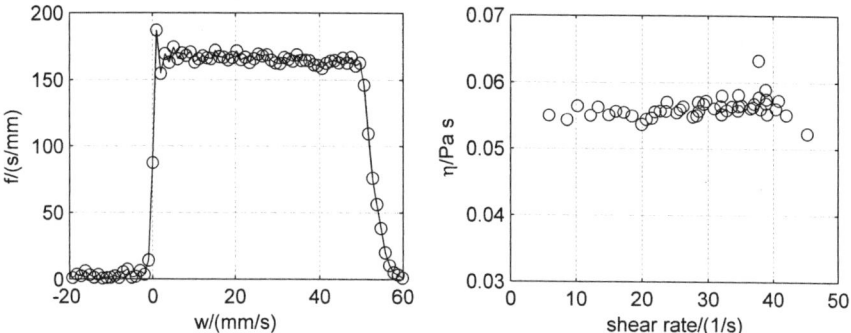

Fig. 4 *Left: VPDF of olive oil flowing through a cylindrical capillary. The probability between zero and the maximum velocity is constant as found for Newtonian fluids. Right: Viscosity function calculated from the NMR-data. As expected for a Newtonian fluid the viscosity shows a constant value in the explored shear rate region.*

4.2.2 Lemon Cake Batter. As another sample we used ready made lemon cake batter. As can be seen from Fig. 5 (left) the flow behaviour differs from the olive oil, high velocities are more probable than low ones. The probability at velocity zero is almost zero which could be interpreted as a wall slip ($f(0) \approx 0$ s/mm, cf. Fig. 5, left). The NMR data and the rotational rheometer results for the batter are in reasonable agreement. The increase in viscosity showing up in the classical measurement may possibly be attributed to drying of the sample. The mean velocities calculated from NMR-data (4.4 mm s^{-1}) and adjusted at the pump (4.3 mm s^{-1}) are in good agreement.

4.2.3 Starch Solution. In Fig. 6 the velocity (cf. Fig. 6, left) as well as the viscosity (cf. Fig. 6, right) data is shown, exhibiting pronounced wall-slip behaviour ($f(0) \approx 0$ s/mm) as there is essentially no contribution of velocities near zero (cf. Fig. 6 left). The calculated average velocity is 2.2 mm/s, evaluation of the NMR-data results in a mean velocity of 2.2 mm/s. NMR and rotational rheometer data are in good agreement, as can be seen in Fig. 6.

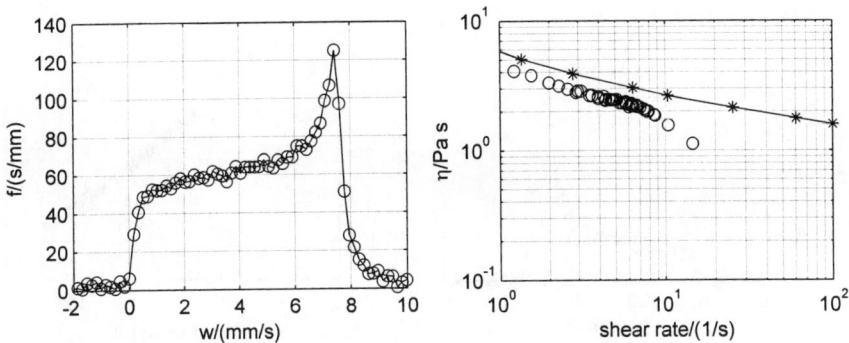

Fig. 5 *Left: VPDF of Lemon cake batter showing increased probability at high velocities. The function shows almost zero contribution at w=0, which is a hint to wall slip behaviour. Right: Viscosity as a function of shear rate exhibiting shear thinning of the cake batter – the viscosity decreases with increasing shear rate. NMR (o) and rotational rheometer (*) data.*

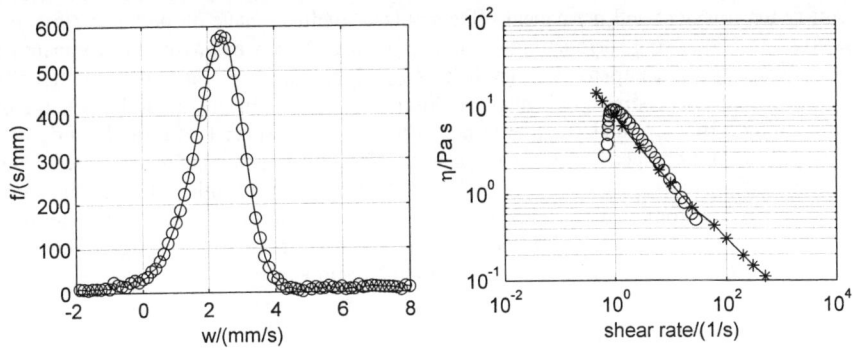

Fig. 6 *Left: VPDF of a starch solution (60 g/l heated to 80 °C). Right: Viscosity as a function of shear rate measured by NMR (o) and a rotational rheometer (*).*

4.2.4 Mayonnaise. The mayonnaise sample was pumped through the NMR-capillary rheometer at a pump flow rate of 0.6 ml/s, resulting in a calculated mean velocity of 34.6 mm/s. Evaluating NMR data a mean velocity of 35.6 mm/s was determined. The NMR-derived viscosity data complies with the rotational rheometer data (cf. Fig. 7) showing pronounced shear thinning (i.e. the viscosity is decreasing with increasing shear rate).

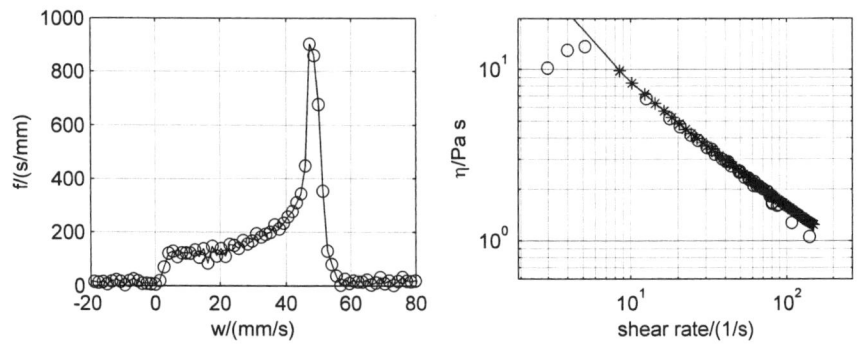

Fig. 7 *Left: VPDF of mayonnaise. Right: Viscosity as a function of shear rate for mayonnaise: NMR measurement (o) and reference measurement (*).*

4   CONCLUSIONS

We have shown that dedicated low-field NMR analyzers have a great potential in the field of flow measurement, even if unshielded gradient systems are used. As we were able to quantify the transient and permanent effects pulsed gradients have on permanent magnet systems and conducting surroundings of the gradient coils, it was possible to compensate both effects by a compensation gradient G_{read} (permanent effect) and intentionally mismatched flow encoding gradient pulses (transient effect). In this study knowledge of the magnitude of G_{perm} and G_{mism} was not necessary, so a special case of the proposed method allows for compensation of the transient effects by saturation, a method used. Both methods enable the measurement of high quality velocity probability density functions which were, in combination with pressure drop measurements, used to determine information on the flow curve of different fluids. The NMR data is in good agreement with data from a rotational rheometer as well as volumetric flow rates.

Acknowledgements
Financial support by DFG (HA2840/4-2) is gratefully acknowledged.

References
1 J. P. M. van Duynhoven, Gert-Jan W. Goudappel, Elena Trezza, Adrian M. Haiduc, Franck Duval and Wladyslaw P. Weglarz, *New Food*, 2008, **2**, 41-44.
2 Robert L. Powell, James E. Maneval, Joseph D. Seymour, Kathryn L. McCarthy and Michael J. McCarthy, *J. Rheol.*,1994, **38**(5),1465-1470.
3 Darren F. Arola, Geoffrey A. Barral, Robert L. Powell, Kathryn L. McCarthy and Michael J. McCarthy, *Chem. Eng. Sci.*, 1997, **52**(13), 2049-2057.
4 Stephen J. Gibbs, Derek E. Haycook, Wiliam J. Frith, Stephen Ablett and Laurance D. Hall, *J. Magn. Reson.*, 1997, **125**, 43-51.
5 Mirko I. Horvat and Charles G. Wade, *J. Magn. Reson.*, 1981, **44**, 62-75.
6 Mirko I. Horvat and Charles G. Wade, *J. Magn. Reson.*, 1981, **45**, 67-80.
7 E. von Meerwall and M. Kamat, *J. Magn. Reson.*, 1989, **83**, 309-323.
8 P. T. Callaghan, *J. Magn. Reson.*, 1990, **88**, 493-500.
9 Dirk Mertens and Edme H. Hardy, manuscript in preparation.

DEVELOPMENT OF A RHEO-NMR SYSTEM TO STUDY THE CRYSTALLISATION OF BULK LIPIDS UNDER SHEAR FLOW

Gianfranco Mazzanti* & Elizabeth M. Mudge

Department of Process Engineering and Applied Science, Dalhousie University,
D405-1360 Barrington st., Halifax, NS, Canada B3J 2X4,* gianfranco.mazzanti@dal.ca

1 INTRODUCTION

1.1 General background

Fats and oils constitute a very large portion of the food industry. Fats are multicomponent systems containing many different triglycerides (and other lipids in smaller amounts). They crystallize in different polymorphic forms, often termed α, β and β' in order of thermodynamic stability and melting point. The proportions of these phases in final products largely determine basic quality attributes of the material, such as palatability, appearance or shelf life.

The effects of shear flow on crystallizing bulk triglycerides such as cocoa butter, palm oil, milk fat and other materials have been qualitatively known for a long time[1-3]. The typical process of crystallization of margarines is done in a scraped surface heat exchanger. Effects such as the competition between enhanced heat transfer and viscous heat generation are commonly discussed. The observation of phase transition acceleration was also noted early[1], and some speculations on the mechanism causing it were put forth[2,3]. Innumerable studies have been done on crystallization of static fats using x-ray diffraction[4] (XRD) to identify phases, and nuclear magnetic resonance (NMR) to quantify the solid fat content[5] (SFC) equivalent to the crystalline fraction. Both methods offered unprecedented understanding of the crystallization of triglycerides. The long story can be found in any good review[6]. With the availability of synchrotron sources[7] time-resolved studies were made easier, and many studies have been conducted on crystallization of multicomponent lipid systems. However, only recently it has been possible to study the crystallization under shear flow using synchrotron or in house XRD[8-16]. The use of benchtop NMR to follow the crystallization under shear flow was reported also only very recently[17-19]. The XRD studies have shown changes in the onset times for phase changes[8,9,11,16], phase proportions and compositions[10], changes in microstructure[8,11,16]. The NMR studies have shown changes in solid fat content[18-20]. The observation of micro and nano crystalline orientation[11] opened the possibility for the nano-structuring of anisotropic materials[21,22]. These effects of shear produce changes in mechanical[23], thermal[23] and likely diffusional properties that modify the general functionality of the material.

What can we do with the knowledge that we are accumulating on shear effects? The possibility to combine new ingredients while controlling the nanostructuring and microstructuring of the materials offer the possibility of optimizing existing processes, design new products, processes or equipment. However, we are still only at the beginning of the understanding of how the application of shear flow affects the crystallization process. It is still necessary to develop adequate conceptual and mathematical models to predict these effects, and thus use them more effectively. This requires an understanding of the thermodynamics of multicomponent systems and driving forces involved, and the distortions that the introduction of shear can introduce to these systems[24]. It also requires the ability to model kinetics[9,13-15] and pathways of phase transitions[10,12] under shear flow. There is still a lack of understanding on heat, mass and momentum transfer in these systems, due to the complex and changing nature of these systems. The regular theories on nucleation and growth may still apply in some cases, such as the JMAEK (Avrami) model[25], because shear can keep the crystallites separate and mix the fluid[15]. However, much is still unknown due to quantification difficulties described later in this section. The characteristics of resulting materials has been attempted using rotational or bending methods[8,16,20,26,27]. The determination of nanostructural characteristics such as crystalline phases, types and proportions are typically done using in-situ (synchrotron) or post-facto XRD. However there is potential for NMR applications since the proton interactions are characteristic of each phase[28].

As mentioned earlier, the determination of SFC is typically conducted using benchtop NMR. The commercial systems provide a total estimate of the crystalline amount using standards that do not distinguish between the responses of different polymorphs. This can lead to some over/under-estimation of the SFC for a particular polymorph [28]. The commercial benchtop systems currently available do not include options of applying shear to the sample, and are not well equipped for fast cooling rates. The combination of shear flow and NMR, Rheo-NMR has been, however, developed in other fields[29].The determination of general microstructural characteristics can be done using microscopy under shear flow[16] and invaluable information can be gathered. More difficult is the determination of particle shape and size distribution. From the XRD data we know the approximate size of the single domains, between 5 and 60 nm typically. Yet their distribution and the size in other directions is difficult to measure[15]. There is a relationship between the chemical composition of a trygliceride crystal and the lamellar thickness of the material[10,13,14,20]. However, establishing precise correlations is still difficult. Perhaps this is another field in which high resolution NMR may help.

1.2 Information from XRD: value and limitations

The information obtained from two dimensional (2D) synchrotron XRD data can tell us primarily information about the nanostructure (molecular composition, crystalline thickness), as well as some information on the microstructure (integrated intensity, orientation). If we take a radial average of a 2D diffraction pattern, we can fit the radial profile using a multi-peak program. Once the peaks have been obtained and separated from the background (sometimes not an easy task at all), one can find the peak position of each phase, and thus determine the thickness of the lamellar structure composed of triglycerides. These lamellar units are the first supramolecular structures that constitute these materials. Since the molecules are much longer than they are wide, the scattering in the small angle region gives information about the average size of the molecules that are present in the lamellae. Combined with the wide angle reflections[30], it is possible to identify the polymorphic form of each phase. The width of the peak is related to the thickness of the individual crystalline units. The area under the peak, called Integrated Intensity, is directly proportional to the crystalline fraction of each particular phase. However, the proportionality factors between integrated intensity and crystalline fraction are impossible to determine a priori in these complex materials and complicated measuring devices (e.g. Couette cell with cooling devices). Therefore, XRD cannot <u>directly</u> tell us the crystalline fraction of a phase, unless we happen to have a condition where we have that particular phase on its own and fully crystalline, seldom possible. The plots of x-ray integrated intensity as a function of time in Figure 1 were obtained from cocoa butter crystallized at 17.5 °C under shear rates of $90s^{-1}$ (Figure1a) and $720s^{-1}$ (Figure 1b). If all the phases had the same proportionality factor between integrated intensity and SFC, we just needed to sum them at any time and divide by the total to obtain the crystalline fraction of each phase. However this is not the case, since each phase is likely to have a different proportionality factor. The proportionality factors depend mainly on the phase type (α, β or β') and on the order of the specific x-ray reflection chosen for the comparison (001, 002, etc). In the case of mixed crystals the phase composition may have an impact on the integrated intensity, but at present we do not even have a rule of thumb to estimate this effect. The experimental conditions (Sample thickness, absorption by the container wall, air and small angle scattering) affect equally all phases. Variations of background are possible during crystallization due to small angle scattering[12]. Crystalline orientation can also change the integrated intensity, and in a multiphase system, if different phases have different spatial orientation distributions, then the integrated intensities will be affected accordingly.

It is apparent then that we need to measure SFC independently in order to find the proportionality constants. This is naturally done using NMR. Typical benchtop instruments operate at 20MHz, and the free induction decay (FID) times for solids are in the order of 10 μs for Hydrogen in the crystalline fraction, and over 200 ms for the liquid. The estimate of the signals at time zero is done by extrapolation of the signals at later times, due to the dead time of the probe.

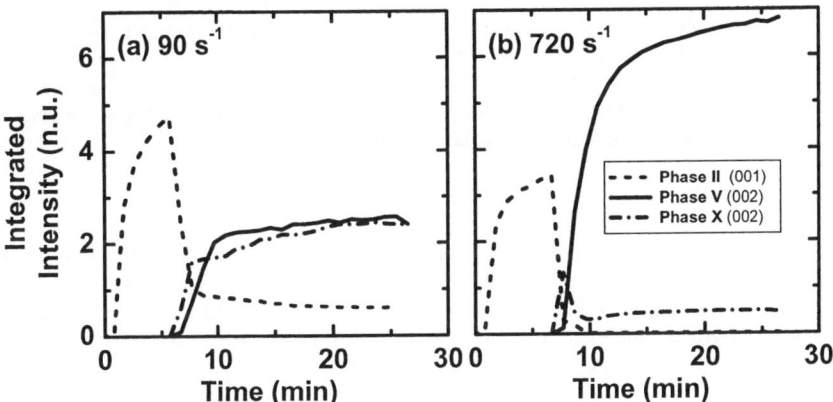

Figure 1 *X-ray integrated intensity (normalized units, n.u.) as a function of time (min) for cocoa butter crystallized at 17.5 °C under shear rates of (a) 90 s⁻¹ and (b) 720 s⁻¹.The phase and x-ray reflection order to which each integrated intensity corresponds is indicated in the graph.*

Though the extrapolation of the liquid is likely precise, extrapolation of the solid can be more difficult due to the short duration of the usable signal. Standards are used to find the calibration constants, sampling at two (or more) times for the solid and liquid signals. These instruments have been of great value for industry and research, but, to the present, they are not equipped to apply shear flow to the sample.

Thus, we endeavoured to develop a system capable of measuring SFC under shear flow.

2. MATERIALS AND METHODS

2.1 Prototype I

We used a Minispec pc20 (Bruker, manufactured in 1983) with a standard ratio-probe for 10mm tubes that had been routinely used for years, without changes any settings. This prototype has been described in a previous publication[18]. In summary, a small motor and a tachometer were used to control the rotational speed of a Teflon® shaft concentric with a 10mm NMR glass tube. The sample size was about 60% of the volume of a regular sample. Since we did not have variable temperature control inside the magnet, we chose a material that could crystallize at 40 °C, which is the regulated temperature of the magnet. The sample was melted by placing the tube with sample and shaft in an external small beaker with water at 80 °C. Then the motor and tube assembly were quickly placed in the NMR probe and let to naturally equilibrate with the magnet temperature. The NMR had been set up in 'mode 0', where each analog FID signal, preceded by a digital trigger, was sent to an external BNC connector. The first experiments were done using a 48 kHz acquisition gadget (National Instruments USB 6009) and could only capture the liquid decay. After installation of a 20 MHz ADLink board in the computer we were able to capture the whole signal. Trials with and without the Teflon shaft, and with the motor turned sequentially on and off showed no significant difference in the FID signals, thus providing clear evidence that the Minispec could be used with samples under shear. Further experiments combining measurements of the liquid and solid signals allowed us to develop a procedure to estimate the temperature of the sample under shear flow. The temperature estimates require some reasonably good idea of the densities of the materials involved. Thus we had the unprecedented option of having an intrinsic NMR temperature probe that does not interfere with the shear experiments[18].

2.2 Prototype II

In order to have temperature control, we changed the 10 mm probe to a probe designed for 13 mm diameter tubes. The end caps that held the glass tube that supports the RF coil were replaced with modified versions that allowed us to create a seal with a 10 mm sample tube, to pump a fluid in the space between the sample tube and the external glass, and to introduce a temperature sensor about 1 cm below the RF coil.

The perfluorinated fluid used was Galden HT135 (Solvay-Solexis) and was pumped using a peristaltic pump. The fluid does not contain significant amounts of hydrogen and is thus suitable for NMR use without interfering with the measurements. The fluid was heated or cooled using a heat exchanger provided with thermoelectric Peltier elements. A controller read the temperature from the thermistor inside the NMR probe and delivered the power needed to adjust this temperature to the setpoint commanded by the computer. The controller produced some signal spikes that were easily identified, but that prevented us from using a larger amplification factor for the signal. Since the volume of the sample remained the same, its relative volume with respect to the volume of the 13 mm probe was reduced. This resulted in a significant decrease of the signal to noise ratio compared to the 10 mm probe. But it was a small price to pay in exchange for the temperature control.

In order to explore if the shear in the mini-Couette cell was in fact affecting phase transitions we made a twin cell compatible with SXRD and designed to fit the motor assembly used for prototype I. We modified the shaft, replacing its bottom part with Lexan of same dimensions as the Teflon shaft. The temperature control was the same used for the NMR modified probe. The cell was successfully tested at the ExxonMobil beamline X10A of the National Synchrotron Light Source (NSLS) at Brookhaven National Laboratory (Upton, NY). The x-ray beam was sent perpendicular to the cell and the diffraction pattern was captured with a single point detector moving in the 2θ angle.

3 RESULTS AND DISCUSSION

3.1 Finding the proportionality constants

From the NMR measurements we can obtain the total SFC. To illustrate this we have plotted in Figure 2 the values of SFC for cocoa butter crystallized at 17.5 °C under two shear rates, 90 and 720 s^{-1} so that we can later compare them with two ways of estimate the SFC using the XRD data. The values of SFC are plotted as squares, and the estimates from XRD as solid lines. The estimating procedure from XRD will be discussed in detail later. In order to make sure that the phases that we were observing in the experiments in the mini-Couette cell in the NMR were the same as those observed in previous XRD experiments, we used the twin (SXRD) mini-Couette cell and conducted experiments with cocoa butter at 17.5 °C. The XRD detector available only allowed us to estimate the peak positions of the XRD diffraction patterns every 5 or 10 minutes. The peak positions clearly showed that a transition from phase II to phase V had indeed occurred, though we could not ascertain the presence of form X[19]. Having confirmed that the mini-Couette cell was in fact modifying the phase transitions in a similar manner as other shear systems[9,10,16], we could attempt to fit the data to get a quantitative relationship[31]. It must be noted that these XRD experiments[10] were conducted with a different shear cell and cooling system and with a different sample of cocoa butter. Two simple models were tested against the data of two of the SFC measurements. The first one, described by Equation (1), was to find just a simple scaling factor for the sum of all integrated intensities, and the result is shown in Figure 2a. It clearly does a poor job, thus each phase needs its own proportionality constant. The second model, Equation (2), uses a proportionality constant for each phase, which yielded a fairly good approximation, as can be seen in Figure 2b.

$$SFC_a = C \cdot \left(II_{II} + II_V + II_X \right) \tag{1}$$

$$SFC_b = C_{II} \cdot II_{II} + C_V \cdot II_V + C_X \cdot II_X \tag{2}$$

We still find variation between the proportionality coefficients computed from experiments between the two shear rates. It is not surprising since we did not have simultaneous acquisition, the samples were different, and so were the surfaces and radiuses. Yet, for so many variations the fit is remarkable. The objective was to prove the potential of a methodology, and offer a preliminary estimate of the proportionally coefficients.

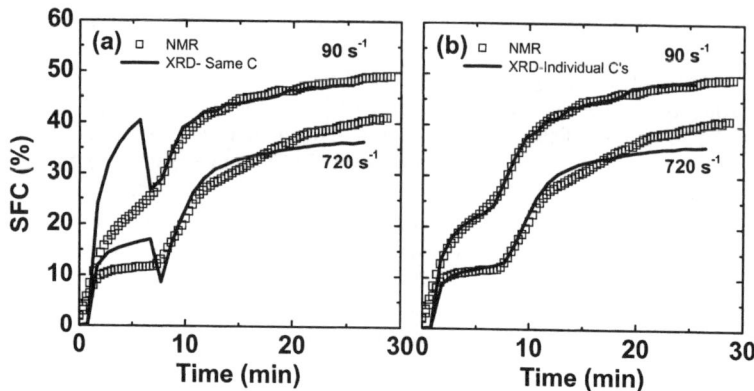

Figure 2 *SFC (%) vs. time (min) for cocoa butter crystallized at 17.5 °C and (a) 90 s⁻¹ and (b) 720 s⁻¹. The squares represent values measured in the Rheo-NMR system, while the solid lines represent the SFC computed using (a) the same factor for all phases, and (b) individual factors for each phase.*

3.2 Effect of crystalline orientation on integrated intensity

Another important example of the need of correlate XRD and NMR data comes from the crystalline orientation induced by the shear flow. In an experiment carried out with milk fat, the material was crystallized under a moderate shear rate of 90 s⁻¹ until the only polymorphic phase present was β' and then allowed to grow for at least another half hour. Then the shear rate was increased above 1000 s⁻¹. As indicated in Figure 3a, in the XRD experiments it was observed that the overall averaged integrated intensity decreased suddenly and then increased progressively as time went by. The vertical (v) and horizontal (h) measurements indicated that the increase in orientation was due both to an increase in the horizontal intensity and a decrease in the vertical intensity. As observed in the first part of the experiment, for a system that exhibits none or little orientation, the SFC is proportional to the XRD integrated intensity. After the sudden increase of the shear rate, the reduction of the integrated intensity was observed as well in the SFC, as shown in Figure 3b. However, the subsequent increase noticed for the XRD integrated intensity was not observed. This means that the increase in the overall average integrated intensity is due to a larger number of crystallites being brought into the Bragg angle by the shear field, and not to an increase in crystallinty. While XRD only samples those crystals that are within the Bragg angle, the solid NMR signal is produced from all crystals. A correct interpretation of the XRD data would be impossible without the corresponding NMR SFC.

4 CONCLUSION

XRD provides identification of phases present, proportions of phases and crystalline orientation, whereas NMR provides total crystallinity. Complementing XRD and NMR measurements is necessary to develop reliable quantitative models of crystallization under shear. The possibility of conducting the Rheo-NMR work in a benchtop machine opens an affordable possibility to perform these types of studies in many laboratories.

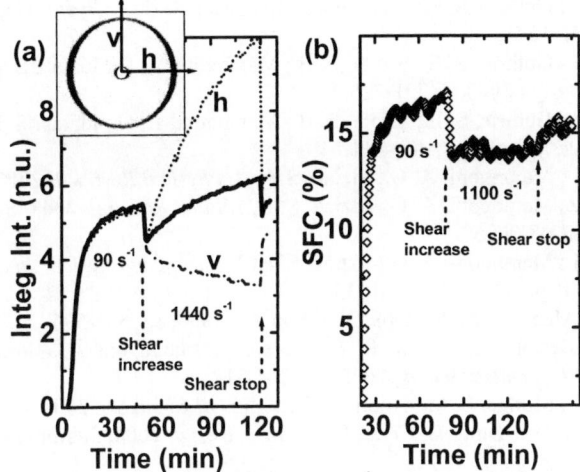

Figure 3 *Milk fat crystallized at 17.5 °C and 90 s^{-1} followed by a shear increase. (a) Integrated intensity (n.u.) vs. time (min). The insert shows a typical diffraction pattern, and the vertical (v) and horizontal (h) directions measured to compare to the overall average (thick line) (b) Rheo-NMR SFC (%) vs. time (min).*

Acknowledgements

We gratefully acknowledge A. Marangoni, S.Bennet, S. Idziak and P. Krygsman; the Dept. of Process Engineering & Applied Science at Dalhousie University; and funding from NSERC Canada. Research carried out in part at the X10A ExxonMobil beamline, NSLS, Brookhaven National Laboratory, NY, US, which is supported by the US DOE, Division of Material Sciences and Division of Chemical Sciences.

References

1   Feuge, R. O.; Landmann, W.; Mitcham, D.; Lovergren, N. V. *J Am Oil Chem Soc* **1962**, *39*, 310-313.
2   Ziegleder, G. *Int. Z. Lebensm. Tech. Verfahrenst.* **1985**, *36*, 412-418.
3   Windhab, E.; Niediek, E., A,; Rolfes, L. *Susswaren* **1993**, *3*, 32-37.
4   Lutton, E. S. *J. Amer. Chem. Soc.* **1951**, *73*, 5595-5598.
5   Van Putte, K.; Van Den Enden, J. *Journal of the American Oil Chemists Society* **1974**, *51*, 316–320.
6   Sato, K. *Chem. Eng. Sci.* **2001**, *56*, 2255-2265.
7   Kellens, M.; Meeussen, W.; Gehrke, R.; Reynaers, H. *Chem Phys Lipids* **1991**, *58*, 131-144.
8   De Graef, V.; Goderis, B.; Van Puyvelde, P.; Foubert, I.; Dewettinck, K. *Eur J Lipid Sci Tech* **2008**, *110*, 521-529.
9   MacMillan, S. D.; Roberts, K. J.; Rossi, A.; Wells, M. A.; Polgreen, M. C.; Smith, I. H. *Crystal Growth & Design* **2002**, *2*, 221-226.

10 Mazzanti, G.; Guthrie, S. E.; Marangoni, A.; Idziak, S. H. J. *Crystal Growth & Design* **2007**, *7*, 1230-1241.
11 Mazzanti, G.; Guthrie, S. E.; Sirota, E. B.; Marangoni, A. G.; Idziak, S. H. J. *Crystal Growth & Design* **2003**, *3*, 721-725.
12 Mazzanti, G.; Guthrie, S. E.; Sirota, E. B.; Marangoni, A. G.; Idziak, S. H. J. *Crystal Growth & Design* **2004**, *4*, 409-411.
13 Mazzanti, G.; Marangoni, A. G.; Idziak, S. H. J. *Physical Review E* **2005**, *71*, 041607.
14 Mazzanti, G.; Marangoni, A. G.; Idziak, S. H. J. *Food Res. Intl.* **2008**, Submission FOODRES-S-08-00960.
15 Mazzanti, G.; Marangoni, A. G.; Idziak, S. H. J. *European Physical Journal E* **2008**, DOI 10.1140/epje/i2007-10359-10350.
16 Sonwai, S.; Mackley, M. R. *J Am Oil Chem Soc* **2006**, *83*, 583-596.
17 Martini, S.; Bertoli, C.; Herrera, M. L.; Neeson, I.; Marangoni, A. *Journal of the American Oil Chemists Society* **2005**, *82*, 305-312.
18 Mazzanti, G.; Mudge, E. M.; Anom, E. Y. *J Am Oil Chem Soc* **2008**, *82*, 405-412.
19 Mudge, E. M.; Mazzanti, G. *Crystal Growth & Design* **2008**, Submitted, cg-2008-00999y.
20 Mazzanti, G.; Li, M.; Qatami, O. A.; Marangoni, A.; Idziak, S. H. J. *Crystal Growth & Design* **2008**, Submission cg-2008-xxxx.
21 Maleky, F.; Marangoni, A. G. *J Food Eng* **2008**, *89*, 399-407.
22 Maleky, F.; Mazzanti, G.; Idziak, S. H. J.; Marangoni, A. G.: US PAT, 2008; Vol. PCT/CA2008/000594.
23 Guthrie, S. E. In *Physics and Astronomy*; University of Waterloo: Waterloo, ON, Canada, 2007.
24 Los, J. H.; van Enckevort, W. J. P.; Vlieg, E.; Floter, E. *J Phys Chem B* **2002**, *106*, 7321-7330.
25 Avrami, M. *J Chem Phys* **1940**, *8*, 212-224.
26 Toro-Vazquez, J. F.; Perez-Martinez, D. B.; Dibildox-Alvarado, E. A.; Charo-Alonso, M. A.; Reyes-Hernandez, J. B. *Journal of the American Oil Chemists Society* **2004**, *81*, 195-202.
27 Guthrie, S. E.; Idziak, S. H. J. *Rev Sci Instrum* **2005**, *76*, 026110.
28 van Duynhoven, J.; Dubourg, I.; Goudappel, G. J.; Roijers, E. *Journal of the American Oil Chemists Society* **2002**, *79*, 383-388.
29 Callaghan, P. T.; Gil, A. M. In *Magnetic resonance in food science: a view to the future.*; Webb, G. A., Belton, P. S., Gil, A. M., Delgadillo, I., Eds.; Cambridge's Royal Society of Chemistry: Aveiro, Portugal, 2001, pp 29–41.
30 Wille, R. L.; Lutton, E. S. *J Am Oil Chem Soc* **1966**, *43*, 491-496.
31 Cisneros, A.; Mazzanti, G.; Campos, R.; Marangoni, A. G. *J Agr Food Chem* **2006**, *54*, 6030-6033.

NMR-based Multi Parametric Quality Control of Fruit Juices

E. Humpfer[1], H. Schaefer[1], B. Schuetz[1], M. Moertter[1], M.Spraul[1] and P. Rinke[2]

[1] Bruker BioSpin GmbH
[2] SGF International e.V.

1 INTRODUCTION

Traditionally, NMR has been perceived as a tool for structure verification, elucidation and purity analysis. However, driven by the needs of the emerging field of Metabonomics/Metabolomics, NMR has rapidly expanded in recent years into the areas of mixture analysis and screening applications[1,2]. These developments were facilitated by high-throughput sample changing technology (either sample tube or flow-injection methods), integrated sample preparation, and the improved quality of digital spectrometers in general.

Today, NMR is an established tool in a wide range of metabonomics-related applications, including drug toxicity and efficacy screening with animal models, clinical diagnosis of inborn errors of metabolism, and general health status screening in the context of epidemiological research, to name but a few examples. In such studies, hundreds of samples may have to be screened per day with regard to the identity and concentration of selected metabolites as well as for between-sample comparison of spectral patterns using multivariate statistics to obtain classification and discrimination information. NMR is a particularly well-suited detector for the analysis of biological fluids, deriving truly quantitative and structural information while featuring high throughput (a 1D spectrum measured in a few minutes), excellent reproducibility and minimal sample preparation (typically only addition of buffer).

In this article we demonstrate that the principles established for NMR applications in metabonomics have been successfully transferred to yet another but closely related field of mixture analysis, i.e. quality control of beverages. A corresponding method for fruit juice analysis, developed in a joint effort by Bruker BioSpin GmbH and SGF International e.V. has been introduced recently under the name Bruker SGF Profiling. The system is fully automated with respect to sample transfer, measurement, data analysis and reporting and is set up on an Avance 400 flow-injection NMR spectrometer. For each juice a multitude of parameters related to quality and authenticity are evaluated simultaneously from a single data set acquired within a few minutes. This multi-marker/multi-aspect NMR screening approach features low cost-per-sample and is highly competitive with conventional and targeted juice quality control based on GC or LC methods, for example.

Initially, the new NMR-based method was developed as a cost- and time-efficient prescreening tool to identify suspicious samples which may have a quality or authenticity issue and hence require more detailed conventional analysis. However, after having established a spectral database, which currently contains spectra from more than 2000 reference juices, including ca. 800 fully authentic samples taken by SGF inspectors on site, one can clearly see that the potential of NMR analysis goes far beyond our original intentions.

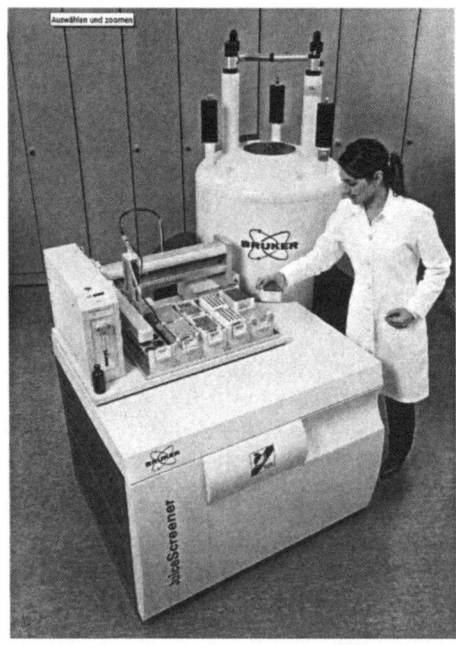

Figure 1: Typical flow injection NMR spectrometer with integrated liquid handler for sample preparation and transfer

2   METHODS AND RESULTS

Bruker SGF Profiling is based on an flow-injection Avance 400 NMR spectrometer which offers low-cost-per-sample capability (figure 1).

Each sample needs minimal preparation effort: for clear juices only the buffer addition (90 % juice / 10 % buffer) is needed, other juices like orange juice need to be centrifuged before. The buffer also contains D_2O for locking and sodium acid to suppress microorganism activity.

The fully automated acquisition procedure includes the adjustment of the temperature, tuning and matching, locking, shimming and the optimization of the pulses and presaturation power for each individual sample. Two NMR experiments are executed: a modified 1D-version of the 2D-NOESY sequence with well-defined water presaturation to allow quantitative evaluation even close to water signal and a fast 2D J-Resolved[3] which is used for safe identification of the NMR signals (figure 3). With modern NMR instruments the baseline correction is obsolete and other processing calculations such as phase corrections and referencing are done in full automation.

This setup yields in excellent spectra quality and reproducibility. Figure 2 shows 50 spectra obtained under the conditions described including sample preparation. It can be seen that phase and baseline are perfectly set in the left part and the signal positions are absolutely stable as seen in the expansion on the right side (signals of malic acid in apple juice).

The whole procedure takes about 15 minutes per sample including evaluation and reporting. Evaluation is done in a twofold mode, namely targeted and non-targeted. Targeted in this case means the identification and quantification of individual compounds, whereas non-targeted approaches apply statistical methods. The following paragraphs are describing these analyses in detail.

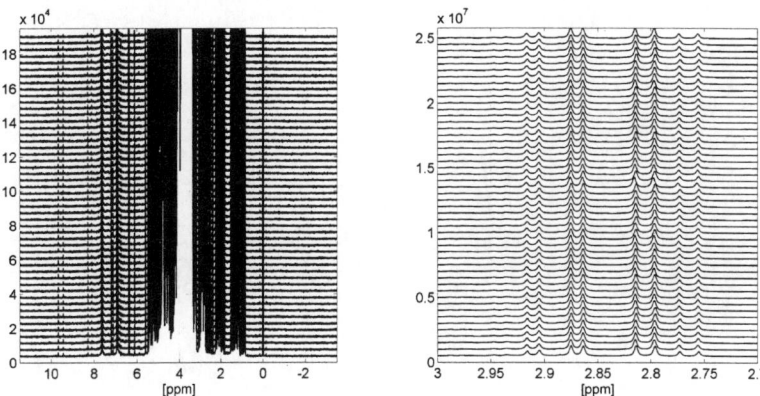

Figure 2: Reproducibility test on one apple juice 50 times injected with automatic preparation, measurement and processing. Left: whole spectrum, right: expansion of malic acid signals.

2.1 Quantification - Targeted Analysis

The primary interest for the food chemists in the classical juice assessment procedure are the concentrations of various ingredients. These values will be compared with reference standards and specific deviations in the concentration of a particular compound or in the profile of a specific combination of compounds may indicate characteristic quality problems, such as the addition of sugar. In this field the usage of NMR spectroscopy provides a clear advantage since it is possible to identify and quantify many compounds in a mixture simultaneously.

The first step in quantification analysis is the safe identification of the ingredients in the mixture spectrum. This is often not possible when using only one-dimensional datasets. The two-dimensional J-RES is a fast and efficient way, since it resolves the couplings in the second dimension. Figure 3 shows the identification of malic acid and citric acid in pear juice by using the J-RES. In the one-dimensional NOESY, the most right citric acid peak (2.78ppm) is here overlapped by the malic acid and a correct assignment in the ppm-region 2.9 to 2.96 is not guaranteed.

The quantification itself is implemented using fitting and deconvolution algorithms. At the moment, our quantification routine provides absolute concentrations for more than 20 different compounds (depending on the type of juice), i.e. sucrose, glucose, fructose, proline, alanine, HMF (5-hydroxymethyl-furfural), ethanol, methanol, acetone, phlorin, and the acids malic, citric, lactic, fumaric, quinic, succinic, citramalic, formic, benzoic, tartaric, acetic, and galacturonic. Furthermore, various useful relationships between compound concentrations are calculated, e.g. the ratio fructose/glucose or the ratio of sucrose to total sugars.

With this huge amount of concentrations it is possible to detect frauds like the addition of sugar, exhaustive enzymatic treatment (galacturonic acid), addition of citric acid or lemon juice (e.g. in apple juice), extraction of orange peel (phlorin) or the usage unripe fruits (quinic acid in apple juice).

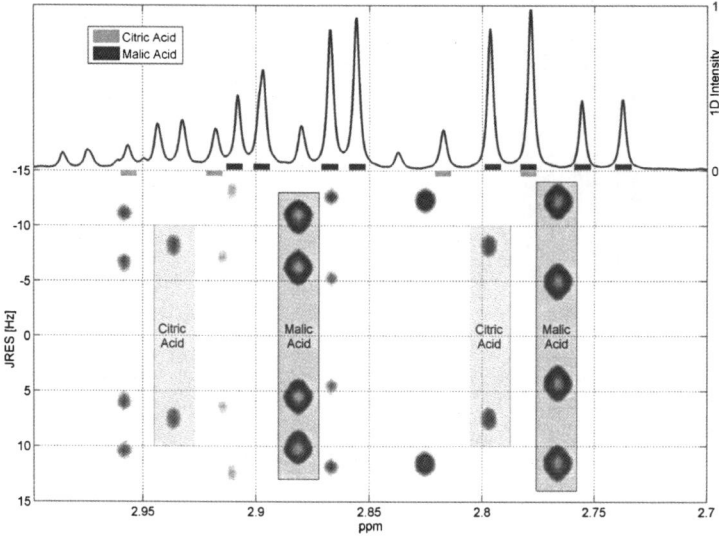

Figure 3: The J-resolved 2D-spectrum shows the region of the overlapping multiplets of citric acid and malic acid in pear juice. The center of gravity of the multiplets is easily visible in the 2D-spectrum and allows the safe assignment of signals which overlap in the 1D spectrum above

2.2 Statistical Models - Non-Targeted Analysis

In addition to compound quantification, an exhaustive statistical analysis is applied to the data. The classification and/or verification of samples are major objectives. Our approach is, first, to exactly identify the sample by cascading classification models and, second, to validate the sample with respect to its most qualificatory group (e.g., rediluted apple juice concentrate from Poland). This reduces the variance of the validation models and, therefore, increases their discriminatory power. The foundation of the statistical analyses is our large reference database of more than 2000 samples of more than 30 different types of fruit juices from more than 50 countries.

The statistical analyses are not based on the quantifiable compounds but on the whole spectroscopical data which also allows the detection of unknown deviations. To reduce the data space, the spectra are bucketed and the resulting data matrix is used as input for the statistical algorithms.

As mentioned before, the first step is the estimation of the type of sample. This is done by cascading models. The first model is able to estimate the type of fruit. This global model can differentiate currently between apple, orange/blood-orange/mandarin, sour cherry, pineapple, black currant, passion fruit, lemon, grapefruit, banana and grape. Of course, this information is usually provided with the sample's metadata and is rarely a reason for reclamation, except when orange juice (Citrus sinensis) is mixed with mandarin juice (Citrus reticulata). The latter is often cheaper so that some companies add it to orange juice without declaration (not allowed in Europe). With conventional analysis the addition of mandarin juice to orange juice is difficult to detect, but with our NMR methods and models we can detect mandarin juice at a level of 10% or more.

More specialized models can distinguish between direct juice and rediluted juice from concentrates and can detect the origin of the fruit. Figure 4 shows the results for the estimation of origin for a particular orange juice sample (left) and an apple juice sample (right). The possible sources or groups included in the models are Spain, Greece, Brazil, Belize/Mexico/Costa Rica, Cuba and USA for orange juice and Poland, Germany, China, Turkey, Brazil and Spain/Italy for apple juice. A 3D projection of the discrimination model space shows the ellipsoids of probability for each source, and the sample of interest represented by a star. Up to now, we have developed detailed classification models for orange juice and apple juice, as shown, sour cherry, lemon and pineapple. The underlying statistical method is a combination of PCA (principal components analysis) and discrimination analysis[6]. The accuracy is checked via cross-validation and Monte Carlo analyses[7].

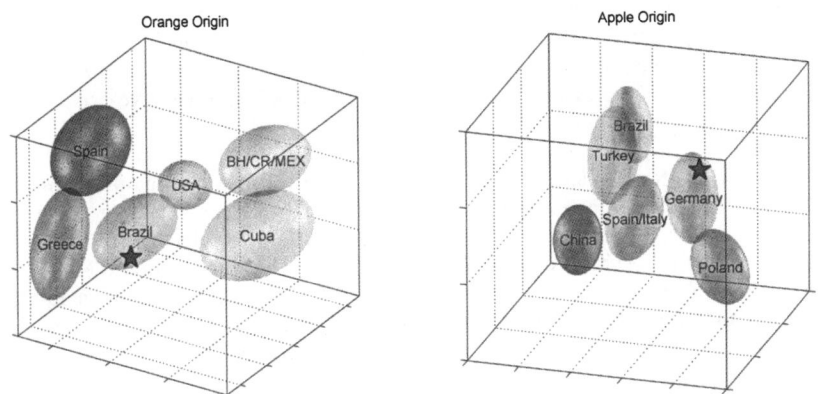

Figure 4: Estimation of the origin of an orange juice (left) and an apple juice (right). The illustrations show the model spaces with the ellipsoids of 95% probability for the various origin groups (BH = Belize, CR = Costa Rica, MEX = Mexico). The stars represent the samples to be tested (left: orange juice sample from Brazil, right: apple juice sample from Germany)

After the determination of the most likely group assignment, the sample is verified in two steps. First, a univariate analysis compares each spectral region of interest with the reference data set and detects deviations in compound concentrations. Figure 5 (left) shows a spectrum expansion near 2 ppm for a rediluted apple juice from Poland overlaid on a quantiles plot of the reference spectra; any unusual component concentrations can be easily detected. The second approach is a multivariate analysis (extended SIMCA analysis, figure 5 right) for detecting deviations which are not apparent in a univariate analysis. If both methods give the same positive result, the sample is considered "representative" and has successfully passed the prescreening trial. In this case there is no need to examine the sample further.

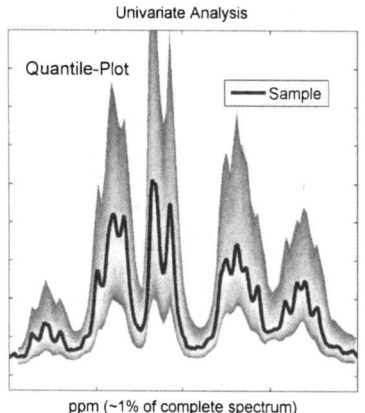
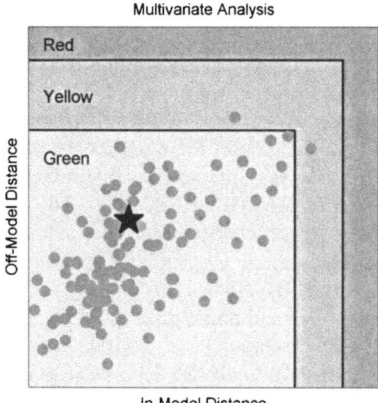

Figure 5: Verification of a fruit juice sample (rediluted apple juice from Poland). **Left:** the 400 MHz ^{1}H spectrum in the region near 2 ppm (black trace) is plotted over a quantiles plot (gray-scale) of the model spectra set (univariate analysis of apple juice at 2 ppm). **Right:** influence plot of a multivariate analysis (circles: reference samples; star: test sample; green region: representative of group; red region: not representative).

Besides the classification and verification, other statistical methods can yield important information about a sample. With a regression analysis, it is possible to quantify compounds indirectly on basis of a training set with known reference concentrations. Figure 6 shows two results of regression analyses by comparing the predicted NMR values with the known reference values of a test set. Here the two parameters titratable acid at pH 7 (left) and the ratio of glucose and fructose (right) are shown. The resulting correlation coefficients larger than 0.98 are indicating a powerful method. The used method is based on ridge regression[4].

Another important index, especially for final market samples, is the fruit content which can be computed by excluding the sugars and main acids (these components are added in products with lower fruit content) and attending the natural variances.

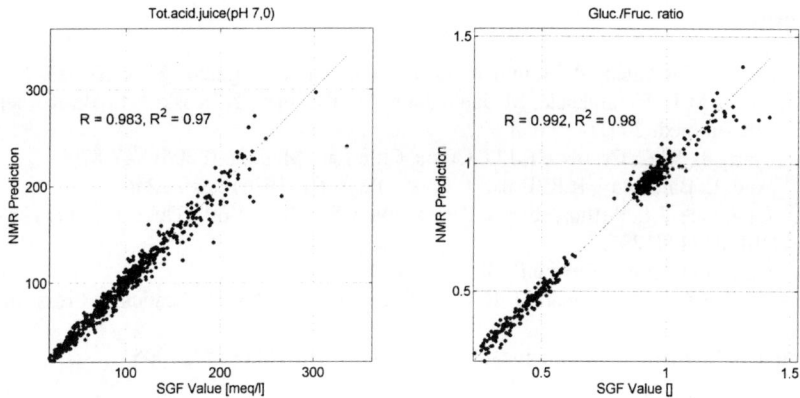

Figure 6: Determination of titratable acid at pH 7 (left) and ratio of glucose and fructose (right) by regression analyses. For a test set, the predicted values by regression (y-axis) are compared with the reference values (x-axis).

3 CONCLUSION AND OUTLOOK

In this article we have introduced the Bruker SGF Profiling method for the authentication, verification and quality control of fruit juices. In addition to the quantification of a large array of characteristic compounds, this fully automated NMR screening technique uses statistical models for the estimation of fruit content or the origin of the juice. This analysis tool can show known and unknown deviations from normality. Currently, routines are under development to identify unknown deviations by constructing spectral patterns which can be compared to an existing reference compound database[5].

The reference juice database is updated regularly, so that additional or improved statistical models will be available in the near future for fruit type, fruit origin or other discriminatory factors. Thus, the predictive power of this screening method will be refined and increased over time.

The examples presented here for the screening of fruit juices can also be seen as proof-of-principle for other upcoming applications. The same workflow (preparation, measurement, processing, reporting) and underlying mathematical methods can be easily transferred to other quality control applications, such as the screening of milk, wine or beer.

Acknowledgement

The methodology described is co-developed by Bruker BioSpin GmbH in Rheinstetten (Germany) and SGF International e. V. in Nieder-Olm (Germany) under the acronym Bruker-SGF-Profiling.

References

1   T. Suna, A. Salminen, P. Soininen, R. Laatikainen, P. Ingman, S. Maškelaš, M.J. Savolainen, M.L. Hannuksela, M. Jauhiainen, M. Taskinen, K. Kaski, M. Ala-Korpela M, NMR Biomed. 20 (2007) 658-672.
2   E.J. Jeyarajah, W.C. Cromwell, J.D. Otvos, Clin. Lab. Med. 26 (2006) 847-870.
3   W.P. Aue, E. Bartholdy, R.R. Ernst, J. Chem. Phys. 64 (1976) 2220-2246.
4   K.D. Lawrence, J.L. Arthur, Robust Regression, CRC, New York (December 11, 1989) ISBN-10: 0824781295.
5   Bruker BioSpin GmbH Germany BBIOREFCODE.
6   Kent, J.T., Bibby, J.M., Mardia, K.V., "Multivariate Analysis", Academic Press Inc., U.S., 1980
7   Eugene S. Edgington, "Randomization Tests", Marcel Dekker Ltd, 1995

SPATIAL MAPPING OF SOLID AND LIQUID LIPID IN CHOCOLATE

A.G. Marangoni[1], B. MacMillan[2], S. Marty[1] and B.J. Balcom[2]

[1]Department of Food Science, University of Guelph, Guelph, Ontario, Canada, N1G 2W1
[2]MRI Centre, Department of Physics, University of New Brunswick, P.O. Box 4400, Fredericton, New Brunswick, Canada E3B 5A3

1 INTRODUCTION

Oil migration is responsible for the poor keeping qualities of composite confectionery products. Quality defects arising from oil migration include softening of the coating, hardening of the filling, deterioration in sensory quality and a greater tendency toward fat bloom formation. For this reason, oil migration has been extensively studied, and yet a clear understanding of the oil migration mechanisms still remains a challenge.

In structures such as dark chocolate, sugar and cocoa solids are embedded in a continuous fat phase which consists of both liquid and crystalline fat[1]. Nuclear Magnetic Resonance (NMR) has been used to determine the extent of migration by measuring the solid fat content of the coatings and fillings[2-4]. It is, however, a bulk measurement, and the result is an average over the entire sample. Magnetic Resonance Imaging (MRI) is a powerful adaptation of NMR which spatially resolves an MR signal. Conventional, or spin-echo, MRI has been used extensively to study the movement of liquid fat within chocolate[4-9]. An inherent limitation in spin-echo MRI, however, is the long echo time required, at least several milliseconds and often more. Such echo times preclude the detection of signals from any solid components, since these have signal lifetimes which are generally tens or hundreds of microseconds[6,10,11]. As well, the liquid signal intensity will be prone to relaxation time weighting, and will not be representative of the true spin density.

Quantitative 3 dimensional imaging in general imposes a significant time penalty. Many physical problems (e.g. transport within materials), however, can be reduced to a 1 dimensional (1D) geometry, and therefore 1D profiling is often the most efficient approach to MRI analysis of materials. We have developed an MRI technique which is particularly well suited to imaging solids and other materials with short lived MR signals. 1D Double half-k (DHK) SPRITE[12] is so called because it is a 1 dimensional SPRITE[13] sequence which covers each half of k-space separately. When coupled with a custom high strength magnetic field gradient and a novel application of zero filling, it enables us to acquire images of very short lived signals such as the solid fat component in chocolate[14].

In this paper we present time resolved 1D images, or profiles, of the solid and liquid lipid components in chocolate during ongoing oil penetration. We began with a simplified model system consisting of hazelnut penetration into chocolate, and proceeded to a more representative two-phase system of cocoa butter and an oil based cream.

2 DHK SPRITE

A centric scan SPRITE acquisition[15] samples the central point of k-space first, such that no T_1 weighting is introduced at the k-space origin. The local image intensity at any point in the image is thus given by

$$S = \rho \exp\left(-\frac{t_p}{T_2^*}\right) \sin \alpha \qquad (1)$$

where ρ is the spin density, α the RF flip angle and t_p the phase encode time. The opportunity offered by Eq. [1] is clear. For centric scan imaging, when $t_p \ll T_2^*$, the resulting image is directly proportional to the local spin density, given that $\sin \alpha$ is a simple geometric factor which is ideally constant over the field of view. Blurring due to the decay of the longitudinal magnetization with repetitive RF pulses[15] limits the choice of flip angle, α.

In the DHK methodology (figure 1), k-space is scanned sequentially from 0 to $+k_z$ and, after a delay equal to $5T_1$, again from 0 to $-k_z$. A single data point is acquired at each gradient step, at a time t_p after the application of the RF pulse. The common 0 points are averaged and the data arranged in a single linear array prior to FFT.

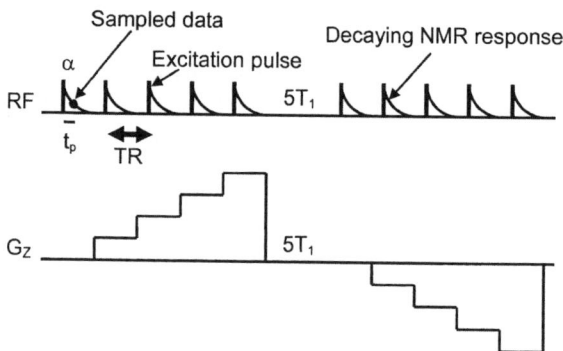

Figure 1   *The one-dimensional DHK-SPRITE sequence.   The magnetic field gradient (G_z) is ramped linearly in steps to $+G_z$ in the 1st half of the measurement and to $-G_z$ in the 2nd half. A single data point is acquired at each gradient step at a time t_p after the application of a radio frequency (RF) pulse of flip angle α. A time delay of $5T_1$ separates the two halves of the acquisition.*

With pure phase encode techniques, the field of view (FOV) is related to the step size of the magnetic field gradient, ΔG, and the encoding time, t_p, via

$$FOV = \frac{1}{\Delta k} = \frac{1}{\frac{1}{2\pi}\gamma\,\Delta G\,t_p} \qquad (2)$$

The MR signal lifetimes (i.e. T_2^*) in solid materials are characteristically short, on the order of µs. In order to successfully image these materials, t_p must be equally short. With a narrow FOV (~1-2 cm), ΔG, and therefore G_{max}, must be correspondingly large, on the order of hundreds of G/cm.

In order to accommodate these requirements, we have designed and built a one dimensional, high power gradient set[12]. With an efficiency of 1.5 G/cm/A, we can achieve a maximum gradient strength of 300 G/cm with a Techron 8710 power supply. Extensive water cooling allows us to maintain the temperature under 17°C with a maximum sustained current of 140 A.

The acquisition of true nuclear density weighted images requires $t_p \ll T_2^*$. However, even with the large gradient values available to us, this is not always possible. When T_2^* is inconveniently short, the true spin density may be recovered by acquiring a series of images with variable t_p, but constant FOV, i.e. a T_2^* map. A fit of pixel intensity to t_p will yield, as a function of position, T_2^* and spin density.

3 MATERIALS AND METHODS

For the first series of measurements, dark chocolate (*Lindt Excellence, Switzerland*) containing 70% cocoa solids was cut into discs 5 mm thick and 3 cm in diameter. The disks were placed on a hazelnut oil soaked filter paper (thickness = 125 µm) in a 9 cm Petri dish. Oil migrated upwards from the filter paper into the chocolate. During absorption, the samples were stored at 23°C and 50% RH in an environmental test chamber (*Caron 6010, Marietta, OH*). Absorption was accompanied by considerable softening of the chocolate, which necessitated minimal handling of the sample. As such, the sample mass was not recorded during this process.

MRI measurements were performed after 0, 17, 35 and 54 hours of contact with the hazelnut oil. 27 profiles were acquired for each exposure time, with each profile having a different phase encode time. Values of t_p from 15.2 to 1083 µs were used. A recovery time, TR, of 2 ms, an RF flip angle of 9° and a delay between successive gradient ramps of 1 s were used in all cases. The maximum gradient strength was 270 G/cm. A total of 32 data points were acquired per profile and zero filled to 64. The field of view was 11.3 mm. 64 signal averages were acquired per profile, giving a total acquisition time per profile of approximately two minutes.

In the second series of experiments, a layer of cocoa butter was placed in contact with a creamy mixture of peanut oil and chemically interesterified fully hydrogenated palm oil blended at a ratio of 60:40 (w/w). Samples were stored under the same conditions used previously. Cocoa butters from six different geographical origins (Brazil, China, Ecuador, Ivory Coast, Malaysia and Nigeria) were analyzed. Samples were tested in both non-tempered and tempered forms to investigate the effectiveness of tempering as a barrier to oil migration.

Profiles were acquired after approximately 0, 1, 4 and 16 weeks of contact. 32 profiles were acquired for each exposure time, with encoding times ranging from 11.2 to 1300 µs. The maximum gradient strength was 125 G/cm. While larger gradient strengths were possible, we were limited in the minimum encoding time by the probe deadtime. A recovery time, TR, of 2 ms, an RF flip angle of 9° and a delay between successive gradient ramps of 2 s were used in all cases. 64 data points were acquired per profile. The field of view was 33 mm. 16 signal averages were acquired, giving a total acquisition time per profile of approximately 66 seconds.

All imaging experiments employed a Narolac (Martinez, CA) 2.4 Tesla, 32 cm i.d. horizontal bore superconducting magnet. The console was a Tecmag (Houston, TX) Apollo. The gradient set was custom built with a single imaging axis (z) and an inside diameter (i.d.) of 7.62 cm, and was powered by a Techron (Elkhart, IN) 8710 amplifier. An eight-rung birdcage ^{1}H probe (Morris Instruments, Ottawa, ON), driven in quadrature by a 2kW AMT (Brea, CA) 3445 RF amplifier was employed for both sample excitation and signal detection.

4 RESULTS AND DISCUSSION

The FID from the chocolate is a superposition of contributions from the solid and liquid phases (see figure 2). The beat pattern is typical of amorphous crystalline materials, and results from the random distribution of crystal orientations[16]. Signal arising from the liquid decays in a simple exponential fashion. Decay of the solid signal is described by a composite function which includes a sinc function with Gaussian broadening.

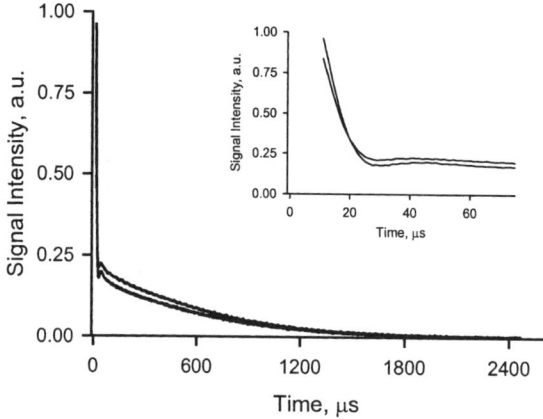

Figure 2   *The bulk FID from a chocolate sample before contact with hazelnut oil (lower line), and after 54 hrs of contact time with the hazelnut oil (upper line). The short component (expanded in inset) decays as a Sinc function with Gaussian broadening, while the long one decays exponentially.*

The time dependence of the FID can therefore be expressed as:

$$S_{obs} = A_S \exp\left(-\frac{1}{2}a^2t^2\right)\frac{\sin bt}{bt} + A_L \exp\left(-\frac{t}{T_{2L}^*}\right) \tag{3}$$

where a is the inverse of the standard deviation of the Gaussian broadening function, b is the angular frequency of the Sinc function, and A_S and A_L are the fractional sizes of the solid and liquid lipid components, respectively. The solid signal decays to less than 5% in 22 μs. Initially it accounts 88% of the signal, decreasing to 84% after 54 hours. The liquid component has a T_2^* of 580 μs.

Profiles of the chocolate collected after 0, 17, 35 and 54 hours of contact time with the hazelnut oil are shown in figure 3 for 27 different encoding times. At longer encoding times (> 80 μs), only the liquid lipid is detected. The peaks at the oil absorbing surface, on the right hand side of the images, correspond to the higher liquid lipid concentration of the infusing oil. Diffusion is slow, with the penetrating front barely moving (< 1mm) in 54 hours. This is a typical characteristic of oil penetration into chocolate[7,17].

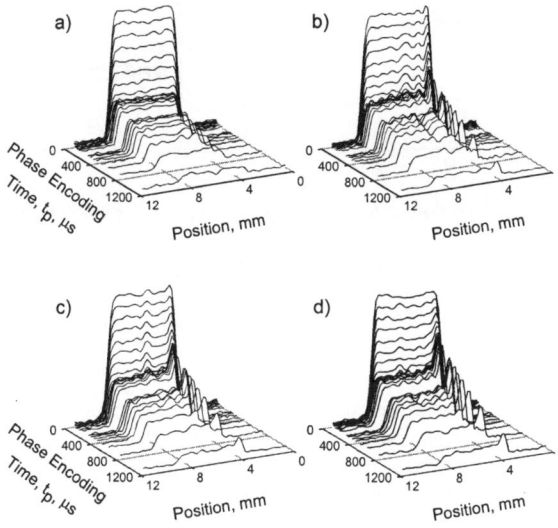

Figure 3 *a) One dimensional T_2^* weighted DHK SPRITE images, as a function of encoding time, of the uniform chocolate sample prior to exposure to oil. The encoding time, t_p, varies from 15.2 μs to 1083 μs. b) Similar images of the chocolate sample after 17 hrs of contact with hazelnut oil. At shorter encoding times, signal from the invading oil is masked by the dominant solid signal. For longer values of t_p, oil ingress is observed at the exposed surface on the right hand side of the image. Increasingly greater penetration is observed after 35 hrs (c) and 54 hrs (d).*

At the shortest encoding times, signal is detected from both the solid and liquid lipid with the solid signal dominating, owing to the greater concentration of solid species in chocolate. Note that these profiles are flat-topped, reflecting the homogeneous distribution of solid within the chocolate.

Analyzing the signal intensity on a pixel by pixel basis allows us to extract the solid and liquid components as a function of position. The same superposition of a Gaussian and exponential decay noted previously was observed here. A plot of the solid and liquid spin densities is given in figure 4. The intensity of the solid signal decreases with prolonged exposure, while that of the liquid lipid increases correspondingly. This is due to changes in the relative concentrations of solid and liquid lipid as the hazelnut oil, which is high in triolein and therefore a good solvent for lipids[18], dissolves the solid lipid as it migrates into the chocolate. This is most pronounced near the oil absorbing surface, where diffusion has progressed several mm. Interestingly, the same effect is seen away from the diffusion front, and is clearly not the result of diffusion mitigated migration. Instead, this is direct evidence of the secondary flow mechanism first reported by Guiheneuf et al.[4] Choi et al.[5] have subsequently identified this as capillary flow. While smaller in capacity, this pathway allows higher speed migration of the oil into the chocolate.

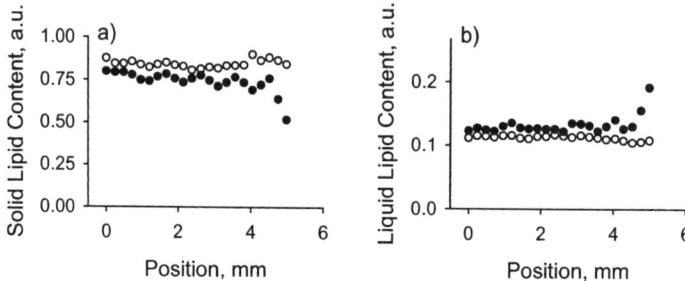

Figure 4   *a) Concentration of the solid lipid, spatially resolved, after 0 (○) and 54 (●) hrs of exposure to hazelnut oil. The exposed surface is on the right hand side of the plot. b) Spatially resolved liquid lipid concentration after 0 (○) and 54(●) hrs of exposure.*

As expected, diffusion in the cocoa butter and cream samples is slower than in the chocolate and oil samples, and requires months, rather than days. The solid and liquid components for non-tempered Cocoa Butter from China are shown in figure 5. An increase in the liquid component in the cocoa butter is accompanied by a decrease in the liquid component in the cream. Unlike the previous case, there is not a corresponding decrease in the solid component of the cocoa butter. This is a reflection of the lower solubility of cocoa butter in peanut oil.

Diffusion in this case is a Fickian process, where the extent of oil uptake is proportional to the square root of the elapsed time. The effective diffusion coefficients for each non-tempered sample are listed in table 1.

The rate of oil ingress is significantly lower in tempered cocoa butter (see figure 6). It is believed the dense and uniform crystalline microstructure obtained during tempering increases the fat matrix tortuosity and thus reduces the oil migration rate.

5 CONCLUSION

We have shown that MRI can be an effective method for quantifying the migration of oil in chocolate. With a newly developed centrically scanned imaging sequence in tandem with a dedicated high strength gradient set, we have acquired density weighted images of both the liquid and solid lipid in chocolate. This is a straightforward, robust and accurate method of acquiring MR images of the solid fat in chocolate, something that has

Table 1 *The effective diffusion coefficients for non-tempered cocoa butter from China, calculated from the liquid component of the 1D profiles.*

Country of origin	Brazil	China	Ecuador	Ivory Coast	Malaysia	Nigeria
D_{eff} (cm^2/s)	1.3×10^{-11}	3.0×10^{-9}	9.2×10^{-11}	5.0×10^{-9}	7.0×10^{-10}	7.0×10^{-11}

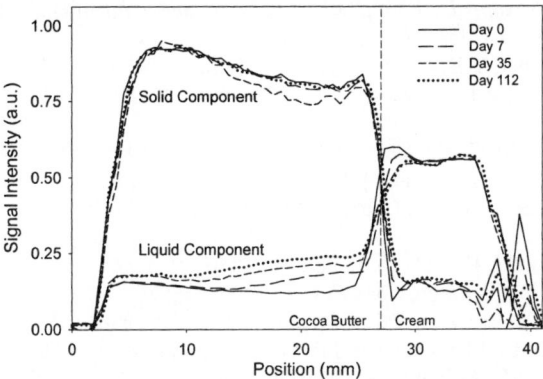

Figure 5 *The solid and liquid components in non-tempered cocoa butter from China. Diffusion is now on the timescale of months rather than days.*

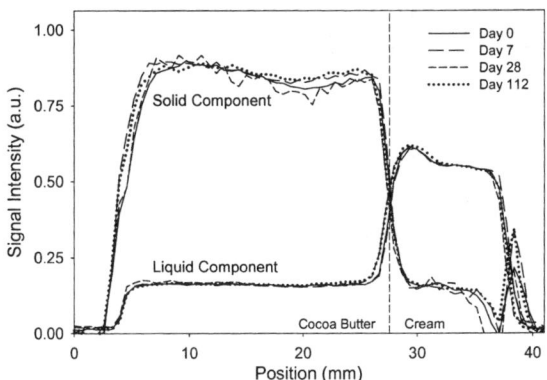

Figure 6 *The solid and liquid components in tempered cocoa butter from China. Tempering markedly reduces the rate of diffusion.*

been, up until now, very difficult, if not impossible. We anticipate that the range of applications in food and other material sciences will be very broad.

The ingress of oil into chocolate is seen to be a complex phenomenon. The dominant mechanism is a slow diffusion. The diffusion rate varies greatly with the type of oil, the phase the oil takes, and the preparation of the chocolate. Our measurements also demonstrate a much faster, albeit more limited, penetration indicative of capillary flow which proceeds in hours or days. The entire process is further complicated by the solubility of the solid lipid in the liquid lipid.

References

1   J.M. Aguilera, D.W. Stanley and K.W. Baker, *Trends Food Sci. Tech.*, 2000, **11**, 3.
2   H. Adenier, H. Chaveron and M. Ollivon, in *Shelf Life Studies of Foods and Beverages: Chemical, Biological, Physical and Nutritional Aspects*, ed. G. Charalambous. Elsevier, Amsterdam, 1993, p 353.
3   A. Ali, J. Selamat, Y.B. Che Man and A.M. Suria, *Food Chem.*, 2001,**72**, 491.
4   T.M. Guiheneuf, P.J. Couzens, H. J. Willie and L.D. Hall, *J. Sci. Food Agr.*, 1997, **73**, 265.
5   Y.J. Choi, K.L. McCarthy and M.J. McCarthy, *J. Food Sci.*, 2005, **70**, E312.
6   S.L. Duce, T.A. Carpenter and L.D. Hall, *Food Sci Technol-Leb.*, 1990, **23**, 545.
7   M.E. Miquel, S. Carli, P.J. Couzens, J.H. Wille and L.D. Hall, *Food Res. Int.*, 2001, **34**, 773.
8   M.E. Miquel and L.D. Hall, *Food Res. Int.*, 2002, **35**, 993.
9   P. Walter and P. Cornillon, *Food Res. Int.*, 2002, **35**, 761.
10 P. Lambelet, C. Desarzens and A. Raemy, *Food Sci Technol-Leb*, 1986, **19**, 77.
11 C. Simoneau, M.J. McCarthy and J.B. German, *Food Res. Int.*, 1993, **26**, 387.
12 K. Deka, M.B. MacMillan, A.V. Ouriadov, I. V., Mastikhin, J.J. Young, P.M. Glover, G.R. Ziegler and B.J. Balcom, *J. Mag. Res.*, 2006, **178**, 25.
13 B.J. Balcom, R.P. MacGregor, S.D. Beyea, D.P. Green, R.L. Armstrong and T.W. Bremner, *J. Mag. Res.*, 1996, **A123**, 131.
14 K. Deka, B. MacMillan, G.R. Ziegler, A.G. Marangoni, B. Newling and B.J. Balcom, *Food Res. Int.*, 2006, **39**, 365.
15 I.V. Mastikhin, H. Mullally, B. MacMillan and B.J. Balcom, *J. Mag. Res.*, 2002, **156**, 122.
16 A. Abragam, *Principles of Nuclear Magnetism*, Clarendon Press, Oxford, 1961.
17 J.M. Aguilera, M. Michel and G. Mayor, *J. Food Sci.*, 2004, **69**, R167.
18 G. Ziegler and K. Szlachetka, *New Food*, 2005, **8**, 45.

Food Systems and Processing

EFFECT OF SOY ADDITION ON MICROWAVABLE PARBAKED FROZEN DOUGHS

L. Serventi[1], J. Sachleben[2] and Y. Vodovotz[1]

[1] The Ohio State University, Department of Food Science and Technology, 2015 Fyffe Court, Columbus, OH, USA, 43210
[2] The Ohio State University, Department of Biochemistry, 484 W. 12th Avenue, Columbus, OH, USA, 43210

1 INTRODUCTION

Microwavable frozen baked goods are used frequently by the food industry to enrobe meat, vegetable and sweet items for convenient meal delivery.[1,2] Frozen doughs suffer from poor texture upon microwave heating. Freezing and storage of dough at -18 °C generate loss in bread quality reflected by a lower loaf volume, longer fermentation time, an increment in the size of gas cells, and less elasticity in bread doughs.[3] Ice crystal formation during freezing has been shown to be a major cause of quality loss, since the migration of water during frozen storage may cause irreversible changes to the structure of the gluten matrix.[1] Microstructure appears altered: less uniform size of air voids and fewer gluten strands are visible around starch granules.[4] Pre-baking the dough prior to freezing prolongs the shelf life of the bread keeping its freshness[5] and improving its quality.[6]

Nonetheless, microwaved doughs' exterior remains tough while the interior is leathery and difficult to chew.[7] These deleterious textural properties may arise from the short baking period of the microwave system[8] when compared to a conventional system. The altered heat and mass transfer patterns and the short baking times associated with microwaved heating cause the development of a crustless product with tougher, coarser and less firm texture compared to oven baked doughs. Some studies report incomplete starch gelatinization as a cause of leathery texture.[9] Other possible explanations involve gluten changes and rapidly generated gas steam.[9] However, Vittadini and Co-workers (1996) proposed that the loss of "freezable" water and networking was responsible for the leathery texture in microwaved pizza while retrogradation of starch was not related.[10]

In fact, microwave heating was found to slightly reduce water content (about 1%);[11] although the water loss was higher in low gluten content doughs (1.7%) because of the low water binding capacity of this protein network. Our preliminary studies[12,13] and literature[10,14] on soy bread doughs showed that their "freezable" water content was slightly higher compared with wheat controls (25% in 40% soy blend versus 23.2% in wheat control). Increased "freezable" water content is potentially related to an increase in water mobility (as may be indirectly inferred by ^{1}H NMR T_1, T_2 experiments) and after microwave heating lead to softer and more homogeneous texture.[12,13,15] However, little information is available on water state and mobility in frozen par-baked doughs.

The objective of this study was therefore to asses the effect of freezing, parbaking and microwaving on water state in wheat and soy containing doughs.

2   METHOD AND RESULTS

2.1   Dough production

Table 1   *Dough Formulations used in this study*

Ingredient (g)	Formulation 1	Formulation 2	Formulation 3	Formulation 4
Wheat Flour	120	100	80	65
HealthyHearth™ Baking Blend*	**0**	**20**	**40**	**55**
Water	72	76	80	88
Yeast	2	2	2	2
Salt	4	4	4	4

* Blend contains defatted soy flour and soy milk powder.

Four dough formulations were developed with 0, 20, 40 and 55 g soy blend (HealthyHearth™ Baking Blend) addition as shown in Table 1. Increasing the soy blend incorporated in doughs required higher amount of water in formulations while other ingredients were kept constant. The following process was used to make the doughs: ingredients were scaled, and then mixed in a high speed mixer until combined for about 1 minute (cuisine mixer, Kitchen aid, K5SS, P.O. Box 218 St. Joseph, MI, 49085). Dough was then allowed to rest for 10 minutes (hydration); rolled in a sheeter (Atlas model 150mm deluxe, Italy), baked at 145 °C for 11 minutes in a convection oven (Frigidaire, Model GLDSM 986, Martinez, GE, USA), cooled on wire racks and finally packaged in polyethylene bags.

2.2   Dough Characterization

Fresh doughs were analyzed before and after microwave heating (60 sec at High power in a convection microwave oven, Sharp Carousel II, Tokyo, Japan). Fresh doughs were also stored in a freezer (-18 °C) for 2 weeks, thawed and analyzed both before and after microwaving. The following methods were used for characterizing the changes in the doughs:

1. Nuclear Magnetic Resonance ^1H T_1 and T_2 experiments were performed using a Bruker NMR DMX 300 MHz (Saturation Recovery[16] and the CPMG sequences,[17,18] respectively, Table 2). For the T_1 experiments, one major ^1H peak was observed and thus this peak intensity was used for the fit using equation 1. For the T_2 experiments, however, 3 different peaks were observed (Figure 1) as documented previously[19] and peak 1 intensity (Figure 1) was used to fit equation 2 since these protons were most likely to represent the water signal. T_1 and T_2 fitted curves for all the formulations and treatments are shown in Figure 5.

Table 2 *^1H NMR T_1 and T_2 settings*

Parameter	T_1	T_2
Spectral Width (MHz)	18	18
Number of Scans	4	8
Acquisition Time (ms)	456	456
Pulse Length (µs)	12.50	12.50

$$T_1 \text{ INTENSITY fit: } I(t) = I(0) + P * \exp\left(\frac{-t}{T1}\right) \tag{1}$$

$$T_2 \text{ INTENSITY fit: } I(t) = P * \exp\left(\frac{-t}{T2}\right) \tag{2}$$

2. Moisture content and thermal transitions were determined by Thermo Gravimetric Analysis (TGA) and Differential Scanning Calorimeter (DSC). For TGA analysis, samples were placed in a TGA Instrument (Q 5000 TA, New Castle, DE) and heated from 25 °C to 150 °C at 5 °C/min. For DSC analysis, samples were placed in a DSC Instrument (Q 100 TA, New Castle, DE) and heated from − 50 °C to 100 °C at 5 °C/min. Transitions were analyzed using the Universal Analysis Software, Version 4.2 (TA Instruments, New Castle, DE).

3. Mechanical properties were performed using an Instron stable Micro System performing a TPA test.[20]

All tests were run in duplicate. Means were calculated with SPSS statistical software (Version 16.0, SPSS Inc., Chicago, Illinois, USA). SPSS was used to perform one-way-analysis of variance ANOVA and Least Significant Difference test (LSD) at a 95% confidence level ($p < 0.05$) to identify differences of evaluated parameters among formulations (soy effect). The same software was used to perform Independent T-test in order to identify differences of evaluated parameters within each formulation (freezing and microwaving effects); Samples were found to be statistically significant at the 95% confidence level and were designated in the statistical tables by different letters.

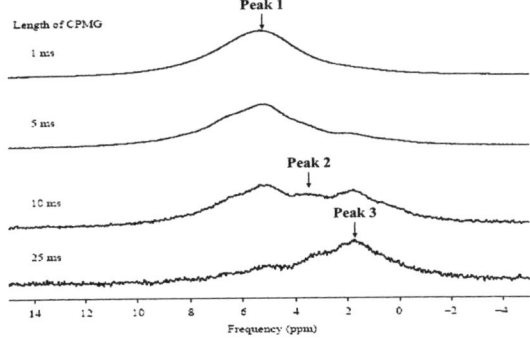

Figure 1 *^1H spectra (obtained using CPMG) used in the calculation of the spin-spin relaxation time of fresh doughs*

2.2.1 Physical properties differences upon microwaving of frozen parbaked doughs.
Soy addition in parbaked dough formulations decreased hardness and chewiness in sample
4 (Figure 2) after frozen storage and subsequent microwaving. Only high amount of
substitution (55 g of soy blend) showed an effect of soy addition on textural properties.

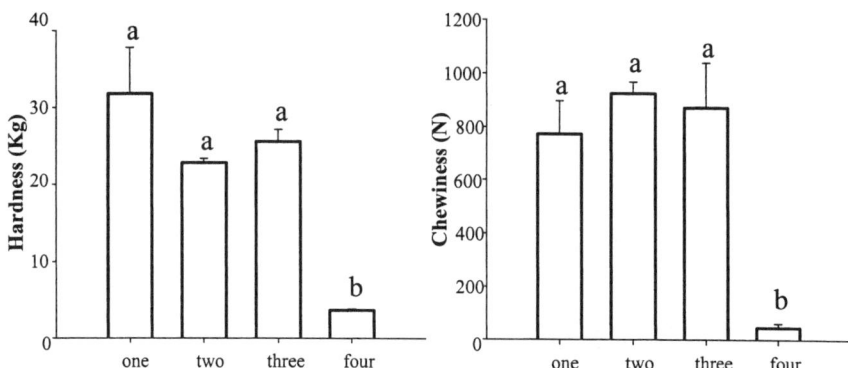

Figure 2   *Effect of soy on textural properties of frozen microwaved doughs*

Figure 3 summarizes the changes in moisture content during low temperature storage and
microwaving of dough samples with varying amounts of soy. No significant differences in
moisture content were observed due to soy addition in fresh samples (Table 3). Similarly,
no significant difference in moisture content was observed during frozen storage (Table 3).
Microwaved fresh samples 2 showed statistically significant lower moisture content
(Figure 3 B, Table 3) compared to the other fresh microwaved samples. This lower
moisture content of formulation 2 samples most likely is due to analysis error since higher
soy containing formulations (3 and 4) do not show this decrease. A previous study[11]
reported moisture decrease during frozen storage of bread samples with differing gluten
contents. In contrast, freezing and microwaving did not affect the moisture contents of the
samples in the present study (Figure 3) potentially due to the lower moisture (about 20%)
of the former compared to about 40%, of the latter.[21] Additionally, samples in the present
study were parbaked affecting their functionality. Nonetheless, moisture content of frozen
microwaved samples (Figure 3 D) could not explain the lower hardness and chewiness of
sample 4 resulted from the TPA test[20] (Figure 2).
Hardness reduction in frozen microwaved doughs has been previously attributed to the
lower amount of gluten and to the higher amounts of fat and emulsifier (soy lecithin) of
soy-containing doughs.[11] Therefore, to explore such change, differential scanning
calorimeter analysis was performed on parbaked doughs after frozen storage and
microwaving. DSC thermograms showed various events in the lower temperature region
(Figure 4). No other thermal transitions were detected. The broad melting peaks in the
range -25 °C/0 °C (Peak °T at about -15 °C) were not affected by soy addition or any of
the treatments conducted (Figure 4). This peak could be attributed to lipid melting[22] since
lipids are present in wheat flour and in soy milk powder.

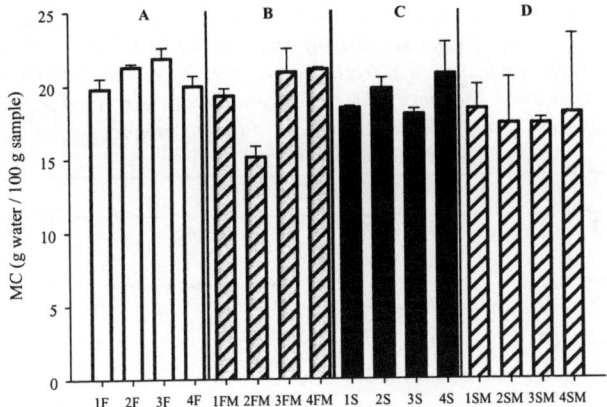

Figure 3 *Moisture content (g water/100g sample) of fresh (A), fresh microwaved (B), frozen (C) and frozen microwaved (D) doughs. 1, 2, 3 and 4 refer to respective formulations in Table 1. F, FM, S and SM refer respectively to fresh, fresh microwaved, frozen and frozen microwaved*

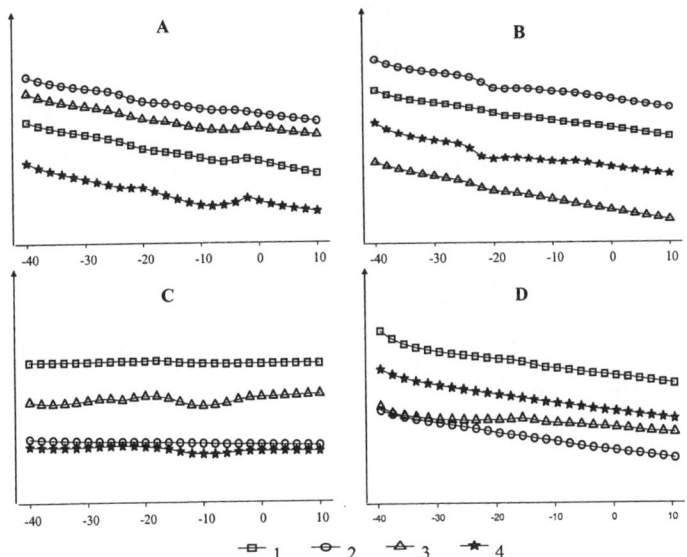

Figure 4 *DSC graphs of fresh (A), fresh microwaved (B), frozen (C) and frozen microwaved (D) doughs: heat flow (W/g) as a function of temperature (°C). 1, 2, 3 and 4 refer to respective formulations in Table 1*

Table 3   *Moisture content statistics: ANOVA one-way-test (soy) and independent T –*
test (freezing, fresh microwaving and stored microwaving). F, S, FM and
SM refer respectively to fresh, frozen, fresh microwaved and frozen
microwaved. Numbers refer to the formulations described in table 1.
Different letters refer to statistically different samples

Moisture Content									
Soy	**1**	**2**	**3**	**4**	*P value*	**Freezing**	**F**	**S**	*P value*
F	a	a	a	a	*0.242*	1	a	a	*0.21*
FM	ab	b	a	a	*0.028*	2	a	a	*0.182*
S	a	a	a	a	*0.414*	3	a	a	*0.175*
SM	a	a	a	a	*0.971*	4	a	a	*0.752*

Fresh Microwaving	**F**	**FM**	*P value*	**Frozen Microwaving**	**S**	**SM**	*P value*
1	a	a	*0.635*	1	a	a	*0.972*
2	a	b	*0.015*	2	a	a	*0.537*
3	a	a	*0.632*	3	a	a	*0.107*
4	a	a	*0.256*	4	a	a	*0.69*

2.2.2   Effect of Soy Addition and Freezing on ^1H mobility of Parbaked Doughs.
^1H NMR spin-lattice relaxation spectra on the different doughs showed only one ^1H peak
while the spin-spin relaxation spectra resulted in 3 different ^1H peaks (Figure 1). Similar
results were observed previously for soy breads[19] where the first ^1H peak (5 ppm) was
attributed to water protons, the second peak (3 ppm) to carbohydrates protons and the third
peak (2 ppm) to lipids protons. Peaks 2 and 3 were clearly visible after about 10 ms of
application of the CPMG sequence, once the intensity of peak 1 had decreased. Only the
main peak (peak 1) was processed since ^1H mobility often attributed to water is most often
related to physico-chemical changes in doughs.[14] For example, a previous study[23] showed
that frozen storage (2 weeks) caused significant shrinkage in doughs resulting in a
decreased porosity and a more viscous crumb. These changes were attributed to damage
caused by ice crystals during storage resulting in structure collapse of the crumb walls
surrounding the air spaces.[23]

T_1 of fresh dough samples showed a reduction in ^1H mobility (potentially water)
proportional to the addition of soy (Figures 5 and 6) indicating a more solid matrix in soy-
containing doughs. However, T_2 of fresh samples decreased slightly upon soy addition but
control formulation was not significantly different until highest soy substitution (formula 4,
Table 4) was reached (Figures 5 and 6). Therefore, addition of soy may have increased
water binding in the dough samples as was found for a higher moisture bread product[24]
thus decreasing the ^1H mobility.

^1H mobility (T_1, T_2) of each soy formulation did not change during frozen storage with the
only exception being the T_2 of the control (sample 1) which decreased during frozen
storage (Table 4). In previous studies water in dough was found to separate upon frozen
storage from the gluten and crystallize.[25] At prolonged storage times (longer than 2 weeks)

large ice crystals also are formed in the gas cells. During thawing of the frozen dough, the water does not return to its original state in the gluten matrix.[23] This phenomenon is particular relevant in yeast-containing doughs. However, in the present system, soy addition to these doughs may have changed the water dynamics ameliorating the effects of the freezing process. Additionally, since the dough was parbaked, the resulting lower moisture content as compared to previous studies resulted in little to no "freezable" water (Figure 4) and thus a decrease in the ^{1}H T_2 in the control samples.

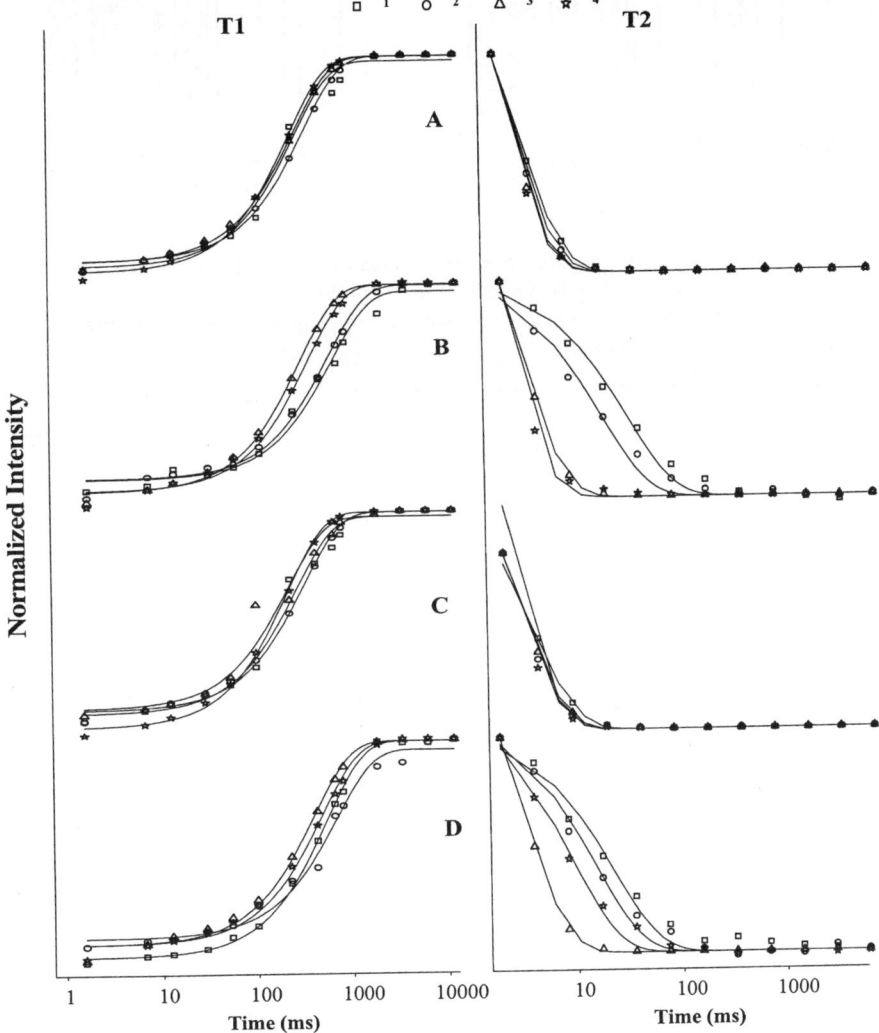

Figure 5   *^{1}H T_1 and T_2 exponential curves of fresh (A), fresh microwaved (B), frozen (C) and frozen microwaved (D) doughs. 1, 2, 3 and 4 refer to respective formulations in Table 1*

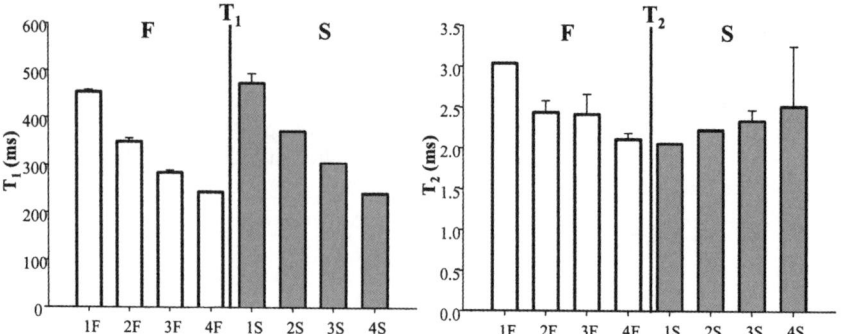

Figure 6   *Effect of freezing on T_1 and T_2 of parbaked doughs. 1, 2, 3 and 4 refer to respective formulations in Table 1. F and S refer respectively to fresh and frozen samples*

Table 4   *^{1}H NMR properties statistics: ANOVA one-way-test (soy) and independent T-test (freezing). F, FM, S and SM refer respectively to fresh, fresh microwaved, frozen and frozen microwaved samples. Different letters refer to statistically different samples*

T_1									
Soy	**1**	**2**	**3**	**4**	**P value**	**Freezing**	**F**	**S**	**P value**
F	a	b	c	d	0.047	1	a	a	0.471
FM	a	a	b	b	0.001	2	a	a	0.103
S	a	b	c	d	0.000	3	a	a	0.072
SM	a	b	c	b	0.000	4	a	a	0.579
T_2									
Soy	**1**	**2**	**3**	**4**	**P value**	**Freezing**	**F**	**S**	**P value**
F	a	ab	ab	b	0.047	1	a	b	0.000
FM	a	b	c	c	0.000	2	a	a	0.277
S	a	a	a	a	0.848	3	a	a	0.817
SM	a	b	c	c	0.000	4	a	a	0.641

2.2.3 Effect of Soy Addition and Microwaving on ^{1}H mobility of Parbaked Doughs.
Effect of microwave heating on [1]H (peak 1) mobility (T_1, T_2) of fresh parbaked samples was studied. [1]H spin lattice relaxation time (T_1) of fresh samples significantly increased after microwave heating (Figures 5 and 7, Table 5). The increase was similar for samples 1 and 2 (about 40%) while in doughs containing high amount of soy (3 and 4) the T_1 increase was lower (about 30%, Figures 5 and 7). T_2 was observed to increase dramatically in the control and low soy addition samples (1 and 2) but not change upon increased soy addition (samples 3 and 4, Figures 5 and 8, Table 5). Increased values of both relaxation times (T_1 and T_2) depict a fast motion of the ^{1}H, likely water molecules[24] potentially due to the microwave heating which liberated water from the gluten network. Soy addition (at greater

levels) had the greatest impact on T_2 where ^1H mobility was kept unchanged. These higher levels of soy blend addition resulted in a lower T_2 value (lower mobility) indicating the observable ^1H (potentially water) had greater association with the solid matrix suggesting the plasticization of polymers.

Microwave heating of frozen samples significantly increased longitudinal relaxation times in all treatments (Figures 5 and 7) and increased transverse relaxation time in samples 1 and 2 (Figures 5 and 8). Again, high level of soy addition (samples 3 and 4) resulted in no significant ^1H T_2 change. It should be noted that the T_2 relaxation was very rapid (Figure 5) and greater care will be employed to obtain more points in the relaxation curve to the reliability of these findings. Nonetheless, the trends are consistent. It therefore appears that the decrease in textural hardness of formula 4 (figure 2) is likely due to the increase of plasticization of the water fraction.

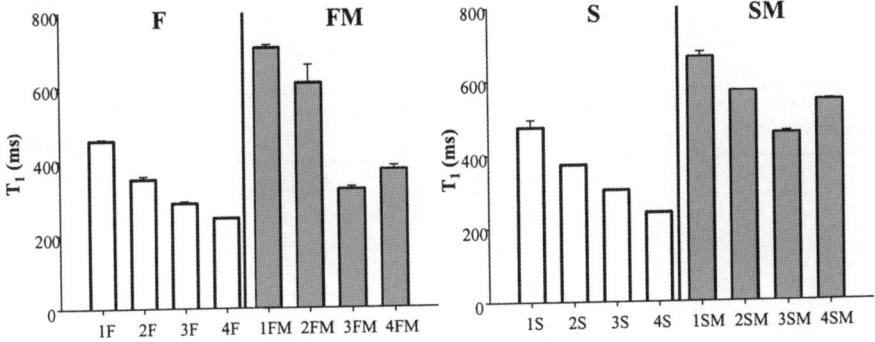

Figure 7   *Effect of microwaving on T_1 of parbaked doughs. 1, 2, 3 and 4 refer to respective formulations in Table 1. F, FM, S and SM refer respectively to fresh, fresh microwaved, frozen and frozen microwaved*

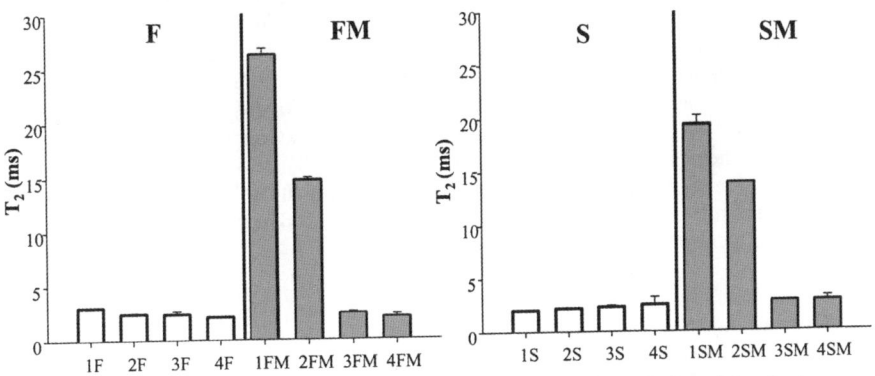

Figure 8   *Effect of microwaving on T_2 of parbaked doughs. 1, 2, 3 and 4 refer to respective formulations in Table 1. F, FM, S and SM refer respectively to fresh, fresh microwaved, frozen and frozen microwaved*

Table 5 ^1H *NMR properties statistics: independent T-test (fresh microwaving and stored microwaving). F, FM, S and SM refer respectively to fresh, fresh microwaved, frozen and frozen microwaved. Different letters refer to statistically different samples*

T_1							
Fresh Microwaving	**F**	**FM**	**P** *value*	**Frozen Microwaving**	**S**	**SM**	**P** *value*
1	a	b	0.004	1	a	b	0.025
2	a	b	0.037	2	a	b	0.002
3	a	b	0.049	3	a	b	0.024
4	a	b	0.01	4	a	b	0.003

T_2							
Fresh Microwaving	**F**	**FM**	**P** *value*	**Frozen Microwaving**	**S**	**SM**	**P** *value*
1	a	b	0.016	1	a	b	0.031
2	a	b	0.002	2	a	b	0.002
3	a	a	0.958	3	a	a	0.085
4	a	a	0.894	4	a	a	0.789

3   CONCLUSION

High amount of soy substitution into parbaked dough formulation (55 g soy blend) reduced hardness and chewiness of frozen microwaved samples resulting in a softer final product. No significant differences were detected in moisture contents of any of the treatments. Increased soy substitution in the dough formulations reduced T_1 and T_2 proportionally to soy addition reflecting the increase in solid-like protons in the soy containing matrix. Formulations 3 and 4 representing 40 and 55g, respectively, of soy blend addition stabilized the ^1H T_2 during frozen storage and microwaving and reduced the ^1H T_1 increases in these doughs indicating the decreased mobility of these ^1H. Such a decrease was likely due to the greater association of the ^1H with the solid matrix suggesting the plasticization role of the water in the high soy formulations that may relate to the softening (Instron) of these products.

References

1. J. Rasanen, H. Harkonen and K. Autio, *Cereal Chem.*, 1995, **72**, 673-642.
2. T.G. Matuda, D.F. Parra, A.B. Lugao and C.C. Tadini, *Lebensmittel-Wissenschaft und -Technologie*, 2002, **38**, 275-280.
3. P.D. Ribotta, A.E. León and M.C. Añón, *J. Agric. Food Chem.*, 2001, **49**, 913-918.
4. S. Zounis, K.J. Quail, M. Wootton and M.R. Dickson, *J. of Cereal Science*, 2002, 35, 135-147.
5. M.E. Barcenas, M. Haros, C. Benedito and C.M. Rosell, *Food Research International*, 2003, **36**, 863-869.

6. N.V. Labutina, L.I. Puchkova, Y.K. Gubiev, S.G. Ilyasov and A.M. Kats, *Khlebopekarnaya Konditerskaya Promyshlennost*, 1981, **8**, 27-28.

7. T.P. Shukla, *Cereal Foods World*, 1993, **38**, 95–96.

8. S. Hegenbert, *Food Prod Des*, 1992, **17**, 29–52.

9. Y. Yin and C.E. Walker, *J Sci Food Agric*, 1995, **67**, 283–291.

10. E. Vittadini, X.J. Chen and P. Chinachoti, *Journal of Food Science*, 1996, **61** (5), 990-994.

11. O. Ozmutlu, G, Sumnu and S. Sahin, *European Food Research and Technology*, 2001, **213** (1), 38-42.

12. K. Smith and Y. Vodovotz, poster presentation, 2003 IFT Annual Meeting Chicago.

13. Y.C. Zhang, J.R. Sachleben and Y. Vodovotz, poster presentation, 2002 Annual Meeting and Food Expo-Anaheim, California.

14. E. Vittadini and Y. Vodovotz, *Journal of Food Science*, 2003, **68**, 2022-2027.

15. M.-Y. Baik and P. Chinachoti, *Cereal Chemistry*, 2000, **77** (4), 484-488.

16. N.K. Goebel, J. Grider, E.A. Davis and J. Gordon, *Food Microstruct*, 1984, **3**, 73–82.

17. A.E. Derome, *Pergamon Press*, 1987, New York.

18. H.Y Carr, E.M. Purcell, *Physical Review*, 1954, **94**, 630-638.

19. A. Lodi, S. Tiziani and Y. Vodovotz, *J. Agric. Food. Chem.*, 2007, **55**, 5850-5857.

20. AACC Method 74-09 Measurement of Bread Firmness by Universal Testing Machine.

21. M.E. Barcenas and C.M. Rosell, *Food Chemistry*, 2006, **95**, 438-445.

22. N. Aktas and M. Kaya, *Journal of Thermal Analysis and Calorimetry*, 2001, **66**, 795-801.

23. I. G. Mandala and K. Sotirakoglou, *Food Hydrocolloids*, 2005, **19**, 709-719.

24. Y. Vodovotz, E. Vittadini and R. Sachleben, *Carbohydrate Research*, 2002, **337**, 147-153.

25. E.F.J. Esselink, H. van Aalst, M. Maliepaard, and J.P.M. van Duynhoven, *Cereal Chemistry*, 2003, **80** (4), 396-403.

IDENTIFICATION AND QUANTIFICATION OF PHOSPHORUS IN CHEESES – METHODOLOGICAL INVESTIGATIONS BY SOLID-STATE ^{31}P NMR SPECTROSCOPY

C. Rondeau-Mouro[1], M. Gobet[2], B. Mietton[3], S. Buchin[4] and C. Moreau[2]

[1] INRA, UR 1268 BIA, Plate-forme BIBS, BP 71627, F-44316 Nantes, France
[2] INRA, UMR 1129 FLAVIC, F-21000 Dijon, France
[3] ENILBIO, BP 49, F-39801 Poligny, France
[4] INRA, SRTAL, BP 89, F-39801 Poligny, France

1 INTRODUCTION

The control over fatty matter and milk protein (casein) content in cheese manufacturing is not sufficient to ensure reproducibility of textural and sensorial properties of the end product. It is difficult, therefore, to guarantee a consistent quality of cheese for the consumer. Milk is a complex fluid in which a mineral equilibrium occurs, mainly a calcium phosphate exchange, between a soluble diffusive phase and an insoluble colloid phase[1-3]. The current lack of information on this equilibrium is due to the complex composition and distribution of phosphates and to the colloidal nature of casein micelles. Phosphorus is present in various forms in milk and cheeses. We can distinguish (i) insoluble or colloidal phosphates (P_{CP}) such as phosphoproteins (P_{Ser}) within the casein micelles and colloidal calcium phosphate (P_{CCP}) associated with casein, from (ii) soluble phosphates as inorganic phosphate (**Pi**) and diesters such as glycerophosphocholine (**GlyPC**) and glycerophosphoethanolamine (**GlyPE**). The different steps in the cheese-making process coupled with the nature of the cheese composition (*e.g.*, moisture, calcium concentration...) all affects the mineral distribution and thus the textural and organoleptic characteristics of cheese[4]. It is clear that a deeper understanding of the calcium phosphate equilibrium in cheese and how it can change under processing conditions would be an important enabling tool for the dairy processing industry.

^{31}P NMR spectroscopy is a useful tool to discriminate between phosphorylated molecules in liquid or amorphous/solid-like sample with respect to their nature and dynamics. The major advantage of the NMR technique is that the sample can be analysed without pretreatment or extraction, and can be recovered since NMR is non-destructive. Phosphates in milk and in isolated casein micelles have been widely investigated using liquid-state ^{31}P NMR spectroscopy [5-8]. As the restricted motion induced by the large colloidal structure of casein micelles does not permit the obtaining of highly resolved spectra, only the mobile phosphates (a part of casein phosphoprotein residues, the dissociated inorganic phosphate and the milk fat phospholipids) found in the soluble phase were detected by liquid-state NMR.

Solid-state NMR with magic angle spinning (MAS) offers new opportunities to characterize the structure and the composition of various dairy materials. Kakalis *et al.*[9] (1994) have suggested the efficiency of ^{31}P solid-state MAS spectroscopy for investigating phosphate composition in cheddar cheese but the low spinning rate chosen in the experiments excluded further molecular identifications. Numerous studies focused on the investigation of colloidal phosphate molecules in native casein micelles. The identification of the colloidal inorganic phosphate was also attempted by analysing a series of calcium phosphates salts[10-13]. Based on the dynamic behaviour and the nature of the various phosphate molecules, the discrimination between ^{31}P NMR signals from mobile ('soluble') and immobile ('insoluble') phosphates as well as organic and inorganic phosphates in native casein micelles was performed[10,12,13].

Herein, we report a solid-state ^{31}P MAS NMR study combining single-pulse excitation (SPE), spin-echo (SE) and cross-polarization (CP) experiments to investigate three 'solid-like' but hydrated phosphorus-containing samples : a milk powder dissolved in water, a lipoproteic matrix composed of raw materials found in cheeses and a home-made semi-hard cheese. The identification of phosphorylated compounds and the determination of their proportion using line shape iterative fitting procedure were investigated in relation with the estimated biochemical composition and structural organisation (making process). Methodological limitations of solid-state ^{31}P MAS NMR techniques in regard to these highly-hydrated samples and fitting procedure for the quantification are discussed.

2 MATERIALS AND METHODS

2.1 Preparation of phosphorus-containing samples

Three different milk samples were analysed by ^{31}P solid-state NMR spectroscopy: a milk powder sample ("PL60" obtained from a tangential microfiltration of skim milk, Triballat, France) rehydrated at 68% (32% PL60 w/w in D_2O, pH=6.37), a lipoproteic matrix as a model cheese sample and a semi-hard cheese.

The lipoproteic matrix was prepared by mixing water (63% w/w), milk powder (29.6%) and anhydrous fatty matter (7.4%) in a blender. The pH was adjusted to 6.20 by an addition of δ-glucono-lactone. Once that pH reached, the coagulation was activated by an addition of rennet. After 3h at 32°C, the preparation was kept at 4°C before NMR analysis. The water/milk powder ratio in the lipoproteic matrix equals 68/32.

The semi-hard cheese, was made-up from cow milk. It is characterized by 61% w/w moisture in non-fat substance (MNFS), 2.3% calcium in non-fat dry matter (Ca/NFDM), 40% fat in the dry matter (FDM) and 4.5% NaCl in moisture rates and a pH of 4.96. For each sample, a cylindrical piece was taken from the core of the sample and inserted in a 4 mm CPMAS rotor.

2.2 Biochemical estimation of the phosphorus content

By measuring the contents of soluble calcium, insoluble calcium and lactose in milk powder and knowing the proportion of minerals in cow's milk we estimated that 90% of the soluble phosphorylated molecules (Pi and phospholids) were eliminated during tangential microfiltration. For the lipoproteic matrix, the acidification of the mix during the making process would lead to a solubilization of about 10% of the colloidal inorganic phosphate. The effect of the coagulation stage was not predictable. For the semi-hard cheese, the elimination of 95% of soluble phosphates during the successive stage of removal of water was estimated but it was not possible to predict the effect of the post-coagulation acidification stage.

2.3 NMR spectroscopy

Solid-state NMR experiments were performed on a Bruker DMX-400 spectrometer operating at a ^{31}P frequency of 161.98 MHz and equipped with a double resonance H/X CPMAS 4mm probe. Each experiment was recorded at a temperature of 298 K (± 1 K). The MAS rate was fixed at 2000 and 7000 Hz for milk powder studies and at 2000 Hz for

real and model cheese samples. The single pulse excitation sequence (SPE) used a 5 μs 90°
^{31}P pulse for an acquisition time of 52 ms during which a 45 kHz dipolar decoupling was
applied. A recycle delay of 9 s was used for milk powder and 30 s for the other samples.
The spin echo experiment (SE) was characterised by an echo time of 16 ms and the same
conditions than previously. The cross polarisation pulse sequence (CPMAS) was applied
with a 5 μs 90° proton pulse, a 2 ms contact time at 62.5 kHz and a 9 s recycle time
preceding an acquisition time of 52 ms during which the same dipolar decoupling than
above was applied. The transients for each experiment varied between 1K (lipoproteic
matrix and real cheese) and 6K (milk powder). Chemical shifts were referenced to 85%
phosphoric acid resonating at 0 ppm. The software used to fit the ^{31}P spectra was based on
the SIMPLEX optimisation (dmfit2003$^{®14}$).

3   RESULTS AND DISCUSSION

3.1 Solid-state NMR of milk powder

Reported studies on native casein micelles have indicated that ^{31}P NMR spectroscopy was
useful to determine the nature of phosphate molecules as phosphoserins and inorganic
calcium phosphate[11,13]. Quantitative assessment of the various micellar components was
also probed by Rasmussen *et al* (1997)[12]. Nevertheless, in the case of cheeses which are
highly-hydrated and heterogeneous samples, investigations by high-field NMR appear as a
true challenge. The NMR technique implies some evident technical limitations as for
instance, the sample preparation (introduction in NMR tubes or rotor), but NMR is also
dependant on the intrinsic sample properties such as heterogeneity and multi-phase
liquid/solid nature which induce some susceptibility effects and the presence of anisotropic
interactions. The feasibility of the ^{31}P NMR spectroscopy to study cheeses has been first
probed on the milk powder, main component of these dairy products.
Figure 1A shows the ^{31}P single pulse excitation spectrum of rehydrated milk powder (32%
w/w in D$_2$O) recorded at ambient temperature. This experiment has been realised in a static
configuration and results in thin signals at -0.12 and 0.40 ppm and broader peaks centred at
0.85 and 2.17 ppm which can be due to fast relaxing phenomenon and/or anisotropic
interactions generally observed in disordered or solid-state samples.

A   B   C

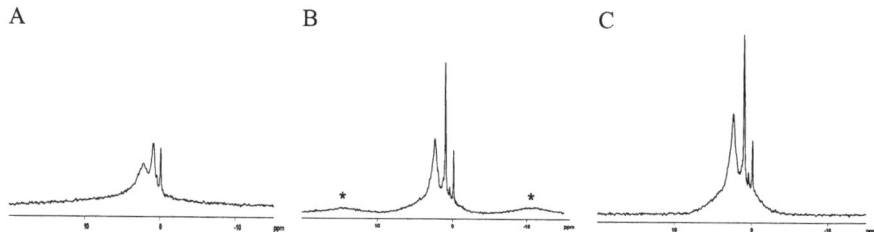

Figure 1 *Single pulse excitation ^{31}P NMR spectra of rehydrated milk powder (32% w/w in
D$_2$O) without (A) and with rotation rates of 2 kHz (B) and 7 kHz (C), * are spinning
sidebands.*

In solid state, the major source of line width for ½ spin nuclei is the dipolar interaction.
While this coupling is rather high for high-γ nuclei as protons (around 100 kHz), the
homonuclear dipolar coupling does not exceed 5 kHz for low-γ nuclei as carbons or
phosphorus. This distance-dependant interaction is accompanied by the orientation

dependency of the chemical shift. The latter phenomenon called chemical-shift anisotropy (CSA) originates from the orientation of the electronic cloud of the nuclei with respect to the static magnetic field, electron density which affects their resonance frequency. In powder samples, the large number of randomly oriented molecules induces a nuclei chemical-shift distribution resulting in the so called "powder pattern".

Based on the NMR theory, the anisotropic nature of the NMR interactions is directly linked to the orientation-dependency of the corresponding couplings with respect to B_0 and suggests that the dipolar and chemical-shift anisotropies become zero when the vector between two interacting nuclei makes an angle $\theta=54.74°$ with the static magnetic field[15]. It has been shown that averaging of these interactions can be achieved by rapidly spinning the sample inclined at $\theta=54.74°$[15,16]. This technique is known as the magic angle spinning MAS. Its application to the rehydrated milk powder sample induces the narrowing of the broad central signal into better-resolved resonances accompanied by the apparition of weak spinning sidebands. These latter signals which appear at frequency distances of integer multiples of the spinning speed, come from the incomplete averaging of the CSA. An illustration of such an experiment is displayed in Figures 1B and 1C for a rotation rate of 2 and 7 kHz using the SPE experiment. Concomitantly to the line narrowing, the intensities of the four thin peaks centred at 2.40, 0.96, 0.40 and -0.12 ppm evolved with the increase of the spinning rate.

A   B

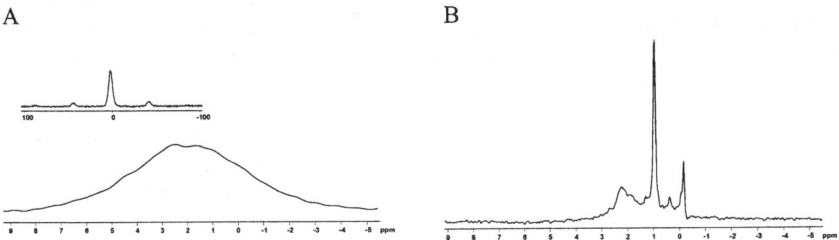

Figure 2 *Milk powder ^{31}P CPMAS (A) and spin echo (B) NMR spectra (32% w/w in D_2O).*

In complement to the SPEMAS ^{31}P spectra, it is interesting to apply other techniques useful for mobility-resolved investigations. In solid state, many NMR experiments imply the transfer of polarisation from the abundant protons to rare nuclei. It is the case for ^{13}C or ^{15}N, which are characterised by a low-γ coefficient reducing their sensibility (low spin polarisation and low signal intensity), in addition to their low abundance. In order to enhance their signal, the rare nuclei are polarised by the magnetisation of abundant and sensitive protons during a contact pulse concomitantly on protons and the rare nuclei[17]. The efficiency of this polarisation transfer is achieved when the radiofrequency amplitudes of the contact pulses on ^1H and the rare nuclei fulfil the Hartman-Hahn matching condition[18]. This process called cross polarisation (CP) has the advantage to depend on the proton relaxation times which are shorter to those of rare nuclei. Moreover, while the SPE experiment provides NMR signal from the entire composition of the studied nuclei whatever their dynamic, the CP technique specifically characterises the "immobile" or "constrained" components. Consequently, for the ^{31}P nucleus, which is abundant but 2.5 times less sensitive than ^1H, recording of a CP spectrum can be interesting for mobility-resolved studies. In complement the phosphorylated "mobile" components can be detected by a spin-echo (SE) experiment using a long echo period to ensure a significant attenuation of the immobile constituents by dephasing and relaxation phenomenon. Figure 2A and 2B display the CPMAS and spin-echo spectra of the rehydrated milk powder. The ^{31}P CPMAS spectrum displays a broad central signal centred at 2 ppm, with weak spinning sidebands indicating that CSA was not averaged to zero in these conditions. The spin-echo spectrum, acquired at 7 KHz, reveals mobile components resonating at 2.28, 0.99, 0.40 and -0.12 ppm. Small variations of the low-field signal chemical shifts are observed between the SPEMAS and SE spectra. Vortex effects due to the sample rotation can produce some pH changes inducing modifications of the chemical shifts for sensitive molecules.

3.2 Quantification of phosphorus-containing molecules using ^{31}P NMR spectroscopy

NMR spectroscopy can be useful for quantifying the amount of material present in a sample. The area under a given resonance signal is indeed proportional to the number of moles of nuclei responsible for that signal. However, great care must be exercised in choosing first the internal calibration standard and secondly the acquisition parameters, more precisely the recycle delay directly proportional to the nuclei spin-lattice relaxation time T_1. While quantification based on NMR studies is feasible on liquid-state samples, its application to solid-state ones should be considered with care. First, the addition into the sample of an internal standard with known concentration in order to calibrate the method is very contaminating and rather difficult to realise. Secondly, the low spectral resolution and the signal overlapping and broadening make the classical signal integration impossible.

The SPEMAS technique has the advantage to be quantitative if the recycle delay is well calibrated as a function of the spin-lattice relaxation times (T_1) previously measured on each signal. However, CP experiment is not much applied for quantitative aspects because the signal intensities are not only dependant of the total nuclei concentrations but also of their dynamics. In order to get quantitative reliable data from CPMAS data, a series of spectra have been recorded in order to exactly determine the match of optimum cross polarization, the proton decoupling power, the pulse width and the delay times. These parameter calibrations are essential before further analyses even if they take a long time. Concerning the spin-echo experiment, the low intensity of signals for few transients excluded precise parameter calibration. Anyway, measurements of spin-spin relaxations times (T_2) is necessary for reliable quantification of spin-echo spectra.

3.2.1 Milk powder

Identification of the phosphorus-containing components of the rehydrated milk powder have been performed using its estimated biochemical composition (Table 1) and assignments of the ^{31}P SPEMAS NMR spectrum based on literature data[6,12,13] (Table 2). As shown above, the SPEMAS milk powder NMR spectrum is characterised by a broad signal even using the magic angle spinning technique with an optimisation of the rotation speed to reduce the spinning sidebands. This large component is clearly detected using the CPMAS technique. This resolution limitation originates from dynamic effects (spin-spin relaxation T_2) but also from the molecule properties. In fact, amorphous and heterogeneous materials, such as polymers, proteins or lipids may display a distribution of isotropic chemical shifts for a single nucleus. Some programs have been developed in order to simulate the feature of individual lines composing large bands. For ½ spin nuclei, each line is characterised by an isotropic chemical shift, amplitude (integrated intensity), line shape (defined by a gaussian/lorentzian ratio) and half-width value. From these parameters and using an optimisation algorithm, the program computes a model spectrum which should be as close as possible to the experimental spectrum. This tool is very interesting in order to identify different components resonating at various frequencies but also to quantify them in term of relative ratio. For the present study, ^{31}P CPMAS has been first iteratively fitted in order to determine the chemical shift and ratio of the "immobile" components characterised by the broad signal centred at 2 ppm. These parameters were reintroduced for the optimisation of the SPEMAS spectrum, known to provide the entire phosphorus composition (Figure 1). SPEMAS spectrum was fitted with taking into account the central CP signal parameters but also the isotropic chemical shifts of signals observed in the spin-echo spectrum (Figure 3). The narrow peaks (1-4) are deconvoluted as Lorentzian line shapes while the contribution of a single broad Gaussian component (884 Hz) was needed to fit well the CP signal. Spinning bands were also matched in the SPE spectrum as a contribution of broad CP signal.

A   B

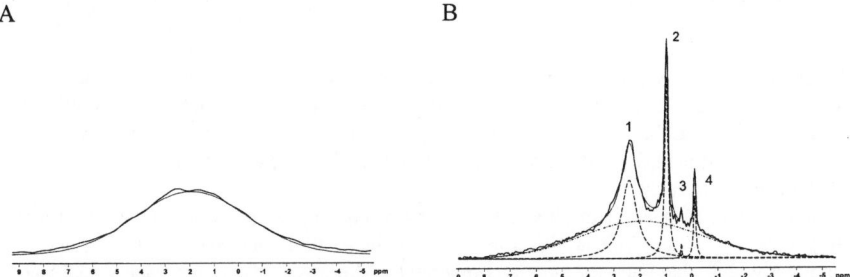

Figure 3 *Deconvoluted (A) CP and (B) SPE MAS ^{31}P spectra of rehydrated milk powder at 7 kHz (32% w/w in D₂O).*

Table 1 *Estimated composition of milk powder, lipoproteic matrix and semi-hard cheese*

Phosphorus type	Proportion (%)		
	milk powder	lipoproteic matrix	semi-hard cheese
Phospholipids	3.1	3.1	2
Soluble inorganic phosphates	6.2	6.2	4
Soluble colloidal phosphates	nd	7.6	56
Insoluble colloidal phosphates	55.6	48	
Phosphoserins	35.1	35.1	38

nd : not determined

Table 2 displays, for various milk-based matrices as the rehydrated milk powder, the assignment of each phosphorus-containing component and their ratio calculated from the NMR signal amplitude.

In the case of milk powder, integration of the NMR signal 2, 3 and 4 agrees with the estimated biochemical composition (Table 1), confirming their assignment to inorganic phosphate (Pi) and phospholipids respectively.

Table 2 *Assignment and quantification of ^{31}P NMR SPEMAS spectra of rehydrated milk powder, lipoproteic matrix and semi-hard cheese*

NMR peak	Assignment	NMR integration (%)		
		milk powder	lipoproteic matrix	semi-hard cheese
3-4	phospholipids	2.7	2.6	--
2	mobile inorganic phosphates	10.4	6.7	11.9
1	mobile colloidal phosphates	20.8	17.8	74.4
CP(*)	immobile colloidal phosphates	66.1	72.9	13.8

(*) with spinning bands

The chemical shift of peak 3 at 0.40 ppm corresponds to glycerophosphoethanolamine (GlyPE) while peak 4 at -0.12 ppm is characteristic of glycerophosphocholine (GlyPC). As already shown by Belton *et al* (1985), the chemical shift of Pi in milk depends on pH. The chemical shift of peak 2 (0.96 ppm) is in accordance with its assignment to Pi in this milk powder sample with a pH 6.37.

The assignment of peak 1, to phosphate groups of phosphoserins is not coherent with its integration and feature. Indeed, this thin signal can be partially detected using the spin echo technique which characterises mobile components. Moreover, peak 1 represents 20.8 % of the total phosphorus signal. It would represent more than the half of the phosphoserins estimated in milk powder (35.1%). It seems difficult to consider that phosphoserins present so many "mobile" phosphate groups since the milk powder is composed of intact caseins. The only "mobile" casein, κ-casein (largely located in the peripherical region of micelles) contains too few phosphoseryl groups to represent such a huge proportion of organic micellar phosphorus. Bak *et al.* (2001)[10] suggested that the colloidal inorganic phosphate exhibits an electronic environment close to hydroxyapatite (HAP) and supposed the presence of a mobile fraction of this colloidal inorganic phosphate (20% of total colloidal P). By analogy, the peak 1 is probably dominated by the contribution of the mobile colloidal inorganic phosphate but one cannot exclude a small contribution from the phosphate groups of phosphoserins. Based on its chemical shift around 2 ppm, the broad signal detected in CPMAS can be assigned to immobile organic P_{Ser} residues from caseins[10, 13]. Nevertheless, its integration (66.1 %) which is too large compared to the expected P_{Ser} content, suggests the contribution of immobile inorganic phosphates from the colloidal phase, as proposed by Thomsen *et al* [13.]

3.2.2 Lipoproteic matrix

In order to identify and quantify phosphorus-containing molecules in a dairy system, a model lipoproteic matrix composed of raw materials found in cheeses (for composition see materials and methods) was first analyzed. As observed for multi-phase highly hydrated samples, the rotor spinning at high rate can produce some vortex effects producing the extraction of the liquid phase by a centrifuge force which packs the solid phase towards the rotor walls. This effect would induce changes in the molecular structural organisation for cheese samples and may influence signal area detection and quantitative reliability of the MAS NMR experiments. To avoid this vortex phenomenon and to preserve the sample structure, the rotation speed was reduced at 2 kHz.

CPMAS and SPEMAS ^{31}P NMR spectra of the lipoproteic matrix are shown in Figure 4. The ^{31}P CPMAS spectrum appears as a broad central signal with spinning sidebands from each side (inset of figure 4A). These sidebands, which are also observed on the SPEMAS spectrum (not shown) arise from the broad resonance of the CP signal and indicate that CSA was not averaged to zero in these conditions.

In analogy with the peak assignment of milk powder signals, the three thin peaks (2, 3 and 4) of SPE spectrum which correspond to mobile phosphates have been assigned to Pi (0.88 ppm), GlyPE (0.35 ppm) and GlyPC (-0.20 ppm), respectively. The broader peak (1) at 2.32 ppm can be assigned to mainly mobile colloidal calcium phosphate ('mobile' P_{CCP}), probably mobile hydroxyapatite[10,13]. These sharp peaks are superposed on the broad signal (around 940 Hz) observed on the CPMAS spectrum. This last is decomposed into two Gaussian peaks centred at 1.70 ppm (53.6 %, bandwidth of 771 Hz) and 4.75 ppm (19.3 %, bandwidth of 716 Hz), corresponding to insoluble micellar phosphates, from caseins phosphoserins (P_{Ser}) and colloidal calcium phosphate (P_{CCP}) in interaction with them[10].

A   B

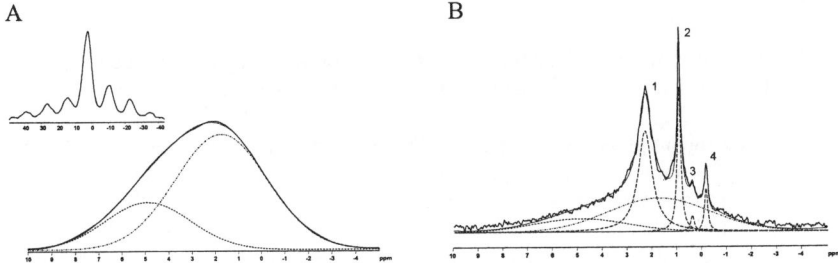

Figure 4 *Deconvoluted (A) CP and (B) SPE MAS ^{31}P spectra of lipoproteic matrix at 2 KHz.*

As expected in such a coagulated system, there is a larger proportion of immobile phosphates in the lipoproteic matrix than in the rehydrated milk powder (Table 2). The setting-up of a three-dimensional network apparently required more cross-linking phosphates, explaining the larger content of immobile colloidal phosphate while the phosphoserin ratio did not change compared to the milk powder sample. As a result, the proportions of mobile inorganic phosphates (peaks 1 and 2) decreased. Moreover, the enzymatic coagulation process seems to give rise to a new type of colloidal phosphate since a second Gaussian component (at 4.75 ppm, not assigned) is needed for the deconvolution of the CP signal of the lipoproteic matrix.

3.2.3 Semi-hard cheese

The feasibility of solid-state ^{31}P NMR in quantifying phosphorus-containing molecules in real cheeses was checked on a semi-hard cheese. Figure 5 shows a SPEMAS spectrum of the semi-hard cheese, acquired and fitted using the same procedures than above. The SPEMAS spectrum is well fitted with three peaks : a narrow (peak 2 at 0.03 ppm) and a broader (peak 1 at 0.57 ppm) Lorentzian peak corresponding to mobile phosphates as previously determined and a broad resonance originating from immobile colloidal phosphates (P_{ser} and P_{CCP}) seen in the CPMAS spectrum (not shown). Contrary to the lipoproteic matrix, the central CP signal at 1.16 ppm for cheese was fitted into only one broad Lorentzian component (199 Hz) indicating that colloid phosphates are in a more homogeneous environment and/or with higher mobility than for those of the lipoproteic matrix. Its chemical shift agrees with an assignment to phosphoserins P_{Ser}.

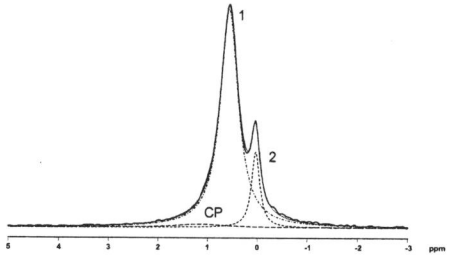

Figure 5 *Experimental and deconvoluted ^{31}P NMR SPE-MAS spectrum of cheese at 2 kHz spinning speed.*

Based on the milk powder and lipoproteic matrix spectrum analyses, we identified and estimated the proportion of mobile and immobile phosphorylated molecules in the semi-hard cheese (Table 2).

A significant change in phosphate proportion can be observed between cheese and the two other samples. The mobile/immobile phosphate ratio in the cheese was about 6 showing that phosphate proportions and dynamics are very different than those of the lipoproteic matrix (mobile/immobile ratio ~0.5) due to the cheese-making process. The higher content of mobile colloidal calcium phosphate in this cheese compared to lipoproteic matrix is in agreement with the fact that a lower pH (4.96 for cheese *vs* 6.2 for lipoproteic matrix) promotes calcium phosphate dissociation from caseins. As a consequence, the inorganic phosphates (Pi) content increases. Moreover, the low content of immobile colloidal phosphates estimated by NMR (13.8 % of the total phosphorus content) compared to the expected ratio of phosphoproteins (38 %), suggests that peak 1 is also composed of mobile phosphoserin residues.

4 CONCLUSION

These preliminary studies demonstrated that MAS ^{31}P NMR could provide a non-invasive method to identify and quantify the various phosphate molecules composing cheeses. Based on multi-sequence investigations and the estimated biochemical composition, it was possible to discriminate between mobile and immobile phosphate molecules in model dairy systems and in a real cheese. The cheese-making process revealed significant changes in phosphate proportion and dynamics. These first results are promising to provide insights into phosphate types and distribution on real cheeses according to their making process and composition.

Acknowledgement
The authors are indebted to the BRUKER BIOSPIN Company, to the European Social Fund and to the Regional Council of Burgundy for financial support (PhD grant to MG). They also acknowledge Anne-Marie Brossard (INRA Nantes) for her help in the solid-state NMR experiments.

References

1. M. A. de la Fuente, *Trends Food Sci. Tech.*, 1998, **9**, 281.
2. J. A. Lucey and P. F. Fox, *J. Dairy Sci.*, 1993, **76**, 1714.
3. F. Salaün, B. Mietton and F. Guacheron, *Int. Dairy J.*, 2005, **15**, 95.
4. R. Kapoor and L. E. Metzger, *Comp. Rev. Food Sci. Food Safety*, 2008, **7**, 194.
5. J. Belloque and M. Ramos, *J. Dairy Sci.*, 2002, **69**, 411.
6. P. Belton, R. Lyster and C. Richards, *J. Dairy Res.*, 1985, **52**, 47.
7. T. Ishii, K. Hiramatsu, T. Ohba and A. Tsutsumi, *J. Dairy Sci.*, 2001, **84**, 2357.
8. M. Wahlgren, T. Drakenberg, H. J. Vogel and P. Dejmek, *J. Dairy Res.*, 1986, **53**, 539.
9. L. T. Kakalis, T. F. Kumosinski and H. M. Farrell, *J. Dairy Sci.*, 1994, **77**, 667.
10. M. Bak, L. K. Rasmussen, T. E. Petersen and N. C. Nielsen, *J. Dairy Sci.*, 2001, **84**, 1310.
11. L. T. Kakalis, T. F. Kumosinski and H. M. Farrell, *Biophys. Chem.*, 1990, **38**, 87.
12. L. K. Rasmussen, E. S. Sorensen, T. E. Petersen, N. C. Nielsen and J. K. Thomsen, *J. Dairy Sci.*, 1997, **80**, 607.
13. J. K. Thomsen, H. J. Jakobsen, N. C. Nielsen, T. E. Petersen and L. K. Rasmussen, *Eur. J. Biochem.*, 1995, **230**, 454.
14. D. Massiot, F. Fayon, M. Capron, I. King, S. Le Calvé, B. Alonso, J.-O. Durand, B. Bujoli, Z. Gan and G. Hoatson, *Magn. Reson. Chem.* , 2002, **40**, 70.
15. I. J. Lowe, *Phys. Rev. Let.*, 1959, **2**, 285.
16. E. R. Andrew, A. Bradbury and R. G. Eades, *Nature*, 1958, **182**, 1659.
17. A. Pines, M. G. Gibby and J. S. Waugh, *J. Chem. Phys.*, 1973, **59**, 569.
18. S. R. Hartmann and E. L. Hahn, *Phys. Rev.*, 1962, **128**, 2042.

USING MRI TO STUDY TOMATO FRUIT

Maja Musse[1,2], Stéphane Quellec[1,2], Marie-Françoise Devaux[3], Mireille Cambert[1,2], Marc Lahaye[3], François Mariette[1,2]

(1) - Cemagref, UR TERE, 17, avenue de Cucillé, F-35044 Rennes Cedex, France
(2) – Université européenne de Bretagne, France.
(3) - UR1268 Biopolymères Interactions Assemblages, INRA, F-44300 Nantes, France

1 INTRODUCTION

Magnetic Resonance Imaging (MRI) is an appropriate technique for studying the internal structure and dynamics of water in plant tissues. Because of its non-invasive and non-destructive character, MRI allows repetitive measurements of the same sample and can thus be applied to study of the growing [1], ripening [2-4] and storage [5] processes. MRI can be used to evaluate various internal quality factors, such as mealiness, [6] worm damage [7] and internal browning [8] and to study changes in internal structure in plants during mechanical compression [9].

It is well known that the MRI signal depends both on the physicochemical properties of samples and on the acquisition parameters. Physicochemical parameters, such as proton density, spin-lattice (T_1) and spin-spin (T_2) relaxation times and diffusion coefficient, provide important physiological information and reflect phenomena such as exchange of water between sub-cellular compartments. Extracting individual physicochemical parameters and interpreting them is therefore often the best way to perform a physiologically meaningful analysis of MRI images. However, the physiological interpretation of relaxation times is not obvious because they can reflect the combined effects of different but simultaneous influences of several chemical and physical properties of the sample and also depend on the acquisition technique. It is therefore useful to combine MRI investigations with an independent experimental technique to complement the interpretation.

Factors determining the sensory texture properties of tomato fruit and their changes during ripening have been investigated in many studies. Tomato texture is affected by both the cellular structure and the biochemical composition of the tissue. Texture also depends on cell turgor, mainly due to the vacuole pressure inside the cell [10, 11]. As in material science, tissue organisation, itself dependent on cell morphology, cell arrangement and cell and tissue properties, also influences the mechanical properties of plants. MRI can provide information at the cell level via measurement of the relaxation times and at the macroscopic level via image analysis. One important advantage of MRI is that it makes it possible to perform repetitive studies on the same sample due to its non-destructive character. Only a few MRI studies of tomato fruit structure have been reported to date and there has been no detailed quantitative investigation of the ripening process.

The aim of this investigation was to perform a quantitative study of tomato fruit and its post-harvest ripening by means of measurement of spatially resolved relaxation times and evaluation of air bubble content. The physical parameters, such as cell size and susceptibility effects which influence relaxation processes, were also investigated in order to contribute to interpretation of the relaxation times.

2 METHODS

2.1 MRI Measurements

Nine Tradiro tomatoes were picked at the same initial green stage and left to ripen at 18°C until a ripe red stage. The MRI experiments were performed on tomatoes at several ripening stages using a 0.2 T electromagnet scanner (Magnetom Open, Siemens, Germany). Images of the tomato median equatorial plane were acquired with a spatial resolution of 1 x 1 x 5 mm^3 and the following tissues (Figure 1) were studied: the core (C), placenta (P), radial pericarp (RP), outer pericarp (OP) and locular tissue (LT).

Spoiled gradient echo (GE) images were acquired with TR=1000ms and TE_1=9 ms and TE_2=40 ms, for the first and second image types, respectively. The gas bubble content was estimated from the relative images obtained by dividing the 40 ms echo time images by corresponding 9 ms echo time images.

T_2 maps were obtain from multi spin echo (MSE) images with 32 consecutive echoes, 30 ms inter-echo time and TR=10 s. T_1 maps were measured using the TOMROP sequence [12] with 32 consecutive echoes, TI=210 ms and TR=10 s. T_2 and T_1 maps were calculated on a pixel-by-pixel basis using corresponding monoexponential functions, via the Levenberg-Marquardt criterion for chi-square minimization

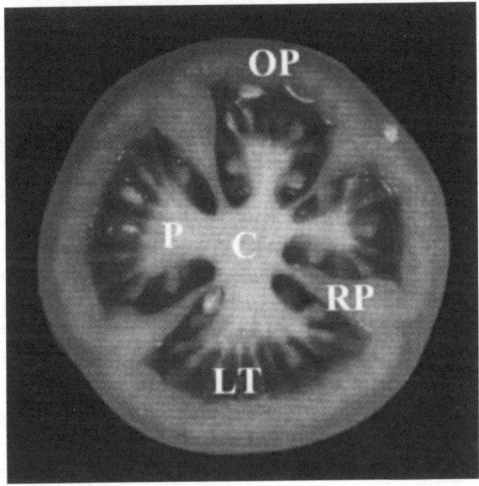

Figure 1 Camera image of the tomato median equatorial plane with tissue annotation: C - core, P - placenta, RP - radial pericarp, OP - outer pericarp and LT - locular tissue.

2.2 MRI Data Analysis

Analysis of variance was performed using the software package Statgraphics Plus (Centurion) to detect significant differences (p< 0.05) between measurements on different days. The two-factor ANOVA test was performed on the results of the MRI measurements corresponding to each measurement day (tissue type and tomato sample) and the F-ratio was calculated to measure how different the means at 95% Least Significant Difference (LSD) confidence level were in relation to the variations within each sample.

2.3 Macro-vision Imaging

Macro-vision Imaging was performed on the samples taken from the placenta and outer pericarp tissues. Tissue sections of 250 μm in thickness were observed using a macro-vision system as described by Devaux et al [13] comprising a CCD camera (Sony XC 8500 CE, Alliance Vision, Montélimar, France) fitted with a 50 mm lens (f 1:1.8 Nikon) and a 20 mm extension tube. Samples were back-lit using a fiber-optic ring-light supplied by Polytec (Pantin, France). The camera and lens were adjusted to observe a 10.7 mm x 14.4 mm area. Images were digitized in 576 x 768 pixels of 18.6^2 mm². After preliminary observation, samples were placed in an ultrasound bath for 15 s and in a dessicator connected to a pump for 30 s to remove air bubbles. Images were compared before and after degassing.

Image resolution did not allow segmentation of cells, and images were considered for visual texture according to both cell morphology and arrangement. Gray level granulometric methods were applied to extract overall information concerning cell size distribution [13].

3 RESULTS

3.1 MRI Macroscopic Analysis and Estimation of Air Bubble Content

An example of the relative GE images of a relatively ripe tomato, obtained by dividing the 40 ms-echo time images by corresponding 9 ms-echo time images, is shown in Figure 2 A, including the tissue nomenclature. The images reflect the morphological features of the tomato. All tomatoes were composed of the major regions shown in Figure 1. On the image of the green tomatoes (not shown), the areas of signal void corresponding to air spaces were observed inside the locular tissue. These areas mainly disappeared on the images of red fruit, indicating that the air spaces filled up during ripening.

Contrast in the relative GE images mainly depended on the presence of gas bubbles, as the T_2 of all tomato tissues were very long compared to the TE of gradient echo images. Tissues rich in air were characterized by lower relative signals than tissues with lower air content. In order to evaluate variations in air bubble content inside the tomato fruit the relative signals were compared between different tissue types. Mean relative signal and the intervals around each mean calculated for an intermediate tomato ripening stage (as in image A) are shown in Figure 2 B. Any pair of intervals that do not overlap correspond vertically to a pair of means that are statistically significantly different, showing that all the tissues were different except that there was no distinction between the core and the placenta. The F-ratio was much higher for the tissue effect than for the tomato sample effect (F-tissue=291, F-tomato=2). The relative signal remained almost constant throughout the ripening process for the core, placenta, radial pericarp and locular tissues of all the tomatoes studied, indicating that in these tissues the air bubble content did not change. However, the relative signal of the outer pericarp in a few tomatoes became grainy as the fruit ripened.

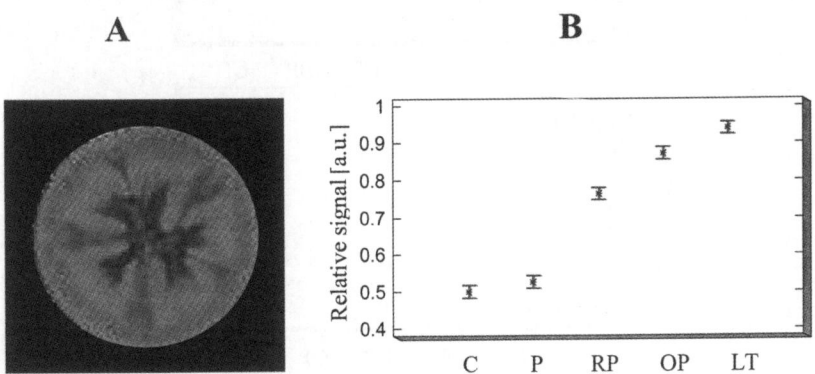

Figure 2 A) Example of relative images obtained by division of the 40 ms echo time GE image by the corresponding 9 ms echo time image. B) Mean relative signal and intervals around each mean at 95% LSD confidence level calculated for each tissue type.

3.2 Relaxation times and proton density measurements

Mean values of T_2 and T_1 for different tomato tissues were extracted from the relaxation maps. The results corresponding to the green and ripe red tomato stages are shown in Figure 3. Relatively short relaxation times were measured in the placenta, core and radial pericarp for all ripening stages compared to the relaxation times in the pericarp and locular tissues. T_2 and T_1 increased gradually through the radial pericarp from the placenta to the outer pericarp. This might be related to progressive changes in the structural aspects of these tissues. In the case of the T_2 measurements, there were three groups of statistically different tissues at the green stage: 1) the core and radial pericarp; 2) the placenta and 3) the outer pericarp and locular tissue (F-tissue=225; F-tomato=4). At the ripe red stage the locular tissue became different from the outer pericarp (F-tissue=176; F-tomato=3). In the case of the T_1 measurements, groups of statistically different tissues were found for the green stage: 1) the core; 2) the placenta and radial pericarp and 3) the outer pericarp and locular tissue (F-tissue=22; F-tomato=5). For the ripe red tomatoes (F-tissue=44; F-tomato=3), the T_1 of the locular tissue was different from the outer pericarp, as for T_2 measurements, but the core, placenta and radial pericarp belonged to the same group. In all cases, the F-ratio was considerably higher for the tissue effect than for the tomato sample effect.

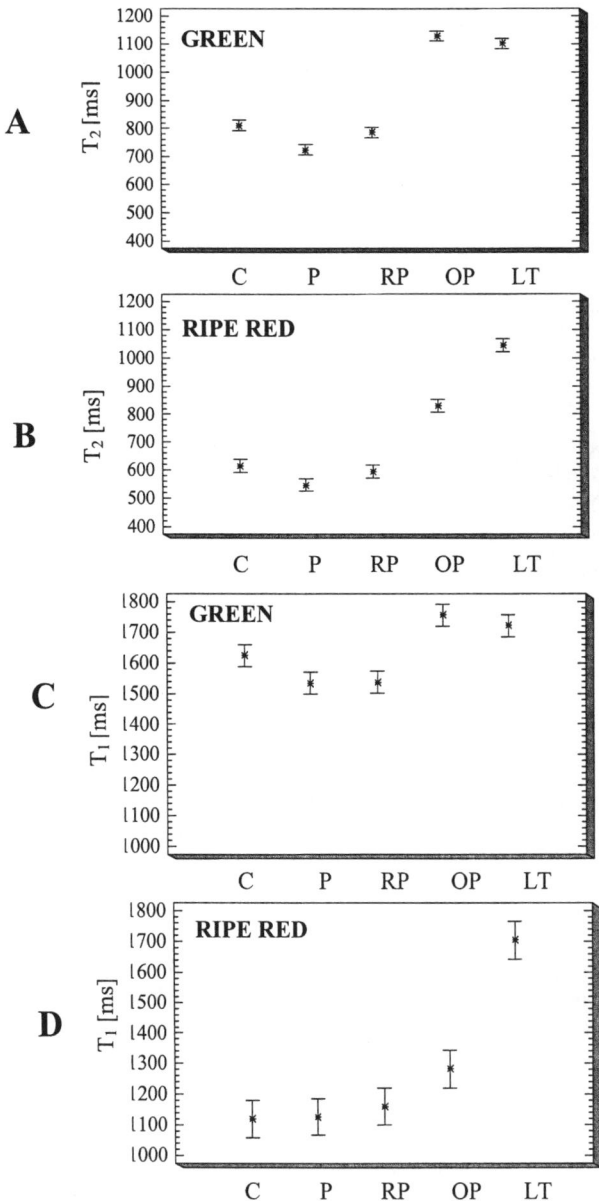

Figure 3 Plot of mean T_2 for each tissue type, and the intervals around each mean at 95% LSD confidence level for the green (A) and ripe red tomato stages. (C) and (D) are the T_1 relaxation times for the same pictures as (A) and (B).

The effects of fruit ripening on the T_2 and T_1 relaxation times were clear. T_2 and T_1 of the core, placenta, radial and outer pericarps decreased as the fruit ripened. These changes were relatively substantial (see Figure 3) and the patterns were similar for all the tomato samples studied (not shown). Relaxation times in the locular tissue remained almost unchanged between the green and the ripe red stages. However, an increase in T_2 and T_1 values occurred during the intermediate ripening stage.

Proton density maps were calculated from the set of MSE and TOMROP images. We assumed that they reflected the density of the water protons although other protons (fat, sugar) may have contributed to the signal intensity. The results showed that the locular tissue had the highest proton density, following by the outer pericarp, radial pericarp, placenta and core. No distinct variations were observed according to ripeness.

3.3 Cell morphology

Macro-vision images of the placenta and outer pericarp are presented in Figure 4. Images were obtained after sample degassing. They represented 8 x 3.2 mm² areas obtained by zooming on the initial 10.7 x 14.4 mm² images. White spots in the tissues corresponded to the vascular bundles (Figure 4).

Cell morphology differed between the placenta and outer pericarp. While cells were large and mainly elongated perpendicular to the cuticle in the outer pericarp, cells in the placenta were much smaller and no specific orientation was observed. The results from the granulometric study showed that cell sizes in the outer pericarp were about 200x600 µm² and about 100x100 µm² in the placenta.

A   **B**

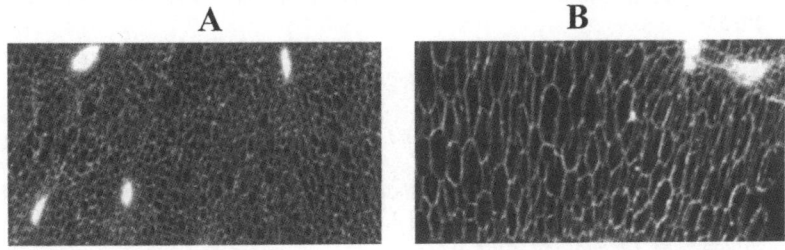

Figure 4 Macro-vision images of placenta (A) and outer pericarp (B) tissues. The FOV of 8 x 3.2 mm² was obtained by zooming on the initial 10.7 x 14.4 mm² images.

DISCUSION AND CONCLUSION

We showed in this study that quantitative evaluation of MRI parameters reflects differences in properties between tomato tissues. Parallel MRI and camera experiments provided greater understanding of the relaxation processes and contributed to the interpretation of origins of contrast in MRI of plant tissues. The core and placenta tissues of the tomato were characterized by lower relaxation times than other tissues, in agreement with the results of the cell size estimations from macro-vision images, which demonstrated that cells of the outer pericarp were significantly larger than cells of the placenta. In addition, proton density maps revealed that the water content was correlated with the relaxation times of tissues. This study also demonstrated significant variations in porosity between tomato tissues that probably have an additional role in T_2 results.

The degree of changes in both T_2 and T_1 suggests that they are sensitive to the ripening process. However, air bubble content, cell size and proton density remained constant throughout the maturation process and the dependency of relaxation time on ripening stage cannot therefore be explained by these properties. Further investigations into the phenomena contributing to the changes in relaxation times are in progress.

References

1   C. J. Clark, A. C. Richardson and K. B. Marsh, *Hortscience*, 1999, **34**, 1071.
2   C. J. Clark and J. S. MacFall, *Magn. Reson. Imaging.*, 2003, **21**, 679.
3   N. Ishida, T. Kobayashi, M. Koizumi and H. Kano, *Agric. Biol. Chem.*, 1989, **53**, 2363.
4   M. E. Saltveit, *Postharvest Biol. Technol.*, 1991, **1**, 153.
5   H. C. W. Donker and H. Van As, *Biochimica Et Biophysica Acta-General Subjects*, 1999, **1427**, 287.
6   P. Barreiro, A. Moya, E. Correa, M. Ruiz-Altisent, M. Fernandez-Valle, A. Peirs, K. M. Wright and B. P. Hills, *Appl. Magn. Reson.*, 2002, **22**, 387.
7   Y. Iwahashi, A. K. Horigane, K. Yoza, T. Nagata and H. Hosoda, *Magnetic Resonance Imaging*, 1999, **17**, 767.
8   N. Hernández-Sánchez, B. P. Hills, P. Barreiro and N. Marigheto, *Postharvest Biology and Technology*, 2007, **44**, 260.
9   J. J. Gonzalez, K. McCarthy and M. J. McCarthy, *Journal of Texture Studies*, 1998, **29**, 537.
10 R. J. Redgwell and M. Fischer, *Fruit texture, cell wall metabolism and consumer perceptions. In: Knee, M. (Ed.), Fruit Quality and its Physiological Basis*, Sheffield Academic Press, 2002.
11 R. L. Jackman and D. W. Stanley, *Trends in Food Science & Technology*, 1995, **6**, 187.
12 S. Pickup, A. K. W. Wood and H. L. Kundel, *J Magn Reson Imaging*, 2004, **19**, 508.
13 M.-F. Devaux, B. Bouchet, D. Legland, F. Guillon and M. Lahaye, *Postharvest Biology and Technology*, 2008, **47**, 199.
14 H. T. Edzes, D. van Dusschoten and H. Van As, *Magn. Reson. Imaging.*, 1998, **16**, 185.

SNIF-NMR AND CHEMOMETRIC METHODS APPLIED TO ^{1}H NMR IN THE STUDY OF BRAZILIAN BRANDY AUTHENTICITY

Elisangela Fabiana Boffo[1], Márcia Miguel Castro Ferreira[2], Antonio Gilberto Ferreira[1]

[1] Departamento de Química, Universidade Federal de São Carlos, São Carlos - SP, Brazil.
[2] Instituto de Química, Universidade Estadual de Campinas, Campinas - SP, Brazil.

1 INTRODUCTION

The terminology *aguardente,* used for Brazilian brandy, designate the alcoholic beverage obtained by fermentation and distillation of sugars from several sources as sugar-cane, honey, pineapple, banana and grape, among others[1]. The sugar source characterises the produced beverage type and shows peculiar chemical composition and sensory profile. The difficulty is to characterise the correct source of material fermentation, since the ethanol produced is chemically the same, and besides, a huge difference in the price and in the products quality is observed.

Site Specific Natural Isotopic Fractionation studied by Nuclear Magnetic Resonance (SNIF-NMR)[2] is proving to be an efficient technique to control the food authenticity and adulteration when the biosynthetic origin of product is in question[3]. This technique provides isotopic criteria to characterise a biochemical transformation such as fermentation and enables measuring the isotopic ratios for the end products which can be correlated with the precursors[4,5]. Therefore, the biosynthetic origin of *aguardente* can be correlated with the sugar, from which it is originated, through the biosynthetic mechanisms C_3, C_4 and CAM that are used for the plants to fix the CO_2.

However, no information about other components in *aguardente* is obtained via this method. ^{1}H NMR spectroscopy is a strong tool for assessing additional constituents, such as aromatics, carbohydrates and acids. However, ^{1}H NMR spectra of simple foods are often complex. For this reason, it is advantageous to analyze the spectra by multivariate methods[6,7]. Chemometric methods, as Principal Component Analysis (PCA)[8] and Hierarchical Clusters Analysis (HCA)[9], allow describing samples clustering and detecting the biochemical compounds responsible for the samples separation[10]. The main use of PCA is to reduce the dimensionality of a data set while retaining as much information as is possible. It generates a compact and optimal description of the data set[11]. The aim of HCA is to reduce the complex data to a minimum and to highlight the natural groupings in the data. The graphical output of HCA is a dendrogram, a tree-like chart which allows visualization of clustering[12].

The aim of this work is demonstrate the application of SNIF-NMR technique and chemometric methods applied to ^{1}H NMR to determine Brazilian *aguardente* authenticity.

2 MATERIALS AND METHODS

2.1 Samples

Forty *aguardentes* obtained from different sources were studied. Some samples were provided by the manufacturer and others were bought in the Brazilian stores, particularly in the S. Paulo state. All samples were collected in the years 2005 and 2006.

2.2 SNIF-NMR analysis

^{2}H and ^{1}H NMR spectra were acquired at room temperature on a Bruker DRX400 9.4 Tesla spectrometer, using a 5 mm direct-detection probe head without fluorine lock device.

The *aguardentes* were distilled and the ethanol was collected with the boiling point in the range of 76 - 78 °C. Samples were prepared, in triplicate, using 600 μL of ethanol and 100 μL of *N,N,N',N'*-tetramethylurea (TMU), 99.0 %, used as an internal standard. Tetramethylsilane (TMS) was used as internal reference (δ 0.0).

^{2}H NMR spectra (61.4 MHz) were acquired using broadband ^{1}H decoupling; spectral width, 983 Hz; 10240 data points; acquisition time, 5.2 s; relaxation delay, 3.0 s; pulse (90°), 17.5 μs and 1024 FIDs were accumulated. Spectra were processed with zero-filling using an exponential multiplication associated to a line broadening (LB) of 1.0 Hz.

^{1}H NMR spectra (400.2 MHz) were acquired using 65563 data points; spectral width, 4664 Hz; acquisition time, 5.2 s; relaxation delay, 3.0 s; pulse (90°), 10.5 μs and 16 FIDs were accumulated. Spectra were processed with zero-filling using a LB of 0.3 Hz. The phase and baseline were corrected in both spectra.

The isotopic ratios were determined to methyl $(^2H/^1H)_I$ and methylene $(^2H/^1H)_{II}$ sites of ethanol using ^{2}H and ^{1}H NMR. Quantitative data were obtained by manual integration of sample and internal standard peaks and using equation 1[13,14]

$$\left(\frac{^2H}{^1H}\right)_i^A = \frac{I_i^A}{I^P} * \frac{P^P}{P_i^A} * \frac{m^P}{m^A} * \frac{M^A}{M^P}\left(\frac{^2H}{^1H}\right)_i^P \tag{1}$$

where I_i^A and I^P are the areas of signal i of A and TMU methyl in the ^{2}H NMR spectrum. P_i^A and P^P are the stoichiometric numbers of hydrogens in site i and in the TMU. M^A, m^A and M^P, m^P are the molecular weight and mass of the A and the TMU, respectively.

2.3 Chemometric analysis applied to ^{1}H NMR spectra

Samples were prepared, in triplicate, using 600 μL of *aguardente* and three drops of D$_2$O. Sodium-3-trimethylsilylpropionate (TMSP-2,2,3,3-d$_4$) was used as internal reference (δ 0.0).

All ^{1}H NMR spectra were obtained with three signals suppression, i. e., water, methyl and methylene of ethanol signals in a 5 mm inverse-detection probe head. Eight FIDs were collected as 65536 data points using a 8.5 μs pulse (90°); spectral width, 8013 Hz; acquisition time, 4.1 s and relaxation delay, 6.4 s. Spectra were processed using 32768 data points by applying an exponential line broadening of 0.3 Hz for sensitivity enhancement before Fourier transformation and were accurately phased, baseline adjusted and converted into JCAMP format to build the data matrix.

All calculations were carried out using the Pirouette® software (v. 3.11, InfoMetrix, Woodinville, Washington, USA). The data matrix was built with 3815 variables (columns) and 78 spectra (lines – 26 samples in triplicate). PCA was applied for exploring the data and for feature selection. Each NMR spectrum was normalized to norm one (the area under the sample profile is set equal to one) and first derivative was taken. The data were also autoscaled, i. e., meancentered and scaled to unit variance, in order to give equal weight to each variable and so, large and small peaks gained the same importance. In HCA, the Euclidean distances among samples are calculated and transformed into similarity indices ranging from 0 to 1 by using the incremental linkage method.

3 RESULTS AND DISCUSSION

3.1   SNIF-NMR analysis

In Figure 1, the ^{1}H and ^{2}H NMR spectra of ethanol from a grape *aguardente* are shown, wherein the ethanol, TMU and water signals can be visualised. The results of SNIF-NMR study are shown in the Table 1.

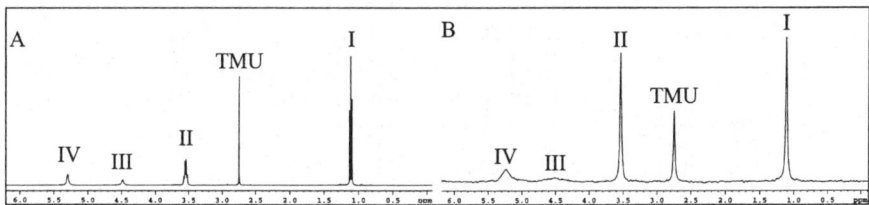

Figure 1   *Natural abundance NMR spectra of A) ^{1}H and B) ^{2}H of ethanol from a grape aguardente, I methyl, II methylene, IV hydroxyl of ethanol and III water signals*

Analysing the isotopic relations (Table 1), it can be observed that the $(^2H/^1H)_{II}$ values are in the range from 126.0 to 129.7 ppm. However, the $(^2H/^1H)_I$ values are useful for the discrimination of sugars produced by the C_3 and the C_4 metabolic pathways[2]. Sugars from C_4 plants have a higher content of heavy isotope than C_3 sugars.

For the banana and grape standard *aguardentes* and two commercial grape *aguardentes* the $(^2H/^1H)_I$ values were observed between 99.7 and 102.8 ppm. These values are characteristic of plants that fix CO_2 using the C_3 mechanism.

$(^2H/^1H)_I$ values of honey *aguardentes* point out that the honeys were probably made from C_3 plants nectars, because they are similar to those of C_3 plants and to those found in honey authenticity studies[15,16].

The cassava *aguardente* (C_3 plant) had a $(^2H/^1H)_I$ value of 99.7 ppm, characteristic of this plant type and confirming its authenticity.

For the sugar-cane and maize *aguardentes* (C_4 plants), $(^2H/^1H)_I$ values are in the range from 111.7 to 113.6 ppm. Therefore, a satisfactory discrimination (approximately 10 ppm) between C_3 and C_4 *aguardentes* has been observed.

For all banana, peach, coconut and pineapple commercial *aguardentes* (C_3 plants), $(^2H/^1H)_I$ values were similar to those of sugar-cane *aguardentes*. For one commercial grape *aguardente* (Grape3), a $(^2H/^1H)_I$ value was found superior to those of standard samples. These results suggest the predominance of C_4 plant in the origin of sugars used in their production.

Table 1 $^2H/^1H$ relations for methyl (I) and methylene (II) sites of ethanol (in ppm)

	Isotopic Relation (ppm)**		Biosynthetic pathway
	$(^2H/^1H)_I$	$(^2H/^1H)_{II}$	
Grape1*	101.3 (± 1.2)	127.5 (± 0.6)	C_3
Grape2*	102.8 (± 0.5)	128. 2 (± 1.6)	C_3
Grape3	106.6 (± 0.2)	128.4 (± 0.2)	C_3
Grape4	101.3 (± 0.2)	128.0 (± 0.1)	C_3
Grape5	102.6 (± 0.1)	127.7 (± 0.4)	C_3
Honey1*	99.7 (± 0.8)	128.8 (± 0.7)	--
Honey2*	100.4 (± 0.7)	129.1 (± 0.7)	--
Honey3*	101.3 (± 0.9)	128.0 (± 0.5)	--
Honey4	103.6 (± 0.2)	126.9 (± 1.1)	--
Honey5	102.1 (± 0.2)	126.9 (± 0.2)	--
Banana1*	101.9 (± 0.3)	127.7 (± 0.3)	C_3
Banana2*	102.2 (± 0.2)	127.4 (± 0.4)	C_3
Banana3*	102.3 (± 0.2)	127.2 (± 0.3)	C_3
Banana4	112.8 (± 0.2)	126.6 (± 0.3)	C_3
Banana5	113.9 (± 0.1)	128.3 (± 0.0)	C_3
Banana6	113.2 (± 0.2)	128.1 (± 0.2)	C_3
Banana7	111.4 (± 0.2)	127.5 (± 0.3)	C_3
Cassava1	99.7 (± 0.3)	128.3 (± 0.4)	C_3
Peach1	113.0 (± 0.3)	126.9 (± 0.1)	C_3
Coconut1	112.2 (± 0.1)	127.2 (± 0.3)	C_3
Pineapple1*	107.8 (± 0.6)	129.0 (± 0.3)	CAM
Pineapple2*	107.9 (± 0.4)	128.0 (± 0.3)	CAM
Pineapple3*	107.1 (± 0.2)	128.6 (± 0.3)	CAM
Pineapple4	113.3 (± 0.2)	128.5 (± 0.4)	CAM
Sugar-cane1	111.7 (± 0.2)	127.8 (± 0.2)	C_4
Sugar-cane2	111.7 (± 0.6)	129.7 (± 0.1)	C_4
Sugar-cane3	112.3 (± 0.4)	126.0 (± 0.5)	C_4
Sugar-cane4	112.3 (± 0.2)	127.5 (± 0.2)	C_4
Sugar-cane5	112.5 (± 0.3)	125.4 (± 0.4)	C_4
Sugar-cane6	112.6 (± 0.2)	128.1 (± 0.2)	C_4
Sugar-cane7	112.7 (± 0.2)	128.2 (± 0.2)	C_4
Sugar-cane8	112.8 (± 0.1)	127.7 (± 0.2)	C_4
Sugar-cane9	112.8 (± 0.4)	125.5 (± 0.4)	C_4
Sugar-cane10	112.8 (± 0.6)	128.5 (± 0.9)	C_4
Sugar-cane11	113.4 (± 1.1)	127.1 (± 1.3)	C_4
Sugar-cane12	113.4 (± 0.4)	127.2 (± 1.0)	C_4
Sugar-cane13	113.4 (± 0.3)	128.4 (± 0.4)	C_4
Sugar-cane14	113.3 (± 0.8)	126.0 (± 0.7)	C_4
Sugar-cane15	113.6 (± 0.7)	127.7 (± 1.1)	C_4
Maize1	113.9 (± 0.3)	127.4 (± 0.3)	C_4

* standard samples   ** triplicate medium values

Pineapple *aguardentes* (CAM plant) has shown $(^2H/^1H)_I$ values between those of C_3 and C_4 plants, because CAM plants fix CO_2 in the photosynthesis using an intermediary mechanism to those used by C_3 and C_4 plants.

The results have shown that a great number of commercial *aguardentes* have undergone for some adulteration process, possibly by using the sugar-cane *aguardentes*, which have a low cost involved in its manufacture process.

3.1 Chemometric analysis applied to ^1H NMR spectra

The ^1H NMR spectra with three signals suppression from *aguardentes* were compared (Figure 2), with the purpose to verify the existence of differences that could be responsible for the samples discrimination. In the Grape4 *aguardente* spectrum the signals more easily identified were the doublets in δ 4.60 and 5.20, with coupling constant *J* of 8.0 and 3.7 Hz, with respect to anomeric hydrogens of β-glucose (H1 – B) and of α-glucose (H1 – A), respectively. In the Sugar-cane1 *aguardente* spectrum, H1 – B and H1 – A signals were also observed, but those of sucrose presented higher intensity, H1 – C was observed as one doublet in δ 5.40 (*J* = 3.8 Hz). In the banana and honey *aguardentes* spectra, the carbohydrates signals were not visualised.

Moreover, acetic acid, ethyl acetate and acetone were identified in Grape4 *aguardente* spectrum, presenting the methyl singlets in δ 2.02, 2.07 and 2.22, respectively. In this sample it was still possible to identify the methanol presence: its methyl was observed as one singlet in δ 3.35. This identification was made by standard addition.

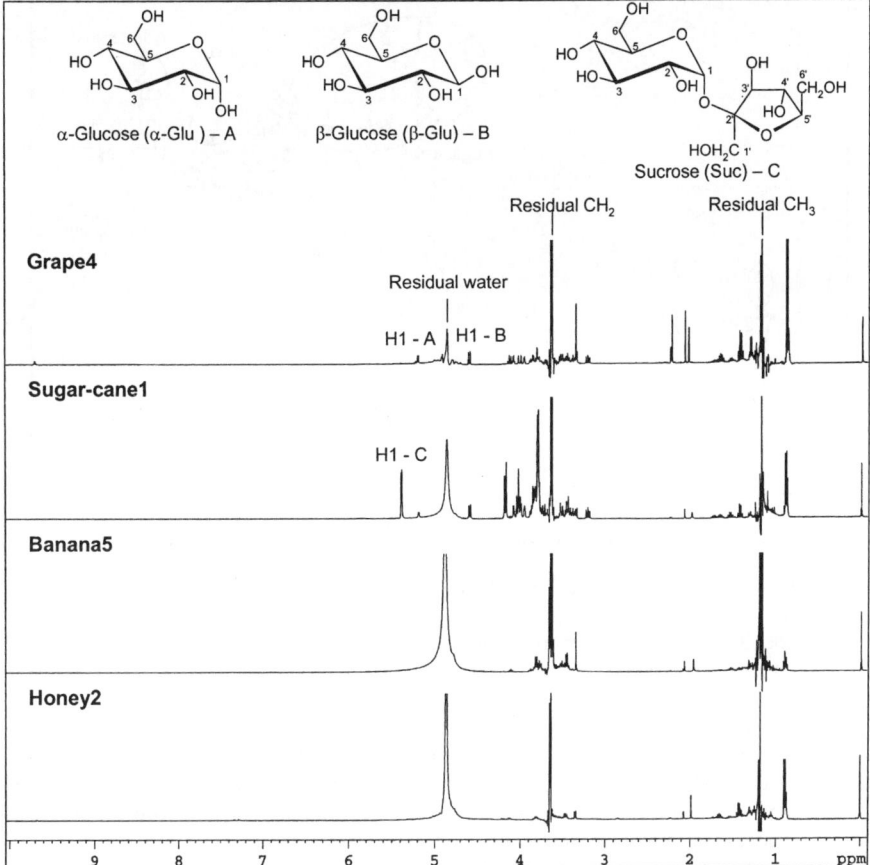

Figure 2   *¹H NMR spectra with three signals suppression of aguardentes (D₂O) and ¹³C satellites decoupling*

Principal component analysis (PCA), applied to ¹H NMR spectra data of *aguardentes* allowed the discrimination of different types of samples and pointed out the compounds related to it. PC1 x PC2 scores plot is shown in the Figure 3. PC1 describes 25.3% of the total variance, while PC2 14.8%, the two PCs together express 40.1% of the original information. In this graph, it is verified that two samples presented different behaviour which can be explained by analysing the loadings graph. Cassava *aguardente* is positioned at positive values of PC1 and negative of PC2 because it contains the largest amount of acetic acid and ethyl acetate. Sugar-cane10 sample is located at negative values in PC1 and PC2 because of two doublets in δ 2.72 and 2.81 (J = 15.5 Hz), not present in other samples from the same group. These two peaks were attributed to the citric acid. Maize, banana and grape samples were grouped near to the sugar-cane *aguardentes* group. A distinct group was obtained for honey *aguardentes*. Adulterated commercial *aguardentes* from banana, peach, coconut and pineapple has shown ¹H NMR spectra quite similar, what suggested that they were made from the same raw material.

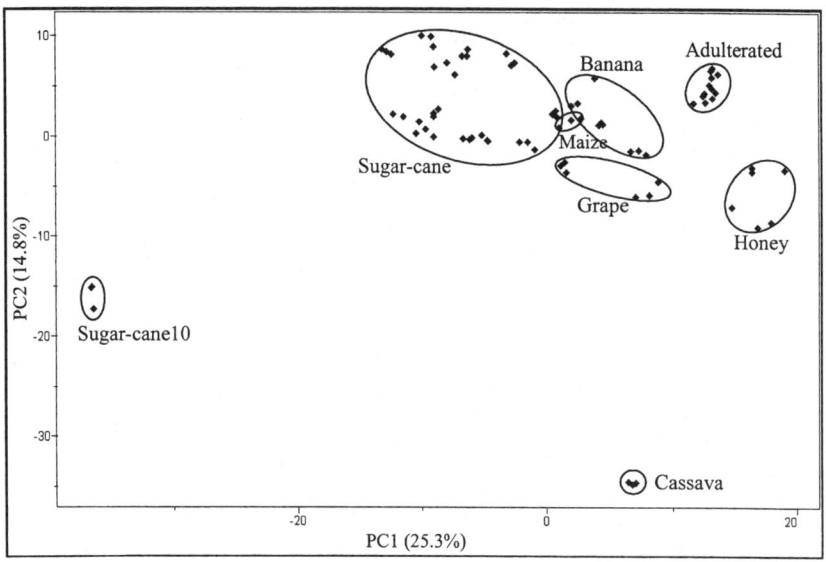

Figure 3   *PC1 x PC2 scores plot of the ¹H NMR data from aguardentes (40.1%)*

The dendrogram obtained from HCA analysis is shown in Figure 4. Using a similarity index of 0.504, eight subgroups can be identified. This result is similar to that obtained in PCA, however, maize *aguardente* is grouped to banana samples and the Grape3 grouped to the sugar-cane samples.

These results have shown that Sugar-cane10 and Cassava *aguardentes* are well distinguished since they have distinct chemical compositions. Therefore, they were excluded from the data set and a new PCA analysis was carried out, with a purpose to obtain a better discrimination among the other *aguardentes*.

The PC1 x PC2 scores plot (Figure 5) represents 39.1 % of the information of the data. PC1 describes 26.6 % of the total variance while PC2 12.5 %. Sugar-cane and maize *aguardentes* form groups at negative values of PC1 because they have the higher quantity of sucrose. On the other hand, distinct groups were observed for honey and adulterated commercial *aguardentes* at positive values of PC1 because of absent carbohydrates signals in the ¹H NMR spectra. Grape4 placed on the inferior right side of the plot for having only α- and β-glucose signals. Grape3 sample is grouped near to the sugar-cane group because of higher sucrose content than Grape4 sample. Moreover, it is observed that banana *aguardentes* are not grouped because they have different carbohydrates concentrations.

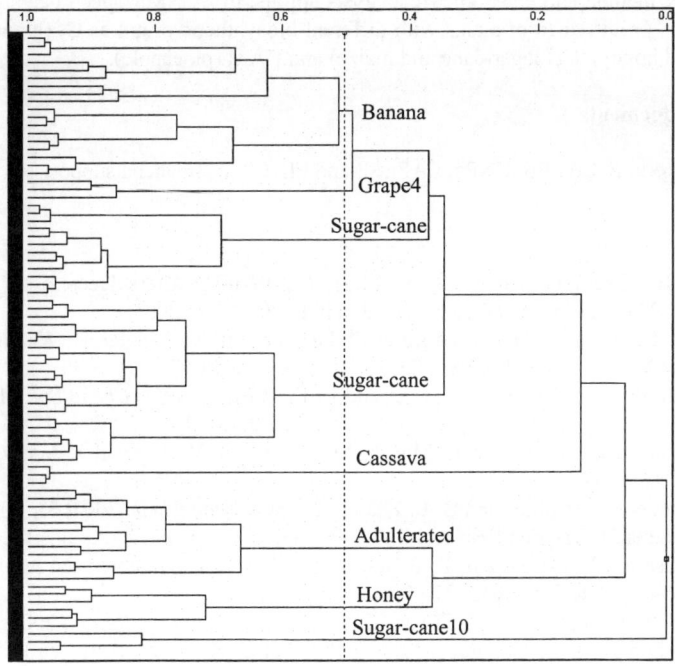

Figure 4 *HCA dendrogram obtained from ^1H NMR spectra from different aguardentes types (similarity index: 0.504)*

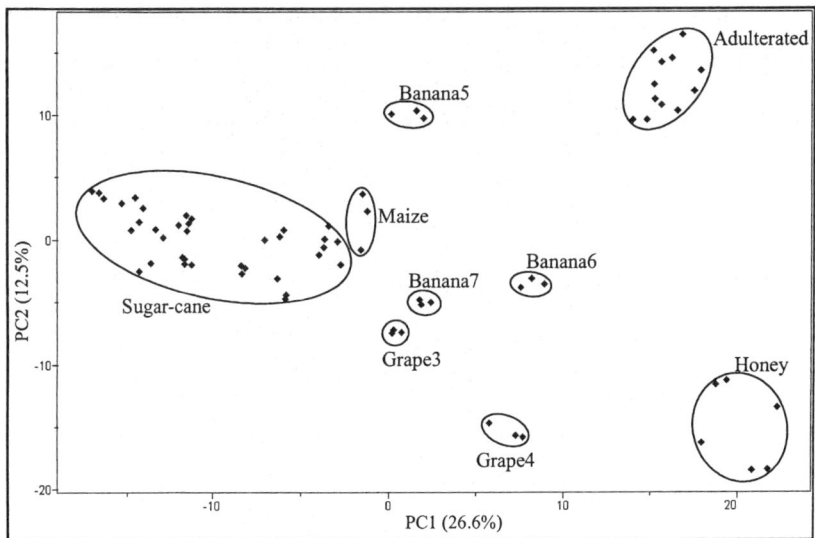

Figure 5 *PC1 x PC2 scores plot of the ^1H NMR data from aguardentes (39.1%)*

4 CONCLUSIONS

SNIF-NMR method and chemometric analysis applied to ^{1}H NMR spectra can distinguish *aguardentes* produced from plants with different biosynthetic origin as C_3 (grape, banana, cassava and honey), C_4 (sugar-cane and maize) and CAM (pineapple).

Acknowledgements

We are grateful to CAPES, CNPq, FAPESP and FINEP for financial support.

References

1   BRASIL, Decreto Presidencial n° 4.851, 02/10/2003. Altera Decreto n$^{\underline{o}}$ 2.314, de 04/09/1997. Diário Oficial da União, Brasília, DF, 03 out 2003.
2   J.F. Cotte, H. Casabianca, J. Lhéritier, C. Perrucchietti, C. Sanglar, H. Waton and M.F. Grenier-Loustalot, *Anal. Chim. Acta*, **582**: 125 – 136, 2007.
3   G.J. Martin, B.L. Zhang, N. Naulet and M.L. Martin, *J. Am. Chem. Soc.,* 1986, **108**, 5116.
4   G.J. Martin, C. Guillou, M.L. Martin, M.T. Cabanis, Y. Tep and J. Aerny, *J. Agric. Food Chem.,* 1988, **36**, 316.
5   S. Pionnier, R.J. Robins and B.-L. Zhang, *J. Agric. Food Chem.* 2003, **51**, 2076.
6   F.H. Larsen, F. van den Berg and S.B. Engelsen, *J. Chemom.*, 2006, **20**, 198.
7   R. Consonni, L.R. Cagliani, F. Benevelli, M. Spraul, E. Humpfer and M. Stocchero, *Anal. Chim. Acta* 2008, **611**, 31.
8   H. Winning, F.H. Larsen, R. Bro and S. B. Engelsen, *J. Magn. Reson.,* 2008, **190**, 26.
9   O. Beckonert, M.E. Bollard, T.M.D. Ebbels, H.C. Keun, H. Antti, E. Holmes, J.C. Lindon, J.K. Nicholson, *Anal. Chim. Acta,* 2003, **490**, 3.
10 B. Lavine and J. Workman, *Anal. Chem.*, 2006, **78**, 4137.
11 M. Jalali-Heravi, S. Masoum and P. Shahbazikhah, *J. Magn. Reson.*, 2004, **171**, 176.
12 J.C. Lindon, E. Holmes and J. K. Nicholson, *Prog. Nucl. Magn. Reson. Spectrosc.* 2001, **39**, 1.
13 G.J. Martin, X.Y. Sun, C. Guillou and M.L. Martin, *Tetrahedron*, 1985, **41**, 3285.
14 M.L. Martin, G.J. Martin and C. Guillou, *Mikrochim. Acta*, 1991, **II**, 81.
15 P. Lindner, E. Bermann, and B. Gamarnik, *J. Agric. Food Chem.,* 1996, **44**, 139.
16 E.F. Boffo, L.A. Tavares, A.G. Ferreira, M.M.C. Ferreira and A.C.T. Tobias, in *Magnetic Resonance in Food Science. From Molecules to Man.* eds. I.A. Farhat, P.S. Belton and G.A. Webb, Royal Society of Chemistry, Cambridge, 2007, p. 105.

IDENTIFICATION AND QUANTIFICATION OF MAJOR TRIACYLGLYCEROLS IN SELECTED SOUTH AFRICAN VEGETABLE OILS BY ^{13}C NMR SPECTROSCOPY

L. Retief[1], J.M. McKenzie[2] and K.R. Koch[1]
[1]Department of Chemistry and Polymer Science, Stellenbosch University, P Bag X1, Matieland, South Africa, 7602.
[2]Central Analytical Facility, Stellenbosch University, P Bag X1, Matieland, South Africa, 7602

1 INTRODUCTION

Vegetable oil production has increased significantly in South Africa in the last decade, due to the use of these oils in the food and cosmetics industry. Apart from olive oils, locally produced oils such marula, apricot kernel, avocado pear, grape seed, macadamia nut, and mango kernel oils are sold for human consumption and widely used in the in the making of handcreams, shampoos and other cosmetic products. Quantitative proton decoupled ^{13}C{^{1}H}NMR has been extensively used for the study of olive oils by several groups in order to determine the geographical origin or cultivar variety of the oil, as well as their authentication (1-4). We have become interested in the use of ^{13}C{^{1}H} NMR for the determination of major components of South African extra-virgin olive oils particularly the possible detection of adulteration and studying variations is these components as a function of geographic and cultivar origin of olive oils (5).

The aim of our project was to identify and determine the fatty acid content of these oils by means of ^{13}C{^{1}H} NMR spectroscopy. The ^{13}C{^{1}H} NMR spectra of locally produced marula, apricot kernel, avocado pear, grape seed, macadamia nut, and mango kernel oils are very similar to those of the well-studied olive oil, we attempted to assign their ^{13}C{^{1}H} NMR spectra by the method used by Mannina et al. for olive oil (1). Mannina et al achieved assignment of the ^{13}C{^{1}H} spectra in part by addition of pure Standard triacylglycerols; we shall refer to as the standard-addition method. For our locally produced oils we found however, that full assignment of the ^{13}C{^{1}H} spectra was complicated by extensive spectral overlap in crowded regions of the spectrum and some concentration dependence of the ^{13}C{^{1}H} resonances of the major components in a given solvent. We here report a ^{13}C{^{1}H} method with which to achieve the reliable full assignment of the major triacylglycerol components of the vegetable oils in question, as well as the preliminary quantitative determination of these triacylglycerols.

2 EXPERIMENTAL

2.1 Materials and sample preparation

Standard triacylglycerols, tripalmitin, tripalmitolein, tristearin, triolein and trilinolein, were purchased from Sigma-Aldrich and used without further purification (≥ 99% purity). Samples of olive oil were provided by Brenn-o-Kem (Wolseley, South Africa). Macadamia nut and avocado pear oil were supplied by Specialized Oils (Paardeneiland, Cape Town, South Africa). All oils were filtered before use. For storage, the oils were flushed with nitrogen gas and refrigerated at -25 °C.

2.2 ^{13}C NMR data collection and processing

Approximately 100 μl of each oil in 700 μl of deuterated chloroform with TMS as reference was used for NMR analysis. NMR spectra were run on a 400 MHz Varian ^{Unity}Inova NMR spectrometer operating at 100 MHz for ^{13}C. Acquisition parameters described by Mannina et al. (10) were used for collecting the ^{13}C NMR spectra: number of points 256 K; spectral width 195 ppm; relaxation delay: 7 s; acquisition time 4.5 s.

2.3 GC analysis of vegetable oils

Methyl esterification of vegetable oils for GC analysis (12) was carried out as follows: Sodium (0.5 g) was dissolved in 100 ml methanol. The sodium methoxide solution (0.3 g) together with 2 g of the specific vegetable oil was placed in a vial and heat sealed. The heat sealed vial was left for 2 hours at 85-90 °C in an oil bath, with occasionally shaking. GC analysis of macadamia nut and avocado pear oil samples was carried out to determine their fatty acid content in order to compare with the ^{13}C NMR spectroscopy data obtained. 20 μl of each sample was diluted with 1 ml of dichloromethane and 1 μl of the solution was inserted in the gas chromatograph inlet. Analysis was performed on a HP 5890 Series 2 Gas Chromatograph. The column employed was a fused silica capillary (30 m x 0.25 mm i.d., 0.2 mm film-thickness) coated with a 100% cyanopropylpolysiloxane non-bonded phase. The column temperature was programmed from 40 to 240 °C at a rate of 4 °C/min.

3 RESULTS

3.1 Assignment of standard-additions method

The idea of the ^{13}C{^{1}H} NMR spectra by standard-additions method involves the spiking of the vegetable oil with a known standard triacylglycerol, such as for example triolein. The ^{13}C{^{1}H} signals representing the fatty acids of triolein that are present in the oil are expected to increase in intensity upon spiking leading to their assignment. We however found some practical disadvantages using this technique. Apart from the standard being expensive and not readily available, the experiments are time-consuming. Our major problem was with the method's lack of reliable use for the 28 – 30 ppm spectral region of the ^{13}C{^{1}H} spectrum, which is very crowded (see to figure 1). More importantly, since it is known that ^{13}C{^{1}H}chemical shifts of the fatty acids in oils are somewhat concentration dependent in a given solvent (1), this renders assignment of the ^{13}C{^{1}H} spectrum particularly in the crowded regions unreliable; for olive oils the 28 – 30 ppm region of the ^{13}C{^{1}H} NMR spectra has not been fully assigned in the literature and we believe it may be due to concentration dependence problem (2). Our own observations confirm this as illustrated for the $^{13}C\{^{1}H\}$ NMR spectrum of olive oil in Figure 1. For this reason for several local vegetable oils we expored an alternative method for the reliable assignment of all peaks in the $^{13}C\{^{1}H\}$ spectrum.

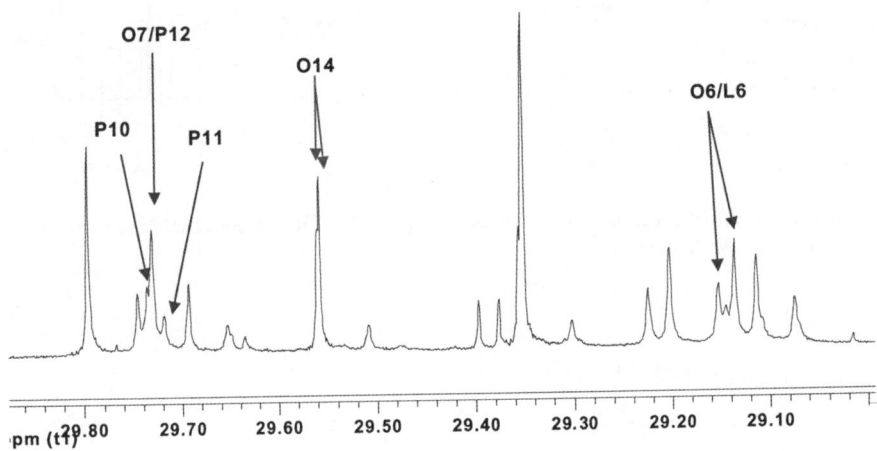

Figure 1 $^{13}C\{^{1}H\}$ NMR spectrum of olive oil in CDCl$_3$ using TMS as reference at 100 MHz at room temperature.

3.2 Linear-graph method

The idea of this method is that the ^{13}C NMR spectrum of the vegetable oil is divided into separate sections (figure 2), namely Section A for the carbonyl carbons, Section B for the olefinic carbons, section C for the glycerol backbone and sections D-J for the aliphatic carbons.

Division of the $^{13}C\{^{1}H\}$ spectrum into these sections is convenient because some of these sections are easily assigned by inspection. It is known that $^{13}C\{^{1}H\}$ signals in the carbonyl region of the spectrum (A) are de-shielded in order of un-saturation, namely the peaks of saturated fragments of the triacylglycerols are most downfield, followed by mono-unsaturated and then polyunsaturated. We also know that the signals due to the carbons in the α position of the glycerol backbone are usually found more downfield than those in the β position. There is no signal representing the saturated carbons in the β position, which is due to the absence of saturated fatty acids in the β position of the naturally occurring triacylglycerols (3). If there a signal is found in the $^{13}C\{^{1}H\}$ spectrum which is ascribable to a β position, this indicates possible adulteration. Our proposed 'linear-graph-method'of assignment is based on the premise that if certain $^{13}C\{^{1}H\}$ peaks in the spectral sections A, B and I can be assigned by inspection for a given component, we expect that these $^{13}C\{^{1}H\}$ shifts should be approximately linearly correlated to all other $^{13}C\{^{1}H\}$ peaks of that molecule. We find that this expectation is largely confirmed from our results, particularly if account is taken of the respective type of carbon atom i.e. whether it is a sp^3 or sp^2 type (*vide infra*).

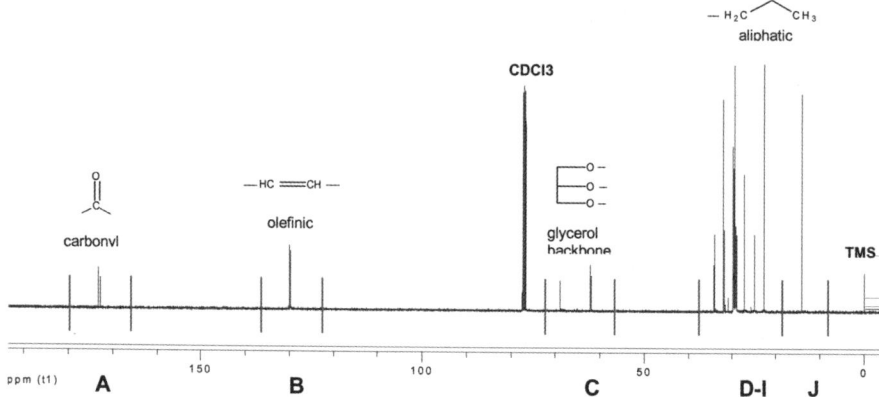

Figure 2 *^{13}C NMR spectrum of macadamia nut oil in CDCl$_3$ using TMS as reference at 100 MHz at room temperature.*

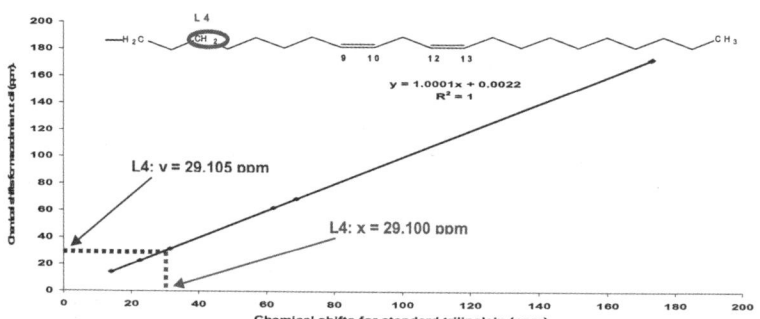

Figure 3 *Linear graph of the chemical shifts in standard trilinolein plotted agains the coreespoinding chemical shifts in macadamia nut oil*

Basically our method works as follows: we construct a graph for which on the x-axis the $^{13}C\{^1H\}$ chemical shift of a standard triacylglycerol e.g. trilinolein is plotted against the corresponding carbon atom's $^{13}C\{^1H\}$ shift of that component in the oil, in the same solvent and at the same temperature and approximately similar concentration. The resulting linear excellent correlation may be used as an aid to assign all other $^{13}C\{^1H\}$ for that component in the oil (Figure 3). By way of example, the assignment of C(4) in the linoleic fatty acid chain, which is difficult by inspection since this peak lies in the crowded 28-30 ppm region, is readily obtained from the linear correlation, which so leads to the assignment of that $^{13}C\{^1H\}$resonance in the vegetable oil spectrum with considerable certainty . We find that this method is remarkably accurate for $^{13}C\{^1H\}$ assignment up to at least one hundredth of a ppm provided that the sp^3 hybridized carbons and sp^2 hybridized carbons should be plotted on separate graphs.

In order to check the robustness of the the linear-graph-method, and in particular to examine a possible concentration dependance of the $^{13}C\{^1H\}$ chemical shifts of the various types of carbon atoms of the triacylglycerols, we measured a series of $^{13}C\{^1H\}$ NMR spectra of different concentration of oil and this is plotted on your x-axis (Figure 4). On the y-axis is the change in chemical shift between each concentration for each type of carbon atom (the aliphatic sp^3 hybridized, carbonyl and olefinic sp^2 hybridized carbon atoms) It can be seen that $^{13}C\{^1H\}$ shifts of sp^3 hybridized carbon atoms show virtually a linear concentration dependence, being shifted downfield with increasing concentration, while the sp^2 carbon atoms undergo a non-linear upfield shift with increasing concentration. This suggests that our method is concentration independant for sp^3 carbon atoms, while for practical oil concentrations ammenable to $^{13}C\{^1H\}$ NMR analysis, even $^{13}C\{^1H\}$ shifts of sp^2 carbon atoms show an approximately linear concentraion dependence.

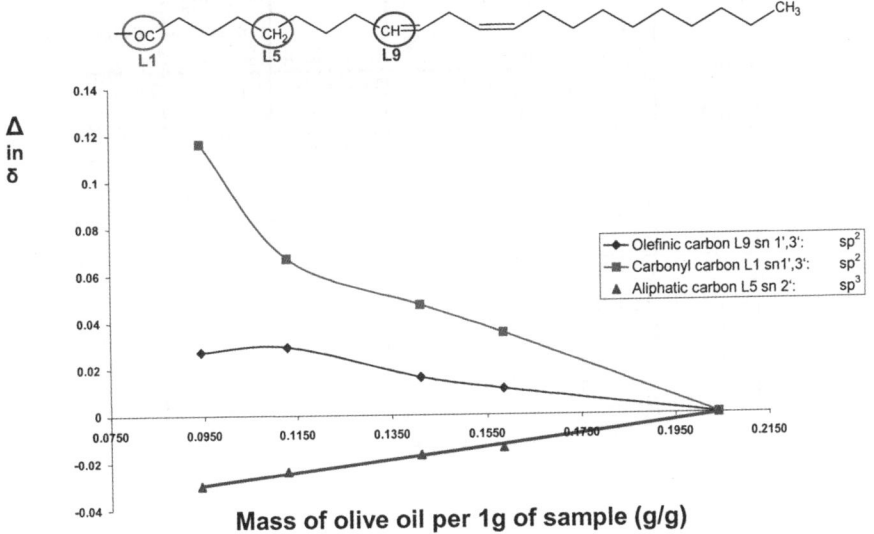

Figure 4 *Graph representing the change in chemical shift between different concentrations of each type of carbon atom in olive oil.*

We validated our linear graph method by the assignment of an $^{13}C\{^1H\}$ NMR spectrum of well studied olive oil in CDCl3, finding that our assignments of $^{13}C\{^1H\}$ chemical shifts compared well to those in the literature (3,4,6) for all the spectral sections (A- J). Moreover it was possible to fully assign the 28 – 30 ppm region of the $^{13}C\{^1H\}$ NMR spectrum of olive oil showm in Figure 1. Thus for the C(14) of the oleic fatty acid we were able to ditstinguish between the α and β postions, as well as suggest assignments of C(7), C(12) and C(11) of the palmitic residue, C(9) and C(6) of oleic residue and C(6) of the linoleic residue, which to the best of our knowledge had not previously been assigned.

With the validation of our method, we were able to apply this the linear-graph-method to the full assignment of the $^{13}C\{^1H\}$ NMR spectra of the major components in marula, apricot kernel, avocado pear, grape seed, macadamia nut, and mango kernel oils which will be published elsewhere (7).

3.3 Quantification of fatty acids

Based of the clear assignment of the major triacylglycerols present in marula, apricot kernel, avocado pear, grape seed, macadamia nut, and mango kernel oils we carried out a preliminary quantitative determination of individual fatty acid using quantitative $^{13}C\{^1H\}$ NMR as applied previously to olive oil (5). For the saturated fatty acids, the carbonyl regions (A) were used, and the $^{13}C\{^1H\}$ were integrated using deconvolution software, which was found to give more reliable integral values for certain signals in the $^{13}C\{^1H\}$ NMR spectrum. For unsaturated fatty acids, the olefinic regions is used (B), and the oleic and linoleic peaks were integrated. These results were compared to those obtained by GC-MS analysis (table 1).

Comparison of fatty acid content as determined by GC with NMR analysis.

		Oleic %	Palmitoleic %	Linoleic %	Vaccenic/ Eicosenoic %	Saturated %
Apricot kernel	GC	63 ± 3.15	1 ± 0.05	31 ± 1.55	< 1	5 ± 0.25
	NMR	61 ± 0.39	nd	31 ± 0.56	nd	8 ± 0.27
Avocado pear	GC	50 ± 2.50	7 ± 0.35	6 ± 0.30	< 1	30 ± 1.50
	NMR	62 ± 0.40		12 ± 0.22	5	21 ± 0.70
Grape seed	GC	12 ± 0.60	< 1	72 ± 3.60	1	14 ± 0.70
	NMR	14 ± 0.09	nd	75 ± 1.36	nd	12 ± 0.40
Macadamia nut	GC	60 ± 3.00	22 ± 1.10	1 ± 0.05	< 1	15 ± 0.75
	NMR	56 ± 0.36	16	nd	7	21 ± 0.70
Mango kernel	GC	50 ± 2.50	< 1	7 ± 0.35	nd	42 ± 2.10
	NMR	56 ± 0.36	nd	10 ± 0.18	nd	34 ± 1.13
Marula	GC	67 ± 3.35	< 1	3 ± 0.15	1 ± 0.05	28 ± 1.70
	NMR	72 ± 0.46	nd	7 ± 0.13	nd	21 ± 0.70

The relative uncertainties for GC analysis were determined based on the maxsimum 5% relative error expected for this technique for oils (8) and those for NMR analysis were determined simsilarly as we reported previously for olive oil by quantitative $^{13}C\{^1H\}$NMR spectroscopy (5). As can be seen within experimantal error the amounts of major fatty acid components in the six vegetable oils obtained by both GC and $^{13}C\{^1H\}$ NMR compare fairly well.

4 CONCLUSION

In conclusion, we believe that the "linear graph method" for the assignment of the $^{13}C\{^1H\}$ chamical shifts of vegetable oil components, particularly in crowded spectral regions is a valuable aid in this regard. We found the method essentially concentration independant for sp^3 carbon atoms at most practical concentrations, while for sp^2 carbon atoms there is only a limited range of concentrations applicable to this method. Nevethelesss it is possible to to fully assign the $^{13}C\{^1H\}$ NMR spectra of these six oils. Quantitative $^{13}C\{^1H\}$ NMR spectroscopy is a suitable method for the rapid determination of the major fatty acids components in six South African vegetable oils with acceptable accuracy and precision.

Acknowledgements

Financial assistance from the NRF and Stellenbosch University is gratefully acknowledged.

References

1. L. Mannina, Sobolev, A.P. and Segre, A. *Spectrosc. Eur.* 2003, **15,** 6-14.
2. L. Mannina, L, and Segre, A. *Grasas y Aceites.* 2002, **53,** 22-33.
3. G. Vlahov, *Magn. Reson. Chem.* 2001, **39,** 689-695.
4. R. Sacchi, R, Addeo, F. and Paolillo, L., *Magn. Res. Chem.* 1997, **35,** S133-S145.
5. J.M. McKenzie, and Koch, K.R., *S. Afr. J. Sci.* 2004, **100**, 349-354
6. A.D. Shaw, di Camillo, A, Vlahov, G. Jones, A. Bianchi, G. Rowland, J. And Kell, D.B, *Anal. Chim. Acta.* 1997, **348,** 357-374.
7. L. Retief, McKenzie, J. and Koch, K.R., *Magn. Res. Chem.* 2008
8. M. Schreiner, *Journal of Chromatography A,* 2005, **1095,** 126-130

POMODORO DI PACHINO: AN AUTHENTICATION STUDY USING ^{1}H-NMR AND CHEMOMETRICS – PROTECTING ITS P.G.I. EUROPEAN CERTIFICATION

F. Savorani,[1,2] F. Capozzi,[2] S.B. Engelsen,[1] M.T. Dell'Abate[3] and P. Sequi[3]

[1] Department of Food Science, University of Bologna, Piazza Goidanich 60, 47023 Cesena (FC), IT
[2] Department of Food Science, Quality & Technology, Faculty of Life Sciences – University of Copenhagen, Rolighedsvej 30, 1958 Frederiksberg C, DK
[3] Agricultural Research Council - Research Centre for the Soil-Plant System (ARC-RPS), via della Navicella 2-4, 00184 Roma, IT

1 INTRODUCTION

In recent years the problem of the objective assessment of authenticity of agro-food products has increased drastically. Due to the global market, the spreading of foodstuffs that were previously localized to well-defined geographical areas is made possible and with this globalization, attempts to imitate high-quality food products are ever increasing. Usually the falsified products are very similar to the original ones in their physical appearance, but they present lower quality and, as a consequence, more affordable prices, causing severe economical damage as well as a reputation damage to the authentic products. In this context, the necessity has emerged for many producers to characterize and to valorize their own products with a legally recognized certification.

Agro-food products that represent a segment of Quality Food Products (QFP's) are an example of the new concept "consumer's choice" in terms of naturalness and authenticity against food globalization. These recent tendencies about food safety aspects, and the need for re-discovering the true values of agriculture strictly related to *terroir*, have led to the creation of quality certification labels that have become a strategic instrument of differentiation that gives the food products a commercial added value.[1] Consumers can nowadays immediately identify products that respect certified quality parameters through quality logos clearly reported on the package. As an example, the European collective marks identifying Protected Designation of Origin (PDO) and Protected Geographical Indication (PGI) have been developed and used for branding those European food products that fulfill the requested and certified characteristics. PDO and PGI were introduced by the EEC Reg. 2081/92 which has been recently upgraded by EC Reg. 510/2006. Figure 1A shows the European PGI label.

Nevertheless, without an adequate analytical technique able to assess the authenticity, it is impossible to restrain attempts of counterfeit. Several studies have been conducted on different food products in order to devise and refine scientific methods able to determine, within an acceptable margin of error, whether or not the analyzed product is consistent with what is declared on its selling package.

Due to the complexity of food matrices several advanced spectroscopic techniques have recently been successfully employed to solve authenticity problems, including near infrared spectroscopy (NIR), mid infrared spectroscopy (MIR), low- and high-field nuclear magnetic resonance (NMR), site-specific natural isotopic fractionation (SNIF-NMR) and isotope ratio mass spectrometry (IRMS), taken stand-alone or hyphenated with chromatographic methods.[2] More recently, also DNA-based technologies have been used to investigate authenticity issues. All the different methods have their own merits and drawbacks with respect to being effective and reliable in coping with the huge variety of foodstuff matrices when an authenticity issue has to be solved.[3,4] Because of the enormous amount of information usually produced by these techniques and because of their intrinsic complexity, chemometric pattern recognition techniques have been exploited in order to extract the information useful to deal with quality and authenticity problems.[5,6,7]

In the present work the authenticity of the Italian cherry tomato "pomodoro di Pachino" in relation to some other Italian cherry tomatoes has been investigated utilizing proton High Resolution Nuclear Magnetic Resonance (^{1}H HR NMR) as the analytical method and chemometrics as a tool for analyzing and interpreting the acquired multivariate data. Thanks to its peculiar characteristics, due to a combination of climate, salt water irrigation and cultivation techniques, the Italian cherry tomato of Pachino was the first tomato to obtain the European PGI certification of quality. Because of its high production costs and the consequent high price of the final product, commercial fakes with lower organoleptic characteristics are present on the Italian as well as on international markets for which reason it is necessary to protect producers as well as consumers against frauds. Consequently, great interest has recently been focused on analytical techniques able to predict the geographical origin of a tomato sample, indicating whether or not it originates from the area of Pachino, Sicily (Italy) (Figure 1B). Recently, an analytical protocol based on the application of magnetic resonance imaging (MRI) was successfully developed by analyzing a set of tomato samples common to a portion of those considered in the present report.[8]

In this study 258 different cherry tomatoes were harvested on selected farms in Pachino and in other Italian geographical regions and then processed and analyzed as aqueous extracts using 400 MHz ^{1}H HR NMR. The NMR data were then preprocessed in order to make them suitable for multivariate data analysis and data-mined by chemometric pattern recognition techniques.

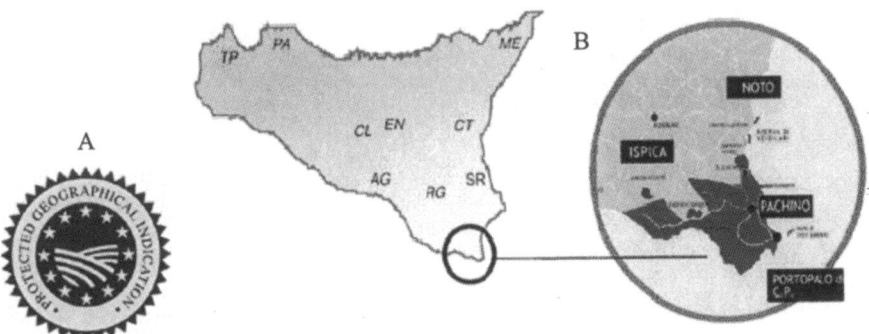

Figure 1 *The European PGI logo (A) and localization of the District of Pachino in the southern part of the island of Sicily (B)*

2   EXPERIMENTAL

2.1   Cherry Tomato Samples: Design of the Experiment

Samples of cherry tomatoes were collected from several farms in the district of Pachino and in other Italian geographical areas using standardized parameters of ripeness and fruit position (fruiting truss) on the selected plants. Since tomatoes are produced all year long, both winter (from November to April) and summer (from May to October) samples were collected over a period of three years (2003-2005). As no winter samples are cultivated outside Sicily for economical reasons, it was decided to collect winter samples in Sicily, but in a district other than Pachino, in order to make a comparison possible and making the authenticity trial more challenging. The area of Licata, located just 100 km from Pachino, was chosen, because it is a very productive area and, indeed, a competitor to Pachino in producing quality cherry tomatoes. Pachino summer samples were compared with cherry tomatoes cultivated and harvested in Sabaudia, located in another Italian region (Lazio). In addition, some random winter and summer cherry tomatoes were bought at different markets in order to compare them with the authentic and standardized ones. The geographical origins and the number of the analyzed and processed samples are summarized In Table 1.

Table 1   *Sampling of the tomatoes*

Origin of Samples	Winter season	Summer season	Total
PACHINO (Naomi)	112	50	162
PACHINO (Shiren)	24	8	32
LICATA (AG)[a]	13	0	13
SABAUDIA (LT)[b]	0	24	24
MARKETS[c]	14	13	27
Total No. of samples analyzed by NMR	163	95	**258**
Samples suitable for chemometrics	153	92	**245**

[a] Adjacent but outside the district of Pachino (very similar pedoclimatic conditions)
[b] Central Italy: no winter cherry tomatoes are produced in Italy outside Sicily
[c] Sicilian cherry tomatoes randomly bought in different markets

2.2   Preparation of samples

The cherry tomato samples were freeze-dried, ground and sealed in labeled glass vials by the ARC-RPS research group. For each sample 100 mg of freeze-dried powder was extracted in a 1.5 ml Eppendorf tube with 1.00 ml of a 100 mM Acetic acid/Acetate buffer solution whose pH was 4.00. The buffer was prepared using D_2O 99.9% and glacial acetic acid in order to avoid increasing the amount of water present in the sample and to allow for acquiring NMR spectra without water suppression pulse cycle. The final pH was adjusted with the proper amount of NaOH. Following 2 hours of shaken extraction the tubes were centrifuged for 10 min at 15000 RPM after which 700 µl of clean extract was transferred to a 5 mm NMR tube for the spectral analysis.

2.3 ¹H-NMR Analysis

All 1D ¹H-NMR spectra were recorded at 298°K using a Varian Mercury AS/400 spectrometer operating at 9.4 T, corresponding to 400.098 MHz ¹H Larmor frequency. D_2O provided deuterium field/frequency lock. The tubes were spun at 20 Hz in the NMR probe, improving the spectral line width. No water presaturation pulse cycle was used. The relaxation delay was 0.5 s, the acquisition time was 3.416 s and the FID's were collected into 32K data points covering a 12 ppm spectral window (4796.2 Hz). After 16 steady scans, a total number of 1024 transients were acquired, requiring 68 min of measurement time for each sample. Following acquisition, each FID was Fourier transformed using a 0.5 Hz Line Broadening apodization. Phase correction and baseline correction as well as the chemical shift calibration were accurately performed manually using MestRe-C 4.9.8.0.[9] The same software was used to export data in ASCII format. Chemical shift referencing was performed by setting the β-D-Glucose downfield signal at 4.650 ppm.

A typical cherry tomato 400 MHz ¹H-NMR spectrum is shown in Figure 2. The strong signal at about 2.03 ppm originates from the methyl protons of acetate, introduced with the acetic-acetate buffer solution.

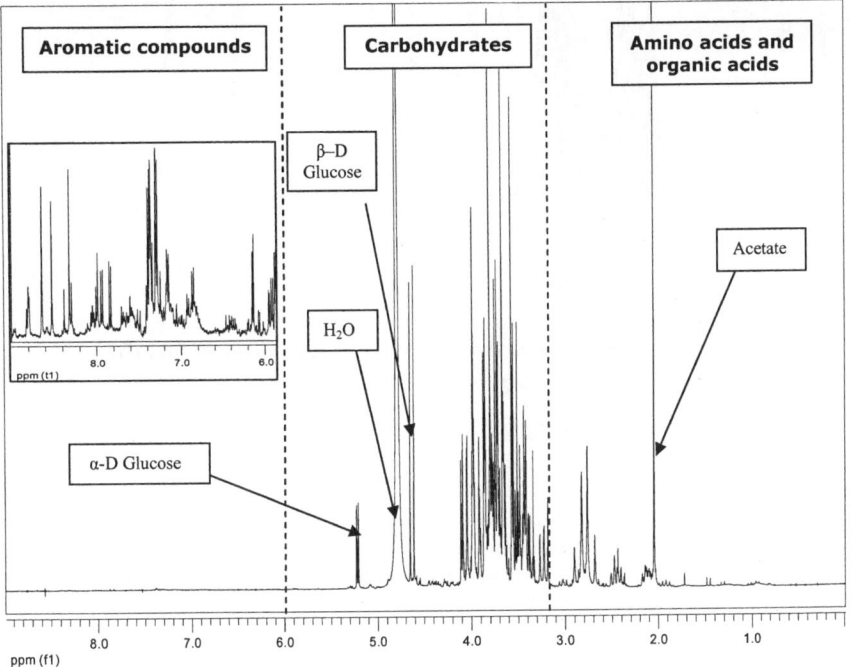

Figure 2 *400 MHz 1D ¹H-NMR spectrum (9.0 to 0.0 ppm) of a cherry tomato extract, Naomi cultivar, buffered at pH 4.0*

2.4   Data Processing and Chemometrics

The spectral area chosen for multivariate analysis was 10.0 to 0.0 ppm, excluding the residual water region (4.9 – 4.7 ppm) and the region containing the acetate peak (2.1 - 2.0 ppm). Both preprocessing and multivariate data analysis were performed using the R software environment.[10] A simple in-house written algorithm was used for correcting the misalignment occurring among the spectra. In order to accomplish this preprocessing step one spectrum was selected among all as the reference on the basis of its property to be the most similar to the average one. All spectra in the dataset were then horizontally aligned toward the β-D-Glucose signal position at 4.65 ppm in the reference spectrum. Vertical scale was standardized using normalization, setting the integral of each spectrum to a common value.[11] The result of these corrective steps is shown in Figure 3. The aligned and normalized dataset was then binned into 200 bins 0.05 ppm wide, containing 68 data points each. As a final step, a total number of six bins, corresponding to the HOD and the acetate signals, were removed from the preprocessed dataset, yielding a binned data matrix having dimensions 258 x 194, suitable for multivariate data analysis and data mining.

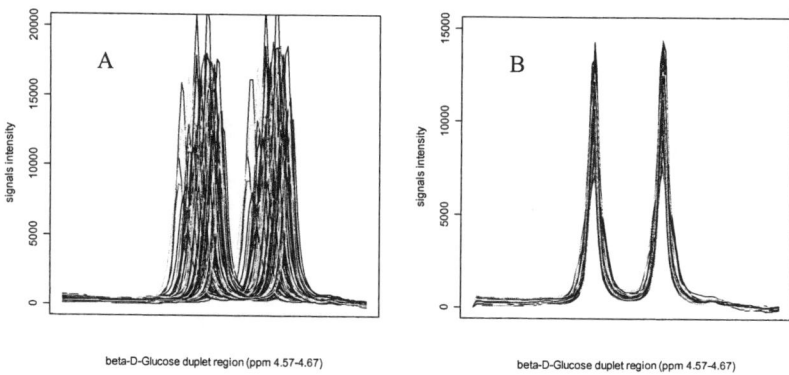

Figure 3   *Superimposed raw spectra within the β-D-Glucose spectral region (A) corrected by means of alignment and normalization algorithms (B)*

3 RESULTS AND DISCUSSION

A first exploratory data analysis was performed on the preprocessed data to investigate if there was a trend to separate among the four different geographical origins of the tomatoes. The result of a principal component analysis (PCA) on the mean-centered data is shown as a score plot in Figure 4A where each geographical origin is labeled with a different symbol. For completeness, a 5th class has been added in order to keep separate the two different cultivars coming from Pachino. In the score plot just principal component No. 1 (PC1) (describing 20.0% of the total variance) and PC2 (16.5%) are shown, but no further components able to discriminate better among the four geographical classes were found. The figure clearly shows that all the classes overlap and it seems that this approach is not effective in achieving separations among the tomato groups.

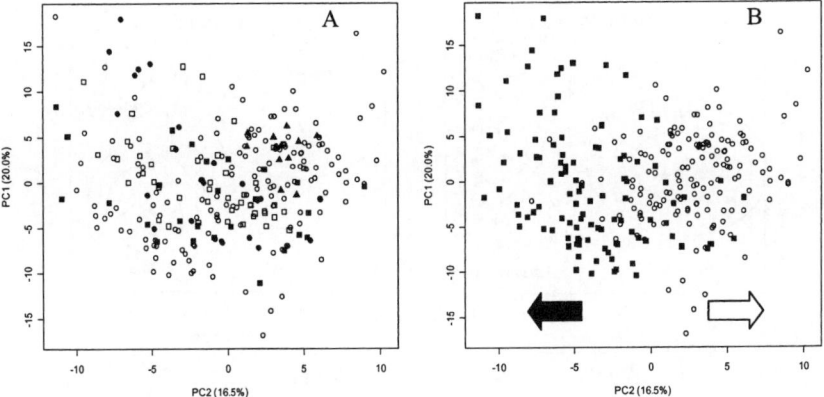

Figure 4 *PCA score plots: in A the samples are colored according to their geographical origin (○Pachino Naomi, □Pachino Shiren, ▲Licata, ●Sabaudia, ■market), whereas in B they are colored according to their harvesting season (○winter, ■summer)*

However, labeling the samples in the same score plot according to their harvesting season (where *winter* refers to tomatoes harvested between the beginning of November and the end of April and *summer* to the remaining part of the year), reveals a strong tendency to separate into these two classes, independent of their geographical origin. Figure 4B clearly shows this tendency along PC2, even though a slight overlap still exists. This is easily explained by the fact that both fall and spring samples have been forced to belong to one of the two extreme seasons. The loadings of PC2 (not shown) indicate that it is primarily the acids region which contains the most discriminative signals. In particular, the doublet of doublets of citric acid (2.9 – 2.6 ppm) seems to be an effective marker for this separation. Since this strong source of variability (seasonality) seems to hide the more relevant discriminative effects (i.e. geographical origin), summer and winter data were separated into two new datasets and analyzed separately.

PCA performed on the two separate datasets reveals that both summer and winter data now tend to provide a clustering according to origin (data not shown). This tendency was much stronger for the summer samples. The difference in behavior probably finds an explanation in the sampling setup: the summers samples come from two different regions with diverse pedoclimatic condition, whereas the winter samples are all from the same region and thus sharing very similar pedoclimatic characteristics. The "random" market samples were spread out in both datasets, but with a clear trend to separate from the others. In light of these results, the data were further investigated using Principal Component Linear Discriminant Analysis (PC-LDA) chemometric pattern recognition technique.[12] In the case of the summer sample, a total of 26 PC's summing up to 90.0% of the explained variance is required for distinguishing among four different classes. Data were cross-validated leaving out 10 of the 92 samples, randomly chosen in order to have at least two samples for each class. The final validation is an average of 50 random repeated cycles and results in a 90.2% correct classification model using three LD canonical variates (CV). The PC-LDA score plot for the summer data is shown in Figure 5A.

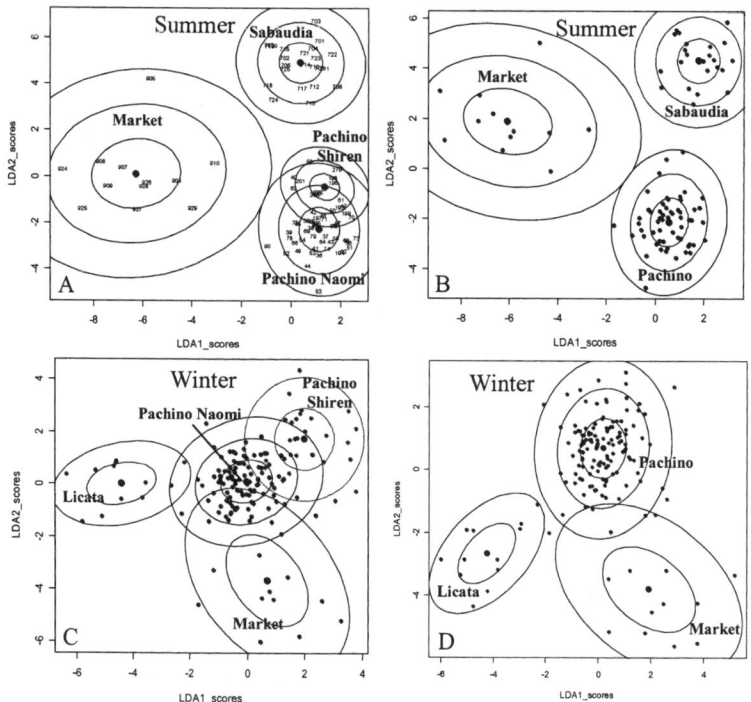

Figure 5   *PC-LDA score plots of summer (A, B) and winter (C, D) samples*

In the PC-LDA score plot the different tomato groups are represented with their Mahalanobis Distances (MD) with ellipses up to three MD. Inspection of the first two PC-LDA scores reveals that the tree different geographic origins are sharply separated. A third dimension was able to distinguish between the two cultivars from Pachino, but it is important to note that the first LDA PC's tend to place them in the same geographical class as they actually are. Indeed, when these samples are considered belonging to the same group, a much better predictivity of the PC-LDA model is reached. In this case the result is 95.6% correct classifications using the same validation procedure (Figure 5B̄).
Surprisingly, the winter samples also gave similar results in the PC-LDA analysis. In this case 31 PC's were required in order to deal with the higher similarity among the samples regarding their origin. The PC-LDA score plot of the winter samples is shown in Figure 5C. In this case the data were cross-validated by leaving out 15 of the 153 samples, randomly chosen in order to have at least two samples for each class. The final validation result is given as an average of 50 random repeated cycles and it yielded a model with 87.8% correct classifications using three LDA CV, despite the fact that the 2D score plot is not effective in representing its discriminative power. As for the summer data, a significant improvement of the model performance is achieved by taking the two cultivars of the winter tomatoes from Pachino as a single group. Figure 5D shows the PC-LDA score plot of that model performing 94.2% of correct classifications obtained using the same validation method.

4 CONCLUSIONS

For a quality certified agricultural product the scientific assessment of its authenticity is a key step in protecting both producers and consumers from attempts of counterfeit which exploit its good name in order to sell fake products. In the present case, the IGP cherry tomato of Pachino has been studied with NMR spectroscopy and chemometrics in order to determine whether tomatoes produced in the Pachino *terroir* are chemically distinguishable from tomatoes produced elsewhere. By taking advantage of the capability of NMR to perform a quick and comprehensive analysis, the fingerprints of aqueous extracts of 258 tomatoes have been acquired and the overwhelming amount of data mined by means of multivariate chemometric techniques for answering the query stated above. While it is credible that tomatoes cultivated in Sabaudia may be qualitatively different from those produced in the area of Pachino, some doubts may arise regarding the legitimacy to discriminate the tomatoes from those originating in the area of Licata, since only slight differences in the pedoclimatic conditions exist between these two production areas.

The results of this research indicate that the tomatoes produced in the Pachino area are indeed different, in terms of chemical quality, from all the non-authentic tomatoes investigated in this study, even including those produced in Licata. It is noteworthy that the NMR method in combination with chemometrics has shown that the quality of production is different according to the harvesting season. This finding holds for the tomatoes produced in Pachino as well as for those produced elsewhere. Once the two seasons of production, summer and winter, are kept separate, non-negligible differences emerge among tomatoes of different origin.

Application of PC-LDA on both the summer and the winter datasets showed that models with a prediction performance better than 90% can be established. The multivariate data analysis also revealed that there is no single molecular component that makes possible distinguishing among the different geographical origins, but rather, it is a pattern of weak and intense signals throughout the NMR spectrum which leads to a reliable model.

Nevertheless, future research should be continued with the chemical identification of those molecular components responsible for the greatest part of the discriminative power evidenced by the multivariate analysis. In the future, hyphenated methods which couple classical methods of separation such as HLPC and GC with NMR will be exploited for identifying those molecular markers.

Acknowledgements

This research has been financed by project "Messa a punto di indici chimici, rilevabili mediante tecnologie analitiche avanzate, per l'identificazione della qualità dei prodotti ortofrutticoli e della loro origine geografica (IGP)" of the Italian Ministry of University and Research (MIUR), 2003-2007.

References

1   G. Belletti, T. Burgassi, A. Marescotti, S. Scaramuzzi, The effects of certification costs on the success of a PDO/PGI, presented at Quality Management and Quality Assurance in Food Chains, University of Göttingen (Germany), May 2, 2005.

2   D. L. Norwood, J. O. Mullis, T. N. Feinberg, Hyphenated techniques, Separation Science and Technology, Academic Press, 2007, Volume 8, HPLC Method Development for Pharmaceuticals, pp. 189-235.

3   L. M. Reid, C. P. O'Donnell, G. Downey, Recent technological advances for the determination of food authenticity, *Trends in Food Science & Technology* (2006), **17** (7), 344-353

4   D. M. A. M. Luykx, S. M. van Ruth, An overview of analytical methods for determining the geographical origin of food products, *Food Chemistry* (2008), **107** (2), 897-911.

5   L. Munck, L. Norgaard, S. B. Engelsen, R. Bro, C. A. Andersson, Chemometrics in food science - a demonstration of the feasibility of a highly exploratory, inductive evaluation strategy of fundamental scientific significance, *Chemometrics and Intelligent Laboratory Systems* (1998), **44** (1-2), 31-60.

6   R. Karoui, J. De Baerdemaeker, A review of the analytical methods coupled with chemometric tools for the determination of the quality and identity of dairy products, *Food Chemistry* (2007), **102** (3), 621-640.

7   H. Winning, F.H. Larsen, R. Bro, S.B. Engelsen, Quantitative analysis of NMR spectra with chemometrics, *Journal of Magnetic Resonance* (2008), **190** (1), 26-32.

8   P. Sequi, M.T. Dell'Abate, M. Valentini, Identification of cherry tomatoes growth origin by means of magnetic resonance imaging. *Journal of the Science of Food and Agriculture* (2007), **87**, 127-132.

9   J. C. Cobas, F. J. Sardina, Nuclear magnetic resonance data processing. MestRe-C: A software package for desktop computers, *Concepts in Magnetic Resonance Part A* (2003), **19A** (2), 80-96.

10 R Development Core Team (2007). R: A language and environment for statistical computing. R Foundation for Statistical Computing, Vienna, Austria. ISBN 3-900051-07-0, URL http://www.R-project.org.

11 A. Craig, O. Cloarec, E. Holmes, J. K. Nicholson, J. C. Lindon, Scaling and Normalization Effects in NMR Spectroscopic Metabonomic Data Sets, *Analytical Chemistry* (2006), **78** (7), 2262 – 2267.

12 R. A. Fisher, The use of multiple measurements in taxonomic problems , *Annals of Eugenics* (1936), **7**, 179

1D AND 2D 1H-NMR ANALYSIS OF TASTE COMPOUNDS EXTRACTED FROM RAW OR FRIED ALLIUM CEPA L. TISSUES. METHODOLOGICAL QUESTIONS

A. Tardieu[1,2], W. de Man[3] and H. This[2]

[1]Mars Chocolat France, Z.I. La Sandlach, 67500 Haguenau, France.
[2]Equipe INRA de Gastronomie Moléculaire, Laboratoire de Chimie, UMR 214 INRA/Institut des sciences et industries du vivant et de l'environnement (AgroParisTech), 16 rue Claude Bernard, 75005 Paris, France.
[3]Mars Food Belgium, Industrielaan 7, B-2250 Olen, Belgique.

1 INTRODUCTION

Molecular Gastronomy (MG) is the scientific discipline looking for new phenomena and mechanisms in culinary transformations. As plant tissues are important in the human diet,[1] one issue of MG is the study of modifications of plant tissues during thermal treatment, including such treatments in aqueous solutions.[2] These modifications can be of physical and chemical nature. Plant tissues, in particular, and food ingredients in general being complex disperse systems,[3] physical and chemical phenomena can be linked.

In order to study material transfers during culinary transformations, including during thermal processing of plant tissues in water, fast and direct analytical methods that could be applied for such systems is needed. Indeed NMR spectroscopy is non-destructive and it can detect simultaneously all proton-bearing compounds of low molecular mass in complex mixtures.[4] Of course, NMR signals are not restricted to chemical shifts, and a wealth of useful information can be drawn from NMR spectra. In particular rapid analysis by 1H NMR can be obtained for unpurified plant extracts, providing information on the major metabolites extracted. A metabolic fingerprint can be obtained,[5-7] metabolic profiling being defined as the identification and quantification of a number of predefined metabolites in a plant sample.

When NMR analysis is performed with complex solutions, the risk of peak overlapping is high, and chemiometric methods can be useful, including multivariate data analysis, such as principal component analysis (PCA) or independent component analysis (ICA).[8] NMR spectra contain the resonances of all relevant spins, but NMR analysis of complex systems is interesting as the method is focusing on components with concentrations higher than the detection threshold. Accordingly, NMR analysis has been used not only for identification[4,9] and structure elucidation but also for quantification.[10] It was formerly used in our team to quantify carbohydrates, amino acids and organic acids in aqueous solutions ("stocks") obtained by thermal processing of carrot (*Daucus carotta* L.) roots and thus to characterize the time course of sugar extraction at different temperatures (55 °C, 75 °C and 100 °C).[11] Direct identification and quantification of the main photosynthetic pigments in green bean crude extracts was performed using NMR methods.[12]

Looking for the mechanisms of plant compounds extraction in aqueous solutions, we wanted to compare the behavior of carrot roots and of onion (*Allium cepa* L.) bulbs, the latter being also a storage organ,[13] where carbohydrates account for a major fraction of their dry matter, contributing as much as 64 to 80 % of the dry weight.[14] More precisely, non-structural carbohydrates are: free glucose, free fructose, sucrose, and fructooligosaccharides, i.e. fructosyl polymers with degrees of polymerization (DP) up to 12, also called fructans.[15-19] Among fructooligosaccharides, the polymers with lowest DP are the more important in terms of quantity,[20] with 1-kestose, nystose, neokestose representing respectively 0.33, 0.16, 0.99 g/100 g of fresh weight.[15] Total fructans represent 3.43 g/100 g of fresh weight.[15] However, the non-structural carbohydrate compositions vary significantly among different cultivars: the mean concentration in fructans was found to be 49.1 and 6.9 mg. g^{-1} of fresh weight for the PLK and the Grano varieties respectively, which indicates a high variability between varieties.[21]

Amino acids and organic acids are also present in onion bulbs. The free amino acid content in onion (tissues) vary considerably,[19] arginine and glutamic acid being both abundant. These amino acids are nitrogen sources; during the growth and the physiological development of onions, arginine and glutamic acid contents increase by 29 % and 7 %, respectively; in contrast, variations in the content of all other free amino acids are small. Sulfur-containing compounds are important in *Allium* plants, but the concentrations in L-cysteine and L-methionine are all relatively low, indicative of their rapid metabolism. These observations are in accordance with another study,[22] with 3.226 and 2.005 µmol/g of fresh weight for arginine and glutamic acid respectively.

In food, onion molecules such as carbohydrates and amino acids are used either for their nutritional or for their flavor interest. In particular, flavor is based on many sensory parameters such as taste, odour, colour, trigeminal effect, consistency, temperature, sound etc.[23] In our studies on onion bulbs, we focused on taste compounds, mainly due to hydrosoluble molecules, including non structural carbohydrates.

To determine the concentrations of such molecules (mainly sugars, amino acids and organic acids) using NMR spectroscopy, some methodological work on NMR parameters was necessary. Temperature, number of scans, pulse angle, recovery time, *p*H, and calibration were investigated in order to get an accurate analytical method by 1H NMR spectroscopy.

2 MATERIALS AND METHODS

Studies were carried out on raw and fried onions. The aim of these investigations was to find an accurate method of quantification of water soluble compounds extracted from two kind of vegetable tissues, either raw or thermally processed in oil (fried), implying possible physical or chemical changes.

2.1 Preparation of Soaking Solutions

Onion bulbs (Armstrong variety) were grown in Poland and stored at 6 °C until experiments. After cutting about 1 cm of the top and the bottom due to different chemical composition[15] and also to reproduce culinary processes,[25] onion bulbs were peeled off by removing the first or the two first dry and brown scales, known as papery scales.

In order to ensure that homogenous material was compared, onion bulbs were cut lengthwise into two halves; one was kept for the experiment on raw onion dice (experiment A), the other was kept for the study on fried onion dice (experiment B); eventual sprouts were removed. For each experiment, precisely determined masses (about 200.00 g) of onion were weighted and cut in parallelepipeds (1 cm x 1 cm x thickness of the scale). For experiment B, the frying step consisted of heating a precise mass of sunflower oil (about 7 g, i.e. enough to fry without having too much remaining oil) in a pan at 140 °C, adding about precisely 200.00 g of raw onion dice until a brown color is obtained (about precisely 15 min with stirring). After frying, most of the adhering oil was removed using absorbing paper and the total mass was weighted. In a 800 mL glass flask, about precisely 200.00 g of raw onion parallelepipeds or 200.00 g of fried onion parallelepipeds were introduced in 600.00 g of MilliQ water with 0.1 % sodium azide (to inhibit bacterial growth).[26] Solutions were stirred at room temperature (around 23 °C). For each experiment, about precisely 15.00 g of soaking water (less than 3 % of the total soaking aqueous solution) was sampled at 5 time points (logarithmic distribution) between 0 and 264 h (11 days). These samples were centrifuged (*Sigma* 3MK) at 10,000 rpm during 10 min at 19 °C, and the liquid part was filtrated (45 μm, 6,2 cm², surfactant-free cellulose acetate membrane and depth filter, Minisart Plus, *Sartorius*). The clear liquid was weighted and frozen at -24 °C.

All experiments were repeated three times; we noticed a possible bacterial contamination at 11 days on one experiment on raw onion dice soaking.

2.2 NMR Acquisition

2.2.1 NMR Sample Preparation. Double freeze-drying was performed to eliminate water which may cause NMR signal saturation. Solids obtained after the first freeze-drying step were dissolved in deuterium oxide (D$_2$O, 99.9 atom % D, *Sigma Aldrich*) and lyophilized again to eliminate residual water. About precisely 15.0 mg of the resulting solids were re-dissolved in 1.0000 g of D$_2$O and the *p*H was adjusted in order to avoid chemical shift drift. Indeed, the influence of *p*H on chemical shifts for different classes of metabolites (carbohydrates/ polyols, amino acids, organic acids, osmolytes, nucleosides derivative, triacylglycerides, etc.) was largely investigated by T. Fan,[4] who observed changes in chemical shifts in function of *p*H. In order to set the *p*H at a value found in literature and to avoid possible chemical reactions (carbohydrate dehydration, hydrolysis, etc.), the *p*H of our samples was adjusted to 7.0 using known amounts of deuterium chloride (DCl) solution in D$_2$O (DCl, 20 % in deuterium oxide, "isotope enrichment 100 %", d = 1.189, *SDS*) at different appropriate concentrations. About precisely 700 μL of each sample were introduced in a NMR tube (5 mm glass, *Wilmad-Aldrich*, USA).

Simultaneously an home made closed capillary tube containing the sodium salt of (trimethyl)propionic-2,2,3,3-d4 acid, TSP (TSP, 98 atom % D, *Sigma Aldrich*) dissolved in D$_2$O (0.2 %) was introduced inside the NMR tube, for chemical shift calibration and for quantification.[11] A coaxial insert tube could have been used in order to avoid possible magnetic heterogeneities that could arise from the presence of the capillary tube. However, many acquisitions made with the same tube (full acquisitions, including removing of the tube, inserting it again, autoshim, etc.) showed a good reproducibility and homogeneity of the TSP peak. Calibration curves of sugars (see § 2.4) show linear shape, with good coefficients of determination ($R^2 > 0.99$).

2.2.2 1D NMR Parameters. A Superconducting Ultrashield 300 MHz (7.05 Tesla) 54 mm magnet system NMR spectrometer 300 MHz BZH 30/300/70 E *Bruker Biospin* (Germany) was used in this study. Four parameters were investigated: temperature, recycle delay (D1), number of scans (NS) and pulse angle (zg).

As this study was focused on carbohydrates, the temperature was an important parameter: glucose and fructose in solution exist in different forms: percentages at equilibrium of α-pyranose, β-pyranose, α-furanose and β-furanose are respectively 38, 62, 0, 0.1 for D-glucose and 2, 70, 5 and 23 for D-fructose.[27] This temperature-dependant equilibrium [28] is reached after about 4 hr. As the room where the samples were stored was thermostated at 21 °C, ^{1}H NMR spectra were recorded at 21 °C in order to avoid different mutarotation equilibrium. Tests were done to check the reliability of this way of doing. Glucose solutions in D_2O at 21 °C were inserted in the NMR spectrometer at different temperatures (21 and 30 °C) and acquisitions were performed after 0, 30 or 60 min. At 21 °C, there was no significant difference in peak area between acquisition at 0 min and acquisition after 30 min; however, significant differences in peak area occurred between an acquisition at 30 °C made directly or after 30 and 60 min.

In former studies on sugars solutions and carrot stocks, the relaxation times T1 (time to recover 63 % of the magnetization) for different sugars were studied and the final D1 was set to 25 s,[29] considering that D1 has to be 5 times bigger than T1.[4,30,31] In this study, we used a D1 value of 25 s.

A pulse angle of 90° was chosen in order to allow complete relaxation and absolute quantification.[32]

In order to obtain a satisfying signal/noise ratio, signal accumulation was performed. The influence of the number of scans (NS) was studied. The same sample was used to do all NMR acquisitions with the same parameters, changing the NS between 8, 16, 32, 64 and 128. For the biggest peaks, i.e. peaks for carbohydrates (Figure 1), a NS of 8 is enough but for smaller peaks such as those from amino acid, a NS of 32 is giving a better signal/noise ratio. In order to get a good resolution for the smallest peaks without increasing too much the acquisition time, a NS of 64 was preferred. As said above, TSP was used for quantification. As this compound has a long relaxation time (over 25 s),[33] NMR analysis would have been too long if full relaxation was performed. However reducing relaxation times induced a reduced apparent quantity of TSP from capillary tube. This is why the number of protons that are used for quantification was calculated by quantifying TSP with a known concentration solution of potassium phthalate, which is a stable and fast relaxing molecule.

To conclude, each spectrum consisted of 64 scans of 32K data points with a spectral width of 6.000 Hz and an acquisition time of 5.3 s, a recycle delay of 25 second per scan and a pulse angle of 90°. The analysis of each solid extract was performed using D_2O as an internal lock. Spectra were acquired under an automation procedure (automatic shimming and automatic sample loading) requiring about 33 min per sample.

2.2.3   2D NMR Parameters.   2D NMR – COSY sequences were recorded with 64 scans, a spectral width (SWH) of 4006.410 Hz, an acquisition time of 0.25 s, and a recycle delay of 2 s.

2.3   Identification of Metabolites

For simple systems such as stocks obtained from plant tissues, NMR spectra include a wealth of peaks.[4] Generally, peak assignment of 1H NMR spectra can be done by addition of pure compounds to solutions, by comparison of published data[4,32,34] and further confirmation can be obtained through 2D (correlation spectroscopy, COSY, and total correlation spectroscopy, TOCSY) experiments. Thus, each major carbohydrate was identified by peak assignment of 1D ^{1}H NMR spectra; and as many peaks overlapped in the amino acids and organic acid region (between 1 and 3.5 ppm), identification was performed using 2D 1H NMR spectra (COSY).

2.4 Quantification of Sugars

The same capillary tube of TSP was introduced in each NMR tube. As said, TSP was used both for chemical shift calibration and for quantification of absolute concentrations of metabolites with calibration curves.[11] Indeed, quantification by NMR spectroscopy can be done either by comparing directly the area of the peak of interest to the area of the referent peak, or by using calibration curves for each molecule. As each molecule has its own chemical and physical environment, and then its own relaxation, calibration curves seemed to be the most accurate method, whereas the direct quantification is faster (and still relevant).

2.4.1 Standard Solutions of a Mix of Three Sugars. As said above, different forms of glucose and fructose exist in solution. However, these proportions vary regarding different publications.[35,36] In our study on sugars, it is then necessary to use calibration curves for glucose and fructose at controlled temperature (see previous §), in order to take care of the different sugar forms. Such issue is not encountered for sucrose, but as explained above and also to take into account most of the possible interactions in complex solutions (that occur even at 6.4 µg/mL),[37] solutions in D_2O of a mixture of the three carbohydrates were studied. Different concentrations (of each carbohydrate) at 8, 3, 0.8 and 0.2 mg/g were achieved by diluting a mother solution at 75 mg/g.

2.4.2 Peaks of interest. Quantification was obtained by integration of one relevant peak for each sugar. Relevant peaks were chosen using two criteria: return to the baseline and no peak overlapping with other molecules.
1D 1H NMR spectra of water where raw and fried onion dice were soaked (Figures 1 and 2 respectively) show that many peaks overlap in a part of the sugar region, between 3.0 and 4.5 ppm. The peaks of interest for sucrose and glucose were easily identified as the chemical shifts of their proton are isolated in the spectrum. The doublet at 5.42 ppm (G1H) was chosen for sucrose.
The glucose doublet at 5.24 ppm (α-C1H) was more convenient than two other peaks. On one hand, the doublet at 4.64 ppm (β-C1H) is sometimes comprised in the water peak and, on the other hand, the doublet of doublet at 3.24 ppm (α–C2H) overlap with the triplet at 3.23 ppm of arginine (δ-CH$_2$). It could have been very interesting to get three possibilities of integration, in order to enhance quantification reliability. As a matter of fact, it could be statistically interesting to work with three different calibration curves, leading to three separated quantifications;hen, an average quantity would be calculated with its standard deviation. This point is now explored.
For fructose, whose peaks are all in the fuzzy part of spectra, picking a relevant peak was more complex. The multiplet at 4.10 - 4.11 ppm (β-C4H, β-C3H, α-C3H) was chosen as there was no theoretical peak overlapping with the present amino acids, organic acids or fructo-oligosaccharides (mainly kestose and nystose). However, in this part of the spectrum the return to the baseline is not completely achieved, so that peak deconvolution was needed (see following § 2.4.3).

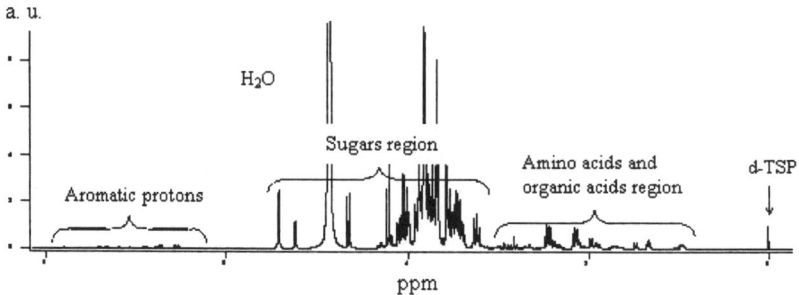

Figure 1 *1D ^1H NMR spectrum of an aqueous solution where raw onions dice where soaked during about 1 day. 0.9 – 3.2 ppm: amino acids and organic acids region, 3.2 – 5.5 ppm: sugars region, 6.0 – 8.0 ppm: aromatic protons.*

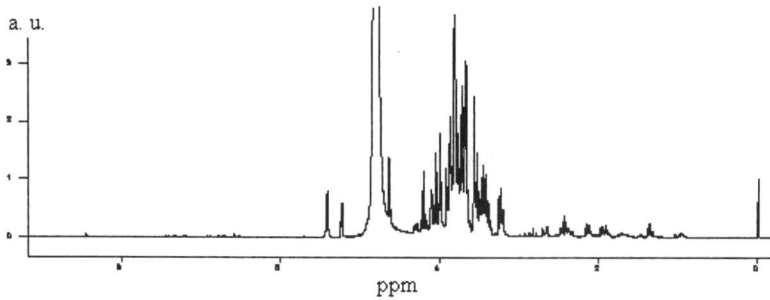

Figure 2 *1D ^1H NMR spectrum of an aqueous solution where fried onions dice where soaked during about 1 day.*

2.4.3 Integration Methods. Of course, baseline is an important step for quantification. After Fourier transformation of the spectra with 0.3 Hz line broadening and phase correction, baseline was corrected automatically and then manually using XWINNMR software (*Bruker Biospin*, Karlsruhe, Germany). The automatic baseline correction of the spectrum is performed by subtracting a polynomial, with the degree of the polynomial set by default at 5. The part of the spectrum that does not contain spectral information is considered by the command as baseline and it is used to fit the polynomial function. If automatic baseline correction does not give satisfactory results (i.e. baseline drift), manual baseline correction can be performed, by arbitrary playing with the parameters A, B, C D and E of the following polynomial function (equation 1). With these corrections, noise is not null.

$$y = A + B*x + C*x^2 + D*x^3 + E*x^4 \tag{1}$$

After spectra acquisition, phase and baseline correction, data treatment for quantification was performed. Four methods of peak integration with the same integration borders (glucose from 5.3 to 5.15 ppm, fructose from the beginning of the multiplet to the end of the shoulder of the multiplet at 4.1 ppm, and sucrose from 5.5 to 5.3 ppm) were tested and compared by the value of R^2 of each calibration curve. The four integration methods tested were: (a) integration of the peaks by *NMR Notebook software*, (b) integration of the peaks by *Igor Pro software*, (c) multipeakfitting and integration of the fit by *Igor Pro software*, and (d) multipeakfitting and integration of the fitted peaks by *Igor Pro software*. This study (to be published) showed no drastic difference between these four methods of integration ($R^2 > 0.99$). In the case of peak overlapping or when problems with baseline occur, the method (d) could be the more accurate. Thus, the carbohydrate concentrations in each sample were calculated from concentrations in the NMR tube and soaking solution volume by integrating the peaks of interest with *Igor Pro software*, using the method (d).

3 RESULTS AND DISCUSSIONS

3.1 Identification

Solutions obtained by soaking raw and fried onion tissues were studied. 1D-1H NMR spectra of all solutions that we prepared had similar shapes (Figures 1 and 2). Three main regions can be recognized: an amino acids and organic acids region (0 – 3.2 ppm), a carbohydrate region (3.2 – 5.5 ppm) and an aromatic protons region (6.0 – 8.0 ppm).

Glucose, fructose and sucrose were identified, and fructo-oligosaccharides peaks may be also observed. However, chemical shifts of the main fructo-oligosaccharides present in onion bulbs, i.e. kestose and nystose, as well as their possible neo-forms, are comprised in the complex sugar region.[34] Except for doublets at about 5.44 ppm and at about 4.27 ppm, corresponding to the proton C1H of the glucose unit and to the proton C3H of the 1F1 fructose unit, all fructo-oligosaccharides peaks overlap with other sugars peaks. This issue was not resolved by COSY experiment, as all the correlation dots for sugars were very broad.

18 massifs/peaks were studied and 14 amino acids were identified, including alanine (Ala), arginine (Arg), aspartic acid (Asp), asparagine (Asn), glutamic acid/glutamate (Glu), glutamine (Gln), histidine (His), isoleucine (Ile), leucine (Leu), lysine (Lys), phenylalanine (Phe), threonine (Thr), tyrosine (Tyr), valine (Val) and 6 organic acids including acetic acid/acetate (Ace), fumaric acid/fumarate (Fum), gamma amino butyric acid (GABA), malic acid/malate (Mal), succinic acid/succinate (Scc) and formic acid/formate (For) (to be published). However, the presence of His has to be confirmed: 1D spectra show some peaks that could be attributed to His, but no correlation peaks were found in 2D spectra. This may be due to a small quantity that could be under the limit of detection, even if the peak is 3 times bigger than the noise.

3.2 Carbohydrate Quantification

Dry matter content of aqueous solutions were onions tissues were soaked was determined. This particular study showed that for soaking of raw onions at different time the coefficients of variation (CV) were small (around 10 %). The frying process is associated with higher CV (> 20 %), because this thermal treatment is difficult to standardize. The total extraction from fried samples is faster (3.6 times) than for raw samples: between 0 and 1.24 h, respectively 24 % and 9 % of the total dry matter was extracted for B and for A. For both experiments, the extraction rates decrease after about 1h, and the variation is not significant afterwards. After about 11 days of soaking, the dry matter extracted during soaking of raw onion tissues is not significantly different from the dry matter extracted during soaking of fried onion samples (analysis of variance showed the results to be not significant at the 10 % level).

The time course of extractions during soaking was similar for glucose, fructose and sucrose. Between 0 and 1.24 h, glucose, fructose and sucrose extraction rates in experiment B_1 are respectively 3.2, 2.2 and 2.2 times more important than the extraction rates for experiment A. After 1.24 h, the extraction rates decrease for both experiments until a similar value of under 0.03 mg/g (f.w.)/h is reached after about 11 days of soaking. Again, CVs are bigger for fried samples. Dry matter extracted by soaking is only composed of water soluble molecules, i.e. mainly carbohydrates (sucrose, glucose, fructose, and also oligo-fructosaccharides that are not quantified by NMR because of their low concentration) and amino acids, organic acids, and some other minor compounds (identified by NMR spectroscopy). The total quantity of glucose, fructose and sucrose extracted after 11 days of soaking is equal to 58.5 ± 6.4 mg/g of f.w. and 74.0 ± 19.9 mg/g of f.w for experiment A and B respectively. This shows that mono- and di-saccharides represents a large fraction of the water soluble molecules extracted by soaking onion tissues. Values of glucose, fructose and sucrose extracted during long-term soaking were compared[38] to O'Donoghue paper[21] in Table 1. Values are in the same order of magnitude, but different, mainly for sucrose.

However some reactions such as sucrose hydrolysis or glucose dehydration (with production of hydroxymethylfurural, HMF)) could occur inside the cells that are exposed to the hot frying oil.[11] NMR spectra of water where fried onion dice were soaked did not show any HMF peaks,[39] but this could be explained by the fact that a brown color can be rapidly induced by a small amount of these reactions products that may be not detectable by NMR spectroscopy.

Table 1 *Comparison of non structural carbohydrate content (in mg/g f.w.) in onion tissues by O'Donoghue[21] (1) for two onion varieties (a: PLK and b: Grano) and in soaking aqueous solutions for A (2) and B_1 (3).*

		Glucose	Fructose	Sucrose
1a	Mean	18.6	3.2	11.2
	s.d.	0.6	0.3	0.7
1b	Mean	19.6	14.4	4.9
	s.d.	1.0	0.6	0.3
2	Mean	17.2	23.2	18.0
	s.d.	2.9	1.2	2.3
3	Mean	19.3	27.7	27.0
	s.d.	5.0	6.6	8.3

4 CONCLUSION

NMR spectroscopy is a useful analytical tool to perform analyses on water soluble compounds extracted during soaking. The fast preparation of samples avoids chemical modifications of the samples and possible bias due to necessary preparation (such as derivatization). Using 1D and 2D NMR studies, 3 sugars (and assumption on 2 fructo-oligosassharides), 14 amino acids and 6 organic acids were identified. A complete quantification of the main sugars was achieved with an internal standard (TSP). Other studies using NMR spectrometry can be investigated, such as relaxometry through time domain NMR with T1-T2 correlations and T1-T2 maps by MRI (Magnetic Resonance Imaging).

References

1　H. D. Belitz and W. Grosh in *Food chemistry*, ISBN 3-540-15043-9, Springer-Verlag Berlin Heidelberg, 2004 (3rd revised edition).
2　H. This, *International Journal of Pharmaceutics*, 2007, **344**, 4.
3　H This, *Br. J. Nutr.,* 2005, **93**, S139.
4　T. Fan, *Prog. Nucl. Magn. Reson. Spectrosc.*, 1996, **28**, 161.
5　A. M. Gil, I. F. Duarte, I. Delgadillo, I. J. Colquhoun, F. Casuscelli, E. Humpfer and M. Spraul, *J. Agric. Food Chem.*, 2000, **48**, 1524.
6　I. F. Duarte, A. Barros, P. S. Belton, R. Righelato, M. Spraul, E. Humpfer and A. M. Gil, *J. Agric. Food Chem.*, 2002, **50**, 2475.
7　G. Le Gall, I. J. Colquhoun, A. L. Davies, G. J. Collins and M. E. Verhoeyen, *J. Agric. Food Chem.*, 2003, **51**, 2447.
8　D. Jouan-Rimbaud Bouveresse, H. Benabid and D. N. Rutledge, *Analytica Chimica Acta*, 2007, **589**, 216.
9　A. P. Sobolev, E. Brosio, R. Gianferri and A. L. Segre, *Magn. Reson. Chem.*, 2005, **43**, 625.
10　G. F. Pauli, B. U. Jaki and D. C. Lankin. *J. Nat. Prod.* 2005, **68**, 133.
11　A. Cazor, C. Deborde, A. Moing, D. Rolin and H. This, *J. Agric. Food Chem.*, 2006, **54**, 4681.
12　J. Valverde and H. This, *J. Agric. Food Chem.*, 2008, **56**, 314.
13　H. D. Rabinowitch and J. L. Brewster in *Onions and allied crops*, CRC Press, Boca Raton, Florida (US), 1990, Vol. 1: Botany, physiology, and genetic.
14　N. Shiomi, N. Benkeblia et al., *J. Appl. Glycosci.* 2005, **52**, 121.
15　K. Kaack, L. P. Christensen et al., *Eur Food Res Technol.*, 2004, **218**, 372.
16　N. Benkeblia, S. Onodera et al., *J Sci Food Agric.*, 2005, **85**, 227.
17　N. Benkeblia, N. Takahashi et al., *Tetrahedron: Asymmetry*, 2005, **16**, 33.
18　L. Jaime, M. A. Martin-Cabrejas et al., *J. Agric. Food Chem.*, 2001, **49**, 982.
19　H. D. Rabinowitch and J. L. Brewster in *Onions and allied crops*, CRC Press, Boca Raton, Florida (US), 1990, vol. 3: biochemistry, food science, and minor crops.
20　J. Wang, P. Sporns et al., *J. Agric. Food Chem.,* 1999, **47**, 1549.
21　E. M. O'Donoghue, S. D. Omerfield et al., *J. Agric. Food Chem.*, 2004, **52**, 5383.
22　C. Selby, I. J. Galpin et al., *New Phytol.*, 1979, **83**, 351.
23　A. J. Taylor, *Critical Reviews in Food Science and Nutrition*, 1996, **36**(8), 765.
24　L. Jaime, E. Molla et al., *J. Agric. Food Chem.*, 2002, **50**, 122.
25　M. Maincent in *Cuisine de référence*, BPI : Paris, 1993.
26　M. J. Havey, C. R. Galmarini et al., *Genome*, 2004, **47**, 463.
27　R. Polacek, J. Stenger, U, Kaatze, *Journal of chemical physics*, 2002, *116*, 2973.
28　N. Le Barc'H, J. M. Grossel, P. Looten, M. Mathlouthi, *Food Chem.* 2001, **74**, 119.
29　Anne Cazor, Etude des solutions obtenues par traitement thermique en phase aqueuse de tissus végétaux (racines de *Daucus carota* L.) ou animaux (tissus musculaires, M. Pectoralis major, Gallus domesticus). Recherche des mécanismes responsables de la constitution de ces solutions (« bouillons ») par spectroscopie par résonance magnétique nucléaire quantitative du proton (q 1H RMN) et par électrophorèse (SDS-PAGE) : analyse des modifications microstructurales ou chimiques des tissus traités et

suivi cinétique des transferts des principales molécules sapides (sucres, protéines, acides aminés et acides organiques). EDCM université Paris VI, Paris 18/06/2007.

30   J. M. Nuzillard, Résonance magnétique nucléaire des liquides, FRE CNRS 2715, Université de Reims-Champagne-Ardenne 2006.
31   S. Akoka, L. Barantin, M. Trierweiler, *Anal. Chem.*, 1999, **71**, 2554.
32   A. Moing, M. Maucourt, C. Renaud, G. Gaudillère, R. Brouquisse, B. Lebouteiller, A. Gousset-Dupont, J. Vidal, D. Granot, B. Denoyes-Rothan, E. Lerceteau-Köhler and D. Rolin, Funct. *Plant Biol.*, 2004, **31**, 889.
33   J. P. Munasinghe, L. D. Colebrook, J. J. Attard, T. A. Carpenter and L. D. Hall, *Magnetic Resonance in Chemistry*, 1997, **36**(2), 116.
34   E. Fulushi et al., *Magn. Reson. Chem.*, 2000, **38**, 1005.
35   P. Arnaud in *Chimie organique*, 16ème ed.; Bordas: Paris, 1997.
36   C. Araujo-Andrade, F. Ruiz, J. R. Martinez-Mendoza and H. Terrones, *Journal of Molecular Structure*,2005, **714**, 143.
37   S. L. Kaufman and F. D. Dorman, *Langmuir*, 10.1021/la800177m, Web Release Date: 2008, August 13.
38   F. Davis, L. A. Terry, G. A. Chope and C. F. J. Faul, *J. Agric. Food Chem.*, 2007, **55** (11), 4299.
39   R. Consonni and A. Gatti, *J. Agric. Food Chem.*, 2004, **52**, 3446.

ANALYSIS OF BUTTER AND MARGARINE BY HIGH-RESOLUTION ^{1}H NMR

Jan Schripsema

Grupo Metabolômica, Laboratório de Ciências Quimicas, Centro de Ciência e Tecnologia, Universidade Estadual do Norte Fluminense, Av. Alberto Lamego, 2000, 28015-620 Campos dos Goytacazes, RJ, Brazil. E-mail: jan@uenf.br

1 INTRODUCTION

In Brazilian butter samples a great variability was observed in organoleptic characteristics. To find out what might be the reason for this, the butter samples were investigated by metabolomic techniques, in this case NMR (nuclear magnetic resonance). A method based on ^{1}H NMR has been developed to obtain quantitative profiles of both the aqueous and the fat phase of the butter.[1] To achieve this the butter was separated in a polar (soluble in water) and apolar fraction (soluble in chloroform). From both these fractions the ^{1}H NMR spectra were obtained, providing in this way the profiles. Compounds which can be observed and quantified in these profiles include the preservatives benzoic and sorbic acid, the organic acids formic, acetic lactic, citric and butyric acid, the carbohydrate lactose and the fatty acids rumenic and linoleic acid.[1] In the present paper the results obtained with this method are further explored.

Butter is rather easily obtained from milk, and therefore the invention of butter is thought to have taken place at the early days of keeping animals for milk. This should have been about 10000 years ago in Mesopotamia.[2] In nearly all cultures butter or similar products are known. Chemically butter is a rather complex product, containing a large variety of different fatty acids. Furthermore it is a water in fat emulsion, containing both water- and fat soluble compounds. The taste and smell of butter are determined by a complex mixture of compounds, originating not only from the milk and the production process but also from storage conditions. Exposure of butter to air easily leads to a rancid taste due to oxidation. Also microbiological spoilage is possible due to the relatively high water content.[2]

Butter is entirely obtained from milk; it contains about 80–82% milk fat, 16–17% water, and 1–2% milk solids other than fat, including salt. Margarine, developed at the end of the 19th century as a butter substitute, is similar in composition, but differs from butter in the fats used for its production, which were originally animal fats but currently almost exclusively vegetable fats are used.[2] Over the years, and especially after World War II, margarine has largely replaced butter as table spread,[2] first due to lower prices and afterwards through suggested beneficial health effects of replacing saturated fatty acids by unsaturated ones.

Most of the studies concerning the chemistry of butter have been devoted to the fat composition of butter.[3,4] In most of these studies the fatty acids were analyzed after hydrolysis by GC-MS. About 98% of the fat was found to consist of triglycerides, and the most common fatty acids are palmitic (22–35%), oleic (20–30%), stearic (9–14%), myristic (8–14%), and butyric (2–5%) acid.[3]

In recent years much attention has been paid especially to the so-called conjugated linoleic acids (CLA) for which many interesting biological activities have been reported.[5,6]

In 1985 Pariza and Hargraves reported a compound with anti-mutagenic activity from beef extract.[7] A few years laters this compound was identified and named conjugated linoleic acid (CLA).[8] Food derived from ruminant animals, beef, milk, and derived products, were found to be the major source of these CLA in human food. The *cis*-9,*trans*-11 CLA isomer, known as rumenic acid, is the major isomer found in ruminant fat, normally representing 80-90% of the total CLA in milk fat. The anticarcinogenic effects were especially related to the rumenic acid. Recently, the range of positive health effects associated with CLA in experimental models has been extended to include reduction in body fat accretion and altered nutrient partitioning, antidiabetic effects, reduction in the development of atherosclerosis, enhanced bone mineralization, and modulation of the immune system.[5,9-11] As a result of the interest in CLA, many studies have investigated CLA levels in dairy products and how these can be influenced.[4,6,12]

2 METHOD AND RESULTS

2.1 NMR analysis

The method utilized in this paper has been extensively described in a previous paper.[1] All the butter and margarine samples were obtained from commercial establishments and maintained in the refrigerator until the analysis was performed. Furthermore, the samples were analyzed well before the date of validity of the product. For the analysis a quantity of 120–130 mg of butter or margarine was weighed. This sample was dissolved in a mixture of deuterated water (Cambridge Isotope Laboratories, 99.9% D) containing an exact quantity of the internal standard (tetradeutero-trimethylsilyl-proprionate, sodium salt) and deuterated chloroform (CIL, 99.9%). The quantity of each solvent was 0.80 ml or 0.70 ml. After the butter had dissolved in the mixture and the mixture had been allowed to stand for a few minutes for a good separation of the layers, 0.50 mL of the water layer was transferred to one NMR tube and 0.50 mL from the chloroform layer to another tube. The pH of the water layer was verified, and in all cases it was between 5.5 and 6.0. NMR spectra were obtained on a JEOL Eclipse+ 400 spectrometer, operating at 400 MHz for protons and 100 MHz for carbon-13. All spectra were obtained at 25 °C using previously described parameters.[1] Typically for the ^{1}H NMR spectra in D_2O, 1000 scans were recorded and the water signal was suppressed by presaturation, while for the $CDCl_3$ samples 100 scans were recorded. For the quantitative analysis individual signals were integrated and the quantities of the compounds were calculated through comparison with the signal from the internal standard. As internal standard for quantification the trimethylsilylproprionate derivative at 0 ppm was used in the D_2O samples. In the $CDCl_3$ samples the signals from the glycerol part of the triglycerides at 4.30 and 4.15 ppm were used.

2.2 Compounds detected in the NMR profiles

In this paper the results of the investigation of 22 samples of butter from 14 brands are reported. Furthermore 3 samples of margarine from 3 brands were investigated. All samples were from Brazil, with the exception of one butter sample which was from Norway. A summary of the compounds which were detected in the NMR spectra is displayed in Figure 1. The absolute number of compounds is rather limited, but the compounds are from quite different classes.

In Table 1 the quantities of these compounds in the butter and margarine samples are given. In the D_2O samples absolute quantifications of the compounds were obtained by adding a known amount of internal standard to the solution. As internal standard the TMS derivative (2,2,3,3-d_4)-trimethylsilylproprionate (TMSP) was used.[1]

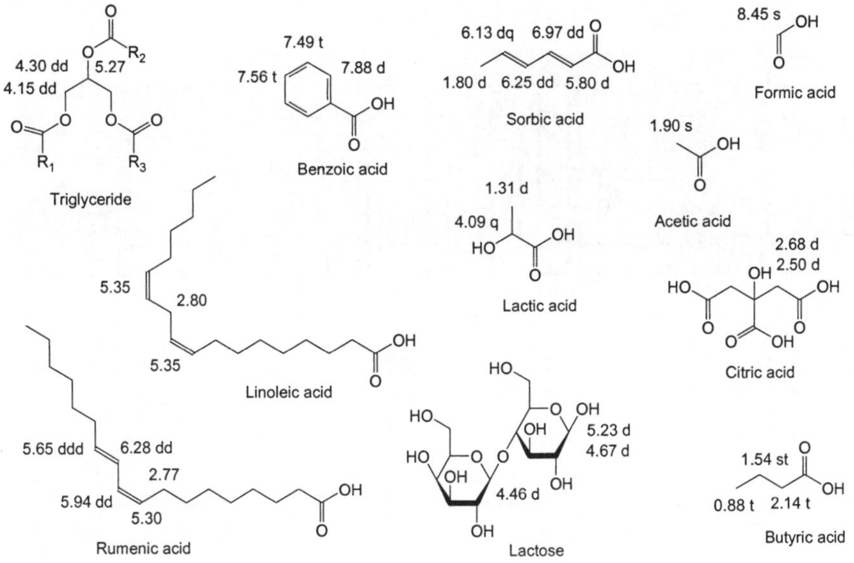

Figure 1 *Structural formulas of compounds observed in the ^1H NMR spectra, with the chemical shifts of the most important signals. (s: singlet, d: doublet, t: triplet, q: quartet, st: sextet).*

2.3 Preservatives in Butter and Margarine

Preservatives were detected in various butter samples as well as in all margarine samples. In six of the thirteen Brazilian butter samples preservatives were found: In five brands sorbic acid and in two brands benzoic acid. The presence of these compounds in butter is not allowed according to the Brazilian legislation.[13] However, in margarine the presence of these compounds is normal and allowed. In three brands of butter the sorbic acid levels were similar to those found in margarine (see Table 1), indicating that these compounds were added to the butter as preservative. In the other three butter samples the low levels of sorbic acid and/or benzoic acid indicated that they probably were the result of the presence of the preservative in the milk from which the butter was prepared.

2.4 Lactose and Lactic Acid Content

Lactose was present in all but one sample of butter and in two of the three samples of margarine (Table 1 and Figure 2). The levels in butter did show large variations, from near to zero up to 12.54 mg/g of butter. Considering that the normal concentration of lactose in milk varies from 3.8 to 5.5 %[14] and that in butter about 18% of residual milk liquid might remain, when after churning the excess of milk is removed and the remainder is directly homogenized, it can be concluded that the high lactose concentrations of about 10 mg/g of butter are the result of a production process without washing steps after the churning (see Scheme 1).

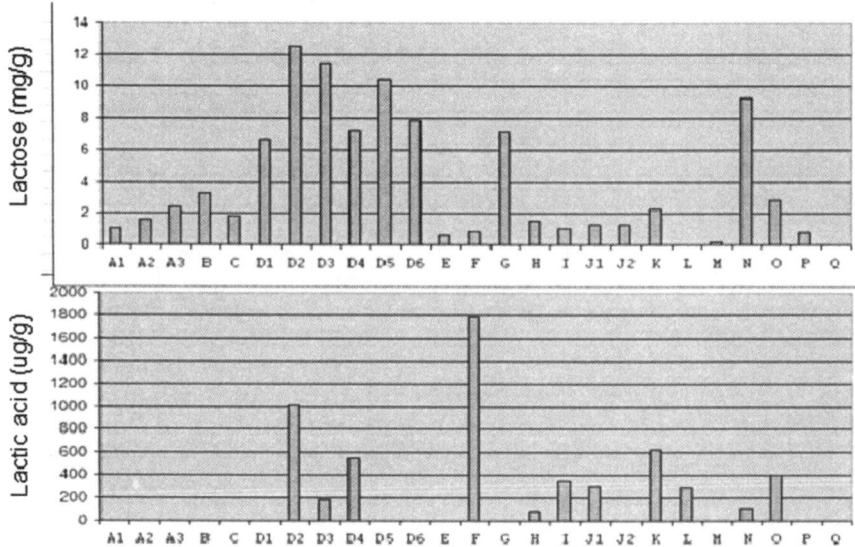

Figure 2   *Comparison of the quantities of lactose and lactic acid in the butter (A-N) and margarine (O-Q) samples. The codes correspond to those in Table 1.*

Lower levels of lactose would then be the result of the inclusion of a washing step of the butter particles after the churning and separation of the buttermilk, in this way leading to a substitution of the residual milk, and an about ten-fold decrease in lactose levels.

Lower lactose concentrations could also be the result of a fermentation process in which the lactose is converted to lactic acid. In a fermentation process also citric acid, normally present in milk in a concentration of about 0.2 %,[15] would be converted, in this case to acetic acid.

The occurrence of rather high levels of lactic acid together with high levels of lactose might indicate that the lactic acid has been added to the butter. Together with the inoculation of butter to obtain cultured butter, often lactic acid is added (see Scheme 1). Also it might have been added for flavour or as a preservative.

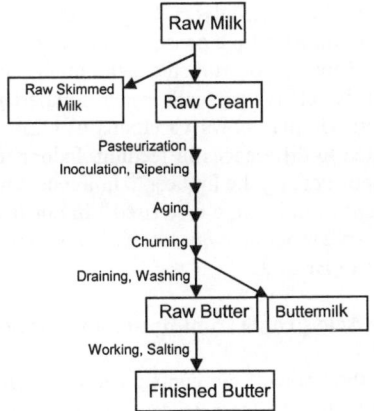

Scheme 1   *Schematic representation of the production process of butter*

2.5 Rumenic Acid in Butter

The fat fraction of butter or margarine dissolved in $CDCl_3$ yields a direct profile of the fats present. Traditionally the fatty acids are analyzed after hydrolysis of the glycerides,[3] but with NMR the glycerides are directly analyzed in the extract. For the quantification of the individual fatty acids (in bound form) specific signals in the NMR spectrum were integrated and the integral was related to the integral of the signal of the *sn*-1,3 glycerol position in the triglycerides at about 4.2 ppm. Rumenic acid shows some characteristic signals in the ^{1}H NMR spectrum (Figure 1), which do not overlap with other signals and permit the quantification of this fatty acid.

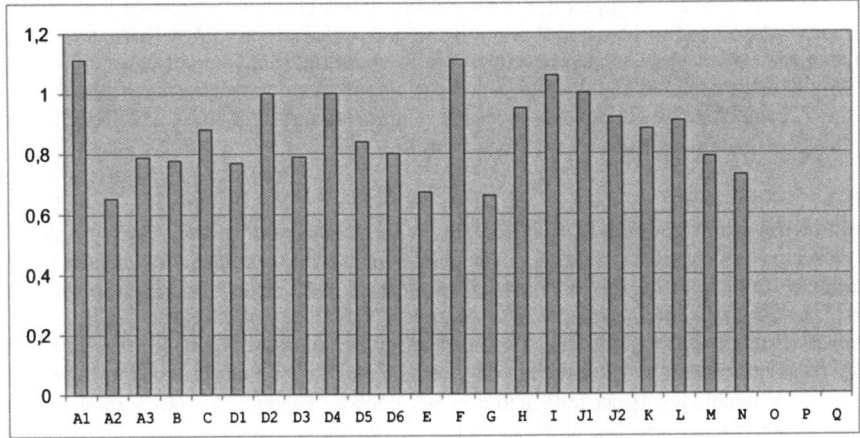

Figure 3   *The relative quantities of rumenic acid in the butter (A-N) and margarine (O-Q) samples. quantities are expressed as percentage relative to the maximum possible number of fatty acids linked to the glycerol of the triglycerides.*

Through the correlation with the *sn*-1,3 glycerol signal, the relative quantity in relation to the total triglycerides is determined, which constitute in butter about 98% of the fat.[3]

In Figure 3 the relative quantities of rumenic acid in the samples is displayed. The variation of the quantities is relatively small, when compared to values related in the literature. E.g. in milk from German cows variations of CLA from 0.26 to 1.14% were related.[16] This was attributed to differences in feeding. Indoor feeding with silage yielded the lowest levels and outdoor grazing the highest.[16] In a recent review the different factors influencing the CLA content of milk were discussed.[6] In our results no differences due to different seasons were observed, but anyway less variation would be expected due to the tropical/subtropical climate of Brazil.

2.6 Analysis of the Fatty Acid Profile from Butter and Margarine

Besides rumenic acid also the linoleic acid content can be determined. Linoleic acid has a specific NMR signal due to the methylene group between the two double bonds. This appears at 2.80 ppm. But also one of the methylene groups next to the double bonds of rumenic acid appears at this position (Figure 1). Therefore to obtain the linoleic acid quantity a correction for the rumenic acid quantity is necessary. In the case of margarine samples one should also count with the fact that other multiple unsaturated fatty acids can be present, e.g. linolenic acid which contains two methylene groups between the double bonds, both appearing at 2.80 ppm. Anyway the signal at 2.80 ppm is a good indicator for the presence of multiple unsaturated fatty acids. For the butter samples linoleic acid quantities were between 1.0 and 2.2 % (Table 1). For margarine much higher levels were found, between 36 and 58%.

Diglycerides with a free hydroxyl at position 3, also show characteristic well separated signals, allowing the quantification of these compounds. For butter samples the levels are typically between 1 and 2%.

The total level of unsaturated fatty acids can be estimated from the signal at 5.3 ppm, which corresponds to the hydrogens linked to double bonds. Because the signal at 5.27 ppm from H-2 of the triglycerides overlaps with this signal, the integral of both signals should be corrected for the contribution of the sn-2 signal of the triglycerides. As a result one obtains the average number of double bonds per fatty acid. In butter samples this value lies around 30%, while in margarine samples values of about 120% were obtained.

2.7 Deterioration of Butter

Some samples of butter were left for a long period (more than 3 years) in a closed tin at room temperature. Analysis of these samples (Table 1, sample D5-o and D6-o) did show a number of modifications. A large quantity of butyric acid was liberated, with estimated levels of 7-8 mg/g. Also high levels of formic acid and acetic acid were formed. Free glycerol was liberated, with levels up to more than 10 mg/g. The diglyceride level increased to about 5-6%. Lactose levels decreased and the level of unsaturated fatty acids decreased dramatically. Rumenic and linoleic acid could not be detected anymore, and the number of double bonds per fatty acid decreased down to about 5%. These samples might not be representative, but they indicate the direction of changes occurring upon deterioration of butter.

Table 1   *Quantities of Polar and Fat Components in Butter and Margarine Samples*[a]

Brand	type	Taste[b]	Formic acid µg/g	Benzoic Acid µg/g	Sorbic acid µg/g	Lactic acid µg/g	Glycerol µg/g	Citric acid µg/g	Acetic acid µg/g	Butyric Acid µg/g	Lactose mg/g	Rumenic acid %FA	Linoleic acid %FA	Diglyc. %	Double Bond No./FA%
A1		++	21	0	0	0	0	0	11	13	1.06	1.11	1.37	1.76	31.0
A2		++	0	0	0	0	0	79	2	5	1.62	0.65	1.92	1.53	28.5
A3	Salted	++	2	0	0	0	0	86	4	4	2.41	0.79	1.72	1.81	29.2
B		++	0	0	0	0	0	219	4	20	3.27	0.78	1.67	2.02	29.3
C		++	1	5	0	0	0	80	3	12	1.85	0.88	2.00	1.98	30.2
D1		++	0	0	0	0	0	238	13	14	6.72	0.77	1.96	1.74	29.7
D2		+++	35	0	0	1017	0	444	159	0	12.54	1.00	1.65	2.06	29.0
D3		++	31	0	0	178	0	437	33	0	11.51	0.79	1.61	1.77	28.7
D4		++	2	0	0	546	0	221	197	0	7.19	1.00	1.45	2.23	30.8
D5	Salted	++	3	0	0	0	0	408	9	0	10.41	0.84	1.82	1.75	29.1
D5-o	Salted		882	0	0	0	4000	0	502	8000	2.04	0	0	5.77	2.6
D6	Salted	++	2	0	0	0	0	298	7	0	7.87	0.80	1.80	1.70	31.5
D6-o	Salted		1311	0	0	0	12000	0	653	7000	3.51	0	0	5.53	6.4
E	Salted	'	24	0	538	0	0	0	8	222	0.60	0.67	1.97	2.01	29.3
F	Salted	'	4	0	0	1788	0	0	271	0	0.84	1.11	0.99	3.50	31.8
G	Salted	++	3	0	0	0	0	241	6	0	7.13	0.66	1.71	1.74	28.0
H	Salted	+	9	0	54	82	0	47	21	0	1.51	0.95	1.59	1.55	32.0
I	Salted	+	49	0	0	349	0	48	29	0	1.06	1.06	1.01	1.65	30.2
J1	Salted	+	29	0	261	295	0	0	42	0	1.28	1.00	2.28	1.81	34.3
J2	Salted	++	0	0	0	0	0	0	2	33	1.29	0.92	1.20	1.58	28.0
K	Salted	'	36	41	41	628	0	0	169	336	2.26	0.88	1.37	2.84	29.0
L	Salted	--	64	0	355	294	0	0	110	23	0	0.91	1.23	2.04	29.8
M	Salted	+	7	0	0	0	0	0	1	2	0.18	0.79	1.21	1.69	30.5
N		++	10	0	0	115	0	618	30	120	9.28	0.73	2.03	1.48	32.9
O	Margarine		3	636	521	402	0	120	19	0	2.81	0	36.79	0.46	116.9
P	Margarine		8	227	200	0	0	156	16	0	0.79	0	36.50	0	113.7
Q	Margarine		0	0	1150	0	0	443	38	0	0	0	57.29	0	132.2

[a] Brands A-M are brands of butter from Brazil. Brand N is a butter sample from Norway, and O-Q are margarine samples from Brazil. The addition –o means the sample was deteriorated. [b] taste: +++ very good, ++ good, + reasonable, - bad, -- bad rancid.

3 CONCLUSION

The ¹H NMR spectra of the water and chloroform soluble components of butter and margarine samples provide characteristic fingerprints, which permit the quantification of a series of compounds, belonging to different classes of compounds. One of the great advantages of this method is the easy preparation of the samples.

The analysis permits the monitoring of forbidden additives, such as the preservatives benzoic acid and sorbic acid. Furthermore, the verification of quality, *e.g.* by determination of the levels of diglycerides, free butyric acid and unsaturated fatty acids. It also provides information about the production process, through the levels of *e.g.* lactose and lactic acid in the butter.

References

1   J. Schripsema, *J. Agric. Food Chem.*, 2008, **56**, 2547.
2   http://webexhibits.org/butter/index.html.
3   R.G. Jensen, *J. Dairy Sci.*, 2002, **85**, 295.
4   M. Ledoux, J.-M. Chardigny, M. Darbois, Y. Soustre, J.-L. Sebedio and L. Laloux, *J. Food Compos. Anal.*, 2005, **18**, 409.
5   Y. Park and M.W. Pariza, *Food Res. Int.*, 2007, **40**, 311.
6   M. Collomb, A. Schmid, R. Sieber, D. Wechsler and E.-L. Ryhanen, *Int. Dairy J.*, 2006, **16**, 1347.
7   M.W. Pariza and W.A.A. Hargraves, *Carcinogenesis*, 1985, **6**, 591.
8   Y.L. Ha, N.K. Grimm and M.W. Pariza, *Carcinogenesis*, 1987, **8**, 1881.
9   M.A. Belury, *Nutr. Rev.* 1995, **53**, 83.
10   C. Ip, S. Banni, E. Angioni, G. Carta, J. McGinley, H.J. Thompson, D. Barbano and D. Bauman, *J. Nutr.* 1999, **129**, 2135.
11   K.L. Houseknecht, J.P. Van den Heuvel, S.Y. Moya-Camarena, C.P. Portocarrero, L.W. Peck, K.P. Nickel and M.A. Belury, *Biochem. Biophys. Res. Commun.*, 1998, **244**, 678.
12   A.K. Seckin, O Gursoy, O. Kinik and N. Akbulut, *Lebensm. Wiss. Technol.* 2005, **38**, 909.
13   Brazilian legislation: Decreto no. 1812 (Feb 8ᵗʰ 1996). *Diario Of. União*, 1996, section 1 (Feb 9), 2241.
14   Alfa Laval/Tetra Pak, 'The Chemistry of Milk' in *Dairy Processing Handbook*, Tetra Pak Processing Systems, Lund, Schweden, 1995, Chapter 2, pp. 13-36.
15   A.W. Bosworth, M.J. Prucha, *J. Biol. Chem.*, 1910, **8**, 479.
16   G. Jahreis, J. Fritsche and H. Steinhart, *Nutr. Res.*, 1997, **17**, 1479.

SPIN-LATTICE RELAXATION TIME MEASUREMENTS AS A PROBE FOR TRIACYLGLYCEROL POLYMORPHISM AND CRYSTAL SIZE

M. Adam-Berret[1, 2, 3], A. Riaublanc[2], C. Rondeau-Mouro[2], F. Mariette[1, 2]

[1] Cemagref, UR TERE, 17 Avenue de Cucillé, CS 64427, F-35044 Rennes, France
[2] UR 1268 INRA-B, Rue de la Géraudière, BP 71627, 44316 Nantes Cedex 3, France
[3] Université européenne de Bretagne, France

1 INTRODUCTION

Food products of high fat content such as butter, margarine and shortening can often be assimilated to a mixture of a solid and a liquid lipid phase. The complex structure formed by the two phases affects physical properties of fats such as quality, texture and spreadability.[1] These properties are dependent on the lipid crystalline microstructure which normally exists as a three-dimensional colloidal fat crystal network.[2] Upon crystallization, fat crystals behave as colloidal gels, thus they aggregate and grow into clusters, flocs and finally form a network. Such a structural hierarchy has been recognized by many groups and further developed to the modelling of the rheological properties of fats.[3, 4] Moreover, knowing the crystalline microstructure is really important because it monitors the physical properties of fats. This microstructure is defined by different parameters such as crystal size, distribution of crystal size, shape and polymorphism which are dependent on processing conditions.[5] However, the crystalline microstructure is also strongly influenced by the presence of a liquid oil phase.[6] Thus solid fat content (SFC) is also an important parameter for the determination of the physical properties of the fat crystal network.

Different methods can be used to determine all these parameters such as X-Ray Diffraction for polymorphism, polarized microscopy for crystal size and NMR for SFC. The last measurement is based on the proton signal intensity of the solid and liquid phases. However, relaxation time measurements recently proved to be useful for the determination of polymorphism. An initial study showed that SFC and polymorphism could be determined through a single measurement combining a Free Induction Decay (FID) and a CPMG (Carr-Purcell-Meiboom-Gill) in one sequence.[7] In a second study based on pure triacylglycerols, NMR relaxation parameters proved to be useful for the determination of polymorphism, independently of temperature and chain length.[8] Second moment (M_2) measurements were sufficiently sensitive to distinguish the α and the β polymorphs, but a greater sensitivity was found for spin-lattice relaxation times (T_1). As low field NMR is available in many food laboratories for the determination of SFC, it is now interesting to determine the potential of the technique concerning the information it can provide on the crystalline microstructure. Determination of polymorphism was proved to be possible, but the technique could also be useful to provide information on crystal size. Different parameters of the fat crystal network are known to evolve as a function of time. There is a decrease in the specific surface area and an increase in the fractal dimension of the system.[9, 10] Such evolutions can be related to the increase in crystal sizes according to the Ostwald Ripening effect. This phenomenon is characterized by the melting of the smaller crystals which recrystallize into larger ones which lead to an increase in the average crystal size. This phenomenon has already been studied by NMR in frozen sugar solutions, and a correlation was found between the crystallization rate and the spin-spin relaxation time (T_2).[11] NMR parameters could thus be sensitive to the evolution in crystal size.

The evolution of the fat crystal network of tricaprin and tristearin mixture as a function of time was followed by NMR and polarized microscopy in order to establish a correlation between crystal size and spin-lattice relaxation time. The T_1 distributions were also measured for two solid mixtures of tricaprin/tristearin and trilaurin/tripalmitin in order to determine crystal size distribution in a solid mixture.

2 EXPERIMENTAL

2.1 Materials and Tempering Procedures

Triacylglycerols were purchased commercially (Sigma, St Louis, MO, USA; >98% purity) and samples were used without any further purification. Before each experiment, triacylglycerols were melted at 80°C to erase all polymorphic memory.

For the effects of storage time, the tricaprin/tristearin blend 50:50 (w/w) was melted at 80°C for 20 min, and then cooled to 5°C by direct immersion in a water bath for 10 min, and the sample was inserted inside the NMR spectrometer heated to 40°C. The sample was kept at this temperature for 14 days. For polarized microscopy, the thermal diagram applied to the sample was the same, but the tricaprin/tristearin blend 75:25 (w/w) was used for a matter of clarity.

Two different systems of triacylglycerols, i.e. tricaprin and tristearin on the one hand and trilaurin and tripalmitin on the other, were studied in the solid state at three different concentrations: 25:75, 50:50, 75:25 (w/w). For the tricaprin/tristearin system, the mixtures were melted at 80°C and then cooled to -50°C for 10 min. The temperature was finally increased to 10°C for 30 min. For the trilaurin/tripalmitin system, the mixtures were melted at 80°C and then cooled to -20°C for 10 min. The sample was first reheated to 0°C for 20 min and then to 20°C for 30 min before NMR measurements.

2.2 Low-field NMR Measurements

Measurements were carried out with a 0.47 T NMR spectrometer (Minispec PC 120, Bruker SA, Wissembourg, France) operating at 20 MHz for protons. The instrument was equipped with a 10 mm probehead. Magnetic field tuning, homogeneity of the magnet, detection angles, receiver gain and pulse lengths were checked before each measurement. Two kinds of NMR sequence were used, i.e. Free Induction Decay (FID) and Fast Saturation Recovery (FSR). For measurements at 10°C and 20°C, FID were recorded for 300 µs with 32 scans and a Recycling delay (Rd) of 10 s, and the FSR were recorded between 30 and 10000 ms with 100 points. For measurements at 40°C, FID were recorded for 300 µs with 16 scans and a Recycling delay (Rd) included between 30 s and 40 s, and the FSR were recorded between 30 and 40000 ms with 100 points.

2.3 Light Polarization Microscopy

Blends of TAG were melted at 80°C for 20 minutes, and then a drop was deposited on a pre-heated glass microscope slide and then rapidly cooled to 0°C by ice contact. The crystallized drop was then observed at 40°C under 90° polarized light using an Eclipse E400 microscope (Nikon, Champigny, France) equipped with a thermostated sample holder (Linkam, Tadworth, UK). Slides were stored in an oven at 40°C between observations.

2.4 Data Analysis

The data from spin-lattice relaxation time measurements were fitted using the Levenberg-Marquardt method according to the monoexponential function presented in Equation (1).

$$s(t) = A \times (1 - \alpha \times e^{-\frac{t}{T_1}}) \tag{1}$$

The α parameter is necessary for low-field spectrometers in order to correct errors from the 90° pulse. Data were also fitted by the MEM (Maximum Entropy Method) algorithm which presents the advantage of providing a continuous distribution of spin-lattice relaxation time components.

3 RESULTS AND DISCUSSION

3.1 Effects of Crystal Size on T_1

The evolution of spin-lattice relaxation times as a function of time was measured for the tricaprin/tristearin 50:50 (w/w) mixture at 40°C. At this temperature, tricaprin was liquid and tristearin was in the β form (checked by XRD measurements, results not shown). The Solid Fat Content (SFC) was 60%. The small deviation from theory (50%) was due to the presence of co-crystals which delay the melting of tricaprin crystals. It is already known that the average crystal size increases as a function of time via the Ostwald Ripening phenomenon according to a power law model.[12] The evolution of T_1 for the present system is shown in Figure 1.

Spin-lattice relaxation time increased continuously as a function of time, passing from 4000 ms to 8500 ms. It was possible to see two different evolutions. T_1 increased rapidly during the two first days and slowed down after this time. Such a curve can be fitted with a power law model. The evolution was the same as for crystal size during Ostwald Ripening. The power law exponent determined for this system was 0.098, which is lower than the 0.2 to 0.3 which can be found for the evolution of crystal size followed by polarized microscopy.[13] The actual deviation was due to the fact that we measured spin-lattice relaxation times and not crystal size directly. However, the fact that we retained the power law model led us to expect a relationship between crystal size and spin-lattice relaxation time.

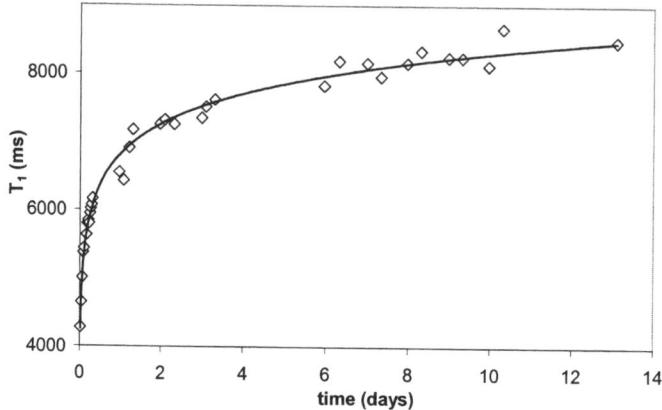

Figure 1   *Evolution of T_1 as a function of time for the tricaprin/tristearin 50:50 (w/w) system at 40°C*

In order to confirm this result, polarized microscopy was used on a mixture of tricaprin/tristearin 75/25 (w/w). This mixture was preferred to the 50/50 one because of the clearer images which can be obtained due to better differentiation of the solid and liquid phases. The images were taken during the three first days, when the increase in T_1 was maximal and are shown in Figure 2. It is worth noting that polarized microscopy is only sensitive to changes in 2D whereas T_1 is sensitive to the changes in the sample volume.

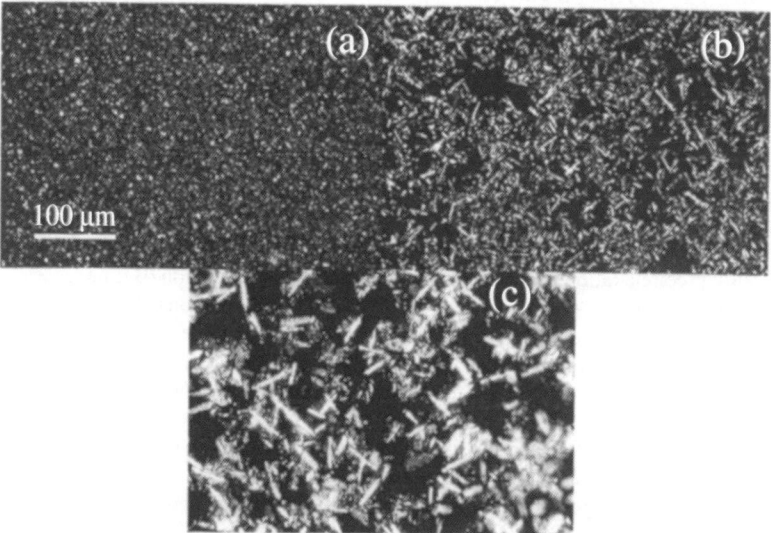

Figure 2   *Polarized light micrographs of tricaprin/tristearin 75:25 (w/w) mixture at 40°C after (a) 5 min, (b) 24 hours and (c) 64 hours. Same scale for all images.*

As expected, it was possible to observe an increase in the average crystal size as a function of time, and the greatest evolution occurred during the 24 first hours. It is worth noting that there were many small crystals of the β polymorphs after 5 min.

Thus from the NMR and microscopy results, it is possible to conclude that T_1 is linked to crystal size. As some physical properties of fats are related to crystal size, this relationship could be useful to obtain information on the properties of the fat crystal network. It should be possible to follow the evolution of the system as a function of storage time and determine the best period for consumption, or follow the evolution of the fat crystal network during heating and cooling cycles.

3.2 Study of Tricaprin/Tristearin and Trilaurin/Tripalmitin Solid Mixtures

The previous results were obtained for crystals of a pure triacylglycerol. In the following part, we investigated mixtures of solid triacylglycerols.

3.2.1 Tricaprin/Tristearin Mixtures. The tricaprin/tristearin mixtures were studied at 10°C. According to melting temperatures, tricaprin is in the β' form and tristearin in the α form at 10°C. However, the polymorphic behaviour depends on the thermal history of the sample. It was confirmed by XRD measurements (data not shown) that in our system, tricaprin was in the β form and tristearin in the α form. We thus had a solid mixture of two different polymorphs, and it was interesting to obtain information on the crystal size distribution profile. The T_1 distributions for the three different ratios are presented in Figure 3.

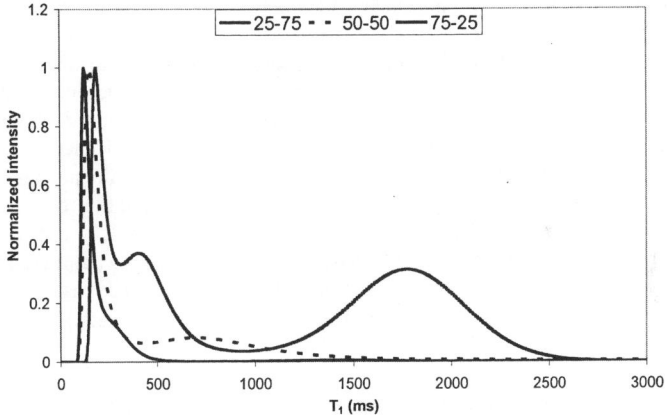

Figure 3   *T_1 distributions of tricaprin/tristearin (% in w/w) mixtures at 10°C.*

The behaviour was different for the three mixtures. There was one peak broadened at the bottom for the 25-75 (w/w) mixture, there were two peaks for the 50-50 (w/w) mixture and three peaks for the 75-25 (w/w) mixture. The first peak was found for all three mixtures around 150 ms and was attributed to T_1 of tristearin in the α form. This peak confirmed the presence of small crystals of tristearin in the systems. This peak shifted a little to a higher value when the tricaprin concentration increased. A second peak was characterized for the three mixtures, and attributed to the β polymorph of tricaprin. It was at 300 ms for the 25-75 (w/w) mixture, 800 ms for the 50-50 (w/w) mixture and 1800 ms for the 75-25 (w/w) mixture. Tricaprin T_1 thus increased as a function of tricaprin concentration. As spin-lattice relaxation time is linked to crystal size, such an increase in T_1 can be related to an increase in the average crystal size of tricaprin. This can be explained by the fact that small tricaprin crystals, firstly crystallized in the α form by quenching at -50°C, were closer and could grow more freely. Indeed, at high tristearin concentration, tricaprin crystals were embedded in the tristearin crystal network and were scattered throughout the sample. When the temperature increased, growth was limited because of the few germs of crystals around the point of growth. When the tricaprin concentration was higher, there were more small crystals and then growth became easier. Thus, there was a large difference between tricaprin and tristearin crystal sizes when there was more tricaprin, and thus it was difficult to discern two peaks for the 25-75 (w/w) mixture whereas the two peaks were easily differentiable for the 75-25 (w/w) mixture. The peak characterizing tricaprin was very broad, which means that there was a wide dispersion of crystal size. Thus it appeared that tricaprin crystal growth was not homogenous in the sample. This was the consequence of the random distribution of the small tristearin crystals. There can be different hypothesis concerning the third peak at 400 ms, present only for the 75-25 (w/w) mixture. The first is the possible presence of tricaprin β' crystals, but this was not confirmed by X-Ray Diffraction. The second involves the presence of co-crystals. The two triacylglycerols can form co-crystals during very fast crystallization due to high supercooling, and the greater the amount of tricaprin present in the sample, the greater the formation of co-crystals. This new species behaves independently from the other two. Thus the last peak may correspond to α or β polymorphic forms of co-crystals. Their polymorphism being the same as tricaprin and tristearin, they cannot be detected by X-Ray Diffraction. However, they have different sizes and they appear on the T_1 distribution.

Spin-lattice relaxation time distributions appeared to be useful to obtain information on crystal size, but they highlighted the limitations of T_1 measurements for the determination of polymorphism. Indeed, for the mixtures with a high tricaprin concentration, there were two well-separated peaks, with a wide difference between the maxima. Therefore, in this case, T_1 measurements effectively determine polymorphism. On the other hand, there was only one peak for the 25-75 (w/w) mixture in spite of the presence of two different polymorphic forms as determined by X-Ray Diffraction. Thus determination of polymorphism was not possible here. Consequently, as determination of polymorphism through T_1 measurements is based on crystal size, it is not possible to achieve this when crystals are of similar size. However, as β crystals are larger than α crystals, the latter situation rarely occurs.

3.2.2 Trilaurin and Tripalmitin Mixtures. The trilaurin/tripalmitin mixtures were studied at 20°C. At this temperature, tripalmitin should be in the α form whereas trilaurin should be in the β' form according to their melting temperatures. It was interesting to determine if the two triacylglycerol crystals could be differentiated by the size of their crystals. It has been already proved that trilaurin crystals grow faster than tripalmitin crystals.[14] Moreover, when the melt was cooled, tripalmitin was submitted to a greater supercooling, which meant smaller crystals.[15] Thus it was interesting to test the sensitivity of the method in order to differentiate each species in the sample. Spin-lattice relaxation time distributions of the trilaurin/tripalmitin mixtures at 20°C are shown in Figure 4.

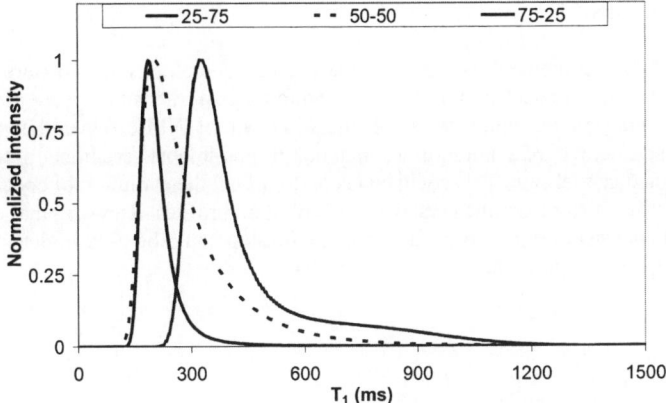

Figure 4 *T₁ distributions of trilaurin/tripalmitin (% in w/w) mixtures at 20°C*

The behaviour of the mixtures appeared to be different according to the trilaurin concentration. Indeed, a single peak was observed for the 25-75 (w/w) mixture with a maximum of 200 ms, whereas there was a broad peak with an enlargement at the base for the 75-25 (w/w) mixture and a maximum at 330 ms. The 50-50 (w/w) mixture presented an intermediate behaviour with a broader peak but still a maximum at 200 ms.

It seemed that there was greater diversity of crystal sizes when there was more trilaurin. The fact that there was only a single peak and not two, as could have been expected from the melting temperatures, can be explained by the formation of co-crystals. Indeed, the two triacylglycerols have only four carbons of difference on their side chains, and as they were in the α form, they tended to form co-crystals. Although trilaurin is more easily trapped in the tripalmitin crystal network, there is consequently one single peak when there is less trilaurin than tripalmitin. The two polymorphs remained in the α form, and the crystal growth was controlled by the tripalmitin growth rate. The α form of trilaurin was stable inside the tripalmitin crystal network for the 25-75 (w/w) mixture. When trilaurin was in excess as in the 75-25 (w/w) system, there were co-crystals and free trilaurin. However, there could not be more than 25% of trilaurin, the amount of tripalmitin in the sample, involved in co-crystallization. From the present measurements for the 75-25 (w/w) system, the intensity of the second peak at 900 ms did not reach the 50% of trilaurin. This can be explained by a slowing down of the polymorphic transformation of the α form to the β' form induced by the presence of tripalmitin. The system did not reach equilibrium before the measurement and was still evolving over time. The increase in T_1 for the first peak proved that co-crystal behaviour was close to the trilaurin behaviour. For the 50-50 mixture, the behaviour remained close to the tripalmitin behaviour. The broadening of this peak was due to free trilaurin crystals which were bigger than tripalmitin crystals.

This experiment thus suggested that trilaurin and tripalmitin tend to form co-crystals, which behave similarly to the triacylglycerol in excess, and that polymorphic modification of trilaurin is slowed down by the presence of tripalmitin.

4 CONCLUSION

Low field NMR equipment is used in many food laboratories to measure Solid Fat Content. This measurement is only based on proton signal intensities. New prospects are opening up for this technique with the measurement of NMR relaxation parameters, especially spin-lattice relaxation times making it possible to establish a relationship between T_1 and crystal size. This could be useful to follow the evolution of crystal size as a function of time or to determine crystal size distribution profiles. Knowing that crystal size monitors the physical properties of fats, this relationship could be used to determine some physical properties of high fat content food products.

References

1   B. S. Ghotra, S. D. Dyal and S. S. Narine, *Food Res. Int.*, 2002, **35**, 1015.
2   D. Tang and A. G. Marangoni, *Trends Food Sci. Technol.*, 2007, **18**, 474.
3   A. G. Marangoni, *Fat Crystal Networks*, Marcel Dekker, New York, 2005.
4   S. S. Narine and A. G. Marangoni, *Food Res. Int.*, 1999, **32**, 227.
5   Y. Shi, B. Liang and R. W. Hartel, *J. Am. Oil Chem. Soc.*, 2005, **82**, 399.
6   K. Sato, *Chem. Eng. Sci.*, 2001, **56**, 2255.
7   E. Trezza, A. M. Haiduc, G. J. W. Goudappel and J. P. M. Van Duynhoven, *Magn. Reson. Chem.*, 2006, **44**, 1023.
8   M. Adam-Berret, C. Rondeau-Mouro, A. Riaublanc and F. Mariette, *Magn. Reson. Chem.*, 2008, **46**, 550.
9   M. Knoester, P. De Bruyne and M. Van Den Tempel, *J. Cryst. Growth*, 1968, **3-4**, 776.
10 R. Vreeker, L. L. Hoekstra, D. C. Denboer and W. G. M. Agterof, *Colloids Surf.*, 1992, **65**, 185.
11 S. Ablett, C. J. Clarke, M. J. Izzard and D. R. Martin, *J. Sci. Food Agric.*, 2002, **82**, 1855.
12 K. Binder, *Phys. Rev. B*, 1977, **15**, 4425.
13 I. M. Lifshitz and V. V. Slyozov, *J. Phys. Chem. Solids*, 1961, **19**, 35.
14 C. Himawan, V. M. Starov and A. G. F. Stapley, *Adv. Colloid Interface Sci.*, 2006, **122**, 3.
15 R. Boistelle in *Crystallization and polymorphism of fats and fatty acids,* ed. N. Garti and K. Sato, Marcel Dekker, Inc, New York, 1988, p 189.

ESR and other techniques

ESR FOR FOOD IRRADIATION DETECTION

Eric MARCHIONI[1]

[1]IPHC-UMR7178, Laboratoire de Chimie Analytique et Sciences de l'Aliment, Faculté de Pharmacie, 74 route du Rhin, 67400 Illkirch - France

1 INTRODUCTION

Food irradiation is considered as a highly effective processing technology to improve and maintain food safety. Indeed this process applied on food products dramatically reduces the populations of pathogens, which are annually responsible for millions of food-borne illnesses worldwide. The World Health Organization and many state agencies around the world have endorsed food irradiation as a major contributor to public health preservation.

The action of ionizing radiation on food results in the formation of free radicals and radiolytic products that are predominantly not radiation specific. It was thus not astonishing that after many years of research before the nineties, it was still not possible to identify any specific radiolytic product that could be used to establish an universal analytical method for the detection of irradiated food. Moreover, the radiation process, when performed at usual absorbed doses (less than 10 kGy), involves many fewer chemical modifications than other treatments such as heating or storage. Indeed, the absorption of the maximal allowed dose for food irradiation in Europe (10 kGy) leads only to an absorption of very few but effective energy and to a temperature rise limited to appproximately 2°C depending on the food composition. This observation pleads certainly in favor of the safety of radiation processing, but represents a major disadvantage when one seeks to identify such a process while studying physical or chemical modifications in the foodstuff itself.

As a result of two concerted actions conducted and funded by the Community Bureau of Reference [1] and by the International Atomic Energy Agency [2] at the beginning of the nineties, no fewer than fifteen analytical methods for the detection of irradiated food were developed, of which ten were standardized by the European Committee for Standardization (CEN). Three of them are reference methods and are based on the analysis of primary radiolytic products, by electron spin resonance spectroscopy (ESR) [3-5].

2 METHOD AND RESULTS

2.1 ESR measurements

The Bruker ESR spectrometer, type ECS 106 (Wissembourg, France), was equipped with a TMH ECS 4108/9105 cylindrical resonator.

2.2 Preparation of the food samples

The irradiated food samples were prepared according to the European protocols [3-5]. The analysis of irradiated ingredients mixed in complex food samples were done accordingly to the protocol of Marchioni *et al* already published [6,7].

2.3 ESR

The energy differences studied in ESR spectroscopy are predominately due to the interaction of unpaired electrons (as radicals in our case) in the sample, with a magnetic field produced by an external magnet. This effect is called the Zeeman effect. The lowest energy of the electron will appear when its magnetic moment ($\vec{\mu}$) is aligned with the external magnetic field. The highest state of energy will come when $\vec{\mu}$ is aligned against the external magnetic field. The difference of energy depends on the magnetic field. A peak in the absorption of the incident microwave energy will occur when the magnetic field tunes the two spin states so that their energy difference matches the energy of the incident radiation. Electron spin resonance spectroscopy is a nondestructive and highly sensitive analytical method that allows the detection of free radicals in matter. It consists in subjecting a test sample to the simultaneous action of a magnetic field (intended to direct the magnetic moments "spin" of the matter and thus those of the free radicals) and of an electromagnetic microwave of very high frequency (X Microwave band, ≈9 GHz).

2.4 Results

Foods having a dry or rigid matrix, or presenting certain dry or rigid parts, are able during a radiation processing to trap free radicals for a period of time that can be longer than the lifetime of the food itself. These free radicals may then act such as magnetic moment ($\vec{\mu}$) and absorbe the electromagnetic energy delivered by the spectrometer. Thus, irradiated foods as meat with bones (poultry, beef, pork, and so on), fish with bones, scales, or teeth, eggs with shells, shellfish, fruits with achenes, nuts with shells, dry fruits containing crystalline sugars, and some seeds and spices can be analyzed by electron spin resonance spectroscopy. The derivative representation of this microvawe absorption, according to the value of the magnetic field, gives the ESR spectrum (Fig. 1b). The integrated intensity of an ESR signal (Fig 1a) is proportional to the concentration of the radical specie. In our work we used the height of the recorded signal (derivative) as a marker of the detected radical concentrations. This value is also known as pic to pic value. External tools such as pellets of dry aminoacid powder (such as alanine) may be used as dosimeters (Fig. 2) as their developed ESR signals present pic to pic values which are proportional to the dose values over a wide range of absorbed doses (from 1 Gy to 100 kGy). It is, up to day, the official international reference for the control of high absorbed doses.

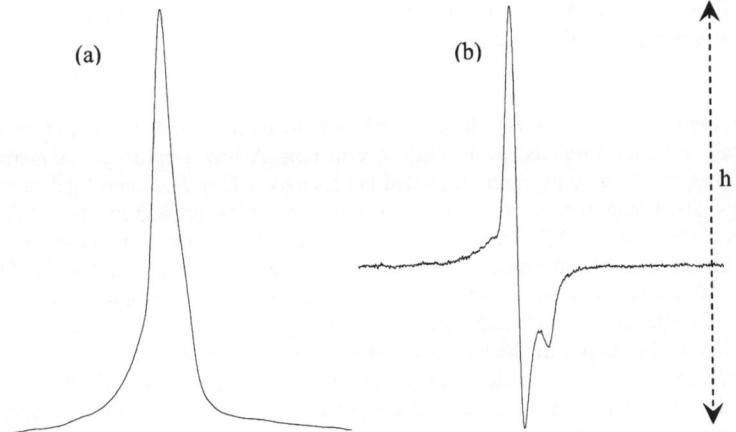

Figure 1 *ESR spectra of a Chicken bone irradiated at 3 kGy. (a) Absorption spectrum, (b) derivative curve of the absorption spectrum. h is the height of the derivative curve*

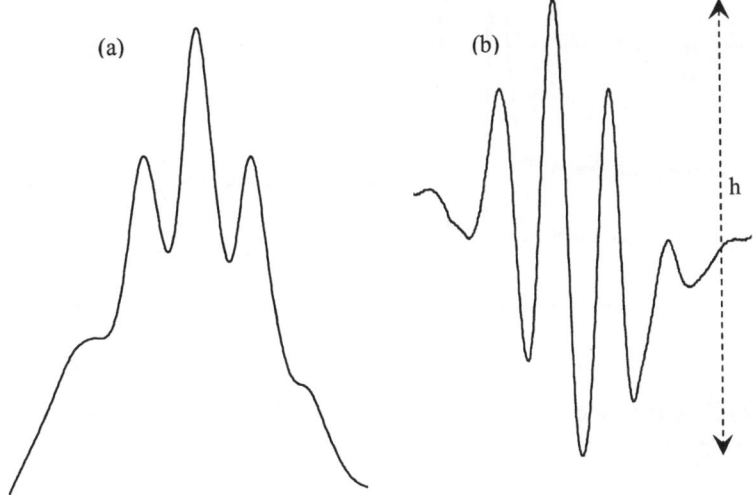

Figure 2 *ESR spectra of a dry powder of alanine (amino acid) irradiated at 100 kGy. (a) Absorption spectrum, (b) derivative curve of the absorption spectrum. h is the height of the derivative curve*

The analytical method is very easy and consists of extracting, out of the food sample, the dry or rigid part which contains the radicals induced by the radiation process (bones in meat, shells in nuts, part of the food in dried fruits, chocolate ....), drying this part (water prevents the ESR analysis because of the O-H dipole, which absorbs the microwave energy) under reduced pressure at 50°C max in order to avoid modification of the food composition (sugars) and the recombination of the radicals, and to record the ESR absorption spectrum as described in the european protocol [3-5].

2.4.1 Limitation. Of course, the presence of radicals in food is not radiation specific as radicals are also produced by heating or crushing. A low amplitude symmetric ESR absorption signal is present in nonirradiated bone samples (Fig 3). Some food samples, as dried figs, dried mangoes, dried papayas and raisins samples, present intense ESR signals when irradiated and no ESR signals when unirradiated (Fig 4), and some other dried foods (dried bananas) do not exhibit any ESR signal even when irradiated at high doses (data not shown). In adition, a high symetric signal is also present in samples containing dry cellulose and increases with a radiation treatment (Fig 5). Nevertheless, the simple visual observation of the shapes of the recorded spectra in case of irradiated samples, as well as their gyromagnetic factors, causes analysts to consider some ESR signals to be radiation specific. It is now widely recognized that the presence of ESR signals (as described in the CEN Normes [3-5]) is radiation specific, but the absence of such ESR signals (except in case of mammal bones) never constitutes proof that the food has not been irradiated.

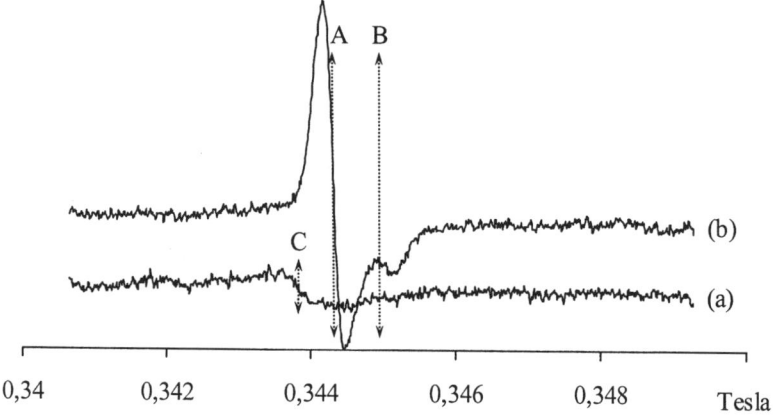

Figure 3 *ESR spectra of a non irradiated (a) and 5 kGy (b) irradiated chicken bone sample. A and B: radiation specific ESR signals and C: symetric non radiation specific signals.*

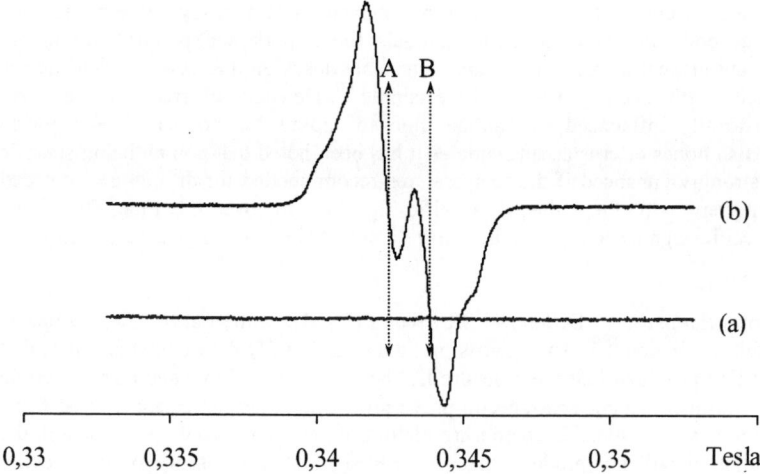

Figure 4 *ESR spectra of non irradiated (a) and 5 kGy (b) irradiated dried mango samples. A and B: radiation specific ESR signals.*

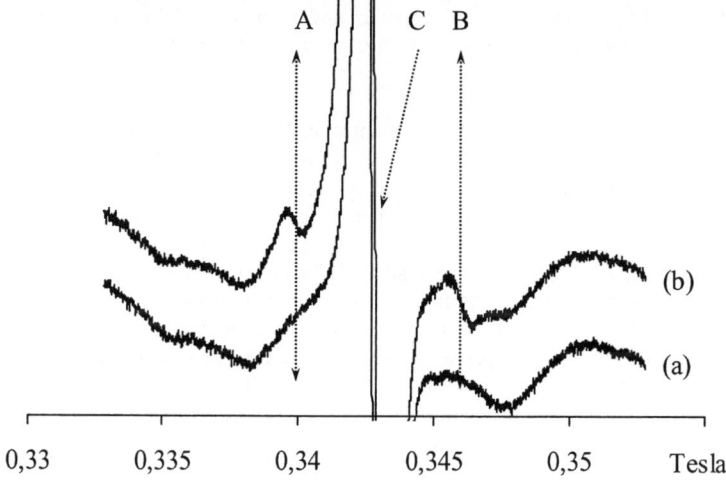

Figure 5 *ESR spectra of non irradiated (a) and 3 kGy irradiated strawberrie achene samples. A and B: radiation specific ESR signals and C: symetric non radiation specific signals.*

2.4.2 Bone Radicals. The recorded ESR signals (Fig 3) correspond [8] to an extremely stable $CO_2^{°-}$ radical trapped in the lattices of hydroxyapatite [$Ca_{10}(PO_4)_6(OH)_2$] which constitutes approximately 60% of the bone composition. It consists of a radiation-specific asymmetrical signal (A and B, g=2.002 g=1.998) superimposed upon a symmetrical endogenous signal (C, g=2.005) of much lower amplitude. Detection of irradiated bone samples is possible above a dose of 0.5 kGy, covering the majority of commercial applications for the foodstuffs considered. Detection limits and stability of the

ESR signals are influenced by the degree of mineralization of hydroxyapatite in the bone sample. In general, the bones of larger animals (beef, pork and poultry) are highly mineralized and present then low minimum detectable doses. In this case, the detection is not significantly influenced by heating of the sample [7]. Detection of irradiation treatment is not significantly influenced by storage times of up to 12 months [3]. For poorly mineralized fish bones or crustaceans cuticles it has been noted that non radiation-specific signals are strongly enhanced if the temperatures recommended for drying are exceeded and may interfere with the radiation specific signals. Moreover the time life of the radiation specific signals may be quite short (a few hours) as soon as the sample is defrosted.

2.4.3 Sugars Radicals. Different ESR spectra (Fig 4), centred at g=2.003, could be produced after irradiation [5,9]. They consist in intense and easily detectable multiplets that can be identified provided that the moisture has been correctly eliminated during sample preparation. Multicomponent ESR spectra prove prior irradiation but as some dried fruits doesn't exhibit any ESR signal even after irradiation (if no sugar crystals are present in the sample, irradiation will not produce specific ESR signals), the absence of the specific spectrum does not constitute evidence that the sample is not irradiated. The limit of detection mainly depends on the crystallinity of the sugars in the sample. The detection of radio-induced ESR signals depends on the presence of sufficient quantities of crystalline sugar in the sample at all stages of handling between irradiation and testing. Detection of irradiated dried figs, dried mangoes, dried papayas and raisins has been validated by international blind tests. Detection of irradiation treatment is not significantly influenced by storage of at least several months.

2.4.4 Cellulose Radicals. The ESR signal presented in Figure 5 is a hyperfine triplet centred at g=2.004 from which only the two outermost peaks (shoulders) A and B (g=2.020 and g=1.985, ΔH≈6 mT) can be used for the detection of food irradiation. The central line cannot be used as irradiation marker because this radiation sensitive triplet is superimposed on an intense central nonradiation-specific singlet C belonging to lignin [10] and being sensitive to inter alia, drying processes [11]. Detection limits and stability are influenced by the crystalline cellulose and the moisture contents of the samples. Positive identification of the cellulose radicals is evidence of irradiation but, as the radiation specific ESR signals are so small compared to the central line and so moisture sensitive, the absence of this signal does not constitute evidence that the sample is unirradiated. Detection of irradiated pistachio nuts, paprika powder and fresh strawberries has been validated by international blind tests. Stability of cellulose radicals in berries depends on storage conditions and may be shorter than the shelf-life of the products. For paprika powder, the stability of cellulose radicals is largely dependent on storage conditions (especially humidity). In case of very dried products such as shells of nuts, stability is not expected to present limitations for detection of irradiation for at least one year after treatment. The special case of liquid whole egg stored in bricks was reported by the official German food control laboratories which detected cellulose radicals in the paper constituting the packing material. Some authors found also these ESR signals in citrus fruits skins, skin components and stalks [12] and fruit cell walls [13,14] in the pulp of citrus fruits after proper elimination of the water, with ethanol. According to these authors, the ESR signals in gamma-irradiated kiwi, papaya and tomato using fruit pulp were clearly detectable even for absorbed doses as low as 200 Gy.

Today, these three analytical ESR methods are CEN and Codex standards used by various EU Member States and also other countries all over the world to exercise control over the international trade of irradiated foods.

2.4 Quantitative assessment

Since the signal height for particular bone types can be related to the dose (linearly proportional up to 14 kGy in the case of frog legs), it is possible to obtain a very rough estimate of the dose absorbed by bones using a 'standard addition' technique, the ESR signal being recorded after each of several successive re-irradiations [15-17] (Fig 6). Nevertheless, it should be noted that ESR signal strengths depend on several parameters such as irradiation temperature, degree of bone mineralization (age of the animal, type of bone analysed, etc.), pre- or post irradiation treatments (cooking or freezing), preparation of the sample (grinding and drying) [18-20]. Also, irradiation of whole pieces of meat including bones may yield higher ESR signals than irradiation of dried excised bones alone [21]. It seems therefore very difficult to get, by this method, an accurate estimation of the dose absorbed by the food.

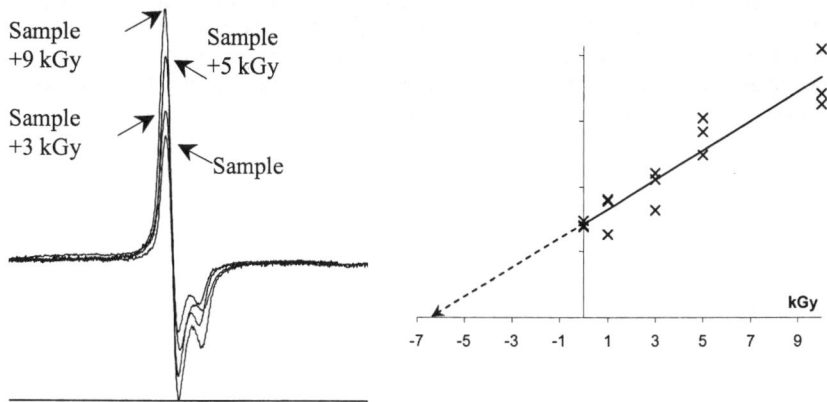

Figure 6 *Standard addition method applied on a sample of bone frog leg. The determined initial dose was 6.3 kGy.*

2.5 Detection of irradiated ingredients

The mechanically recovered meat (MRM) is used as a protein contributor in food and also for its technological qualities (emulsifying and binding). An irradiation process (5 kGy) of this product, in a frozen state, allows its use as an ingredient in culinary preparations (aerobic mesophilic count $< 10^4$ CFU g^{-1}, absence of *Salmonella* in 25 g). During the mechanical separation and freezing steps, some of the proteins are denatured, and the binding capacity of a MRM may then decrease by 70% compared to that of a normal meat. The quantity of MRM added to a food will thus always be higher than the quantity of a fresh meat used for the same objective. In a practical way, the concentrations of MRM used by food industry are always 6-10% (wt/wt) and can even rise to 25% in some cases. A more or less large quantity of bone fragments remains in the produced MRM. The total quantity of bone residues must always remain <1%. The simple application of the CEN protocol for the detection of irradiated meats with radicals induced in bones is not possible when only little bone fragments are included, in small quantities, in a lipidoproteic complex matrix (cheeses, quenelles, etc). It is actually essential to remove as efficiently as possible the food matrix before analysis of bone fragments by ESR spectrometry. We used a protocol comprising an enzymatic hydrolysis (Alcalase) of the proteins, dissolution of the lipids thanks to sodium dodecyl sulfate (SDS) and a purification of the extracts by decantation in an high density (d=2) aqueous solution of sodium polytungstate. This extraction step was associated it with an analysis by ESR spectroscopy to carry out a high-sensitivity detection of irradiated MRM or fish products included in various culinary preparations.

Figure 7 presents the ESR spectrum obtained by the analysis of bones fragments extracted as proposed from frozen commercial irradiated MRM. It seems clear that this spectrum identical to the one obtained with an excised fragment of an irradiated bone analyzed with the CEN protocol. The extraction procedure does interfere neither with the shape nor with the intensity of the signals induced by the radiation process.

The same protocol has been applied to several complex foods containing mixture of ingredients, some being irradiated other not. As example, Figure 8 presents the ESR signal obtained with fish quenelles containing 12 % (m:m) of non irradiated salmon and 7% (m:m) of 5 kGy irradiated salmon. The proposed protocol was able to extract the fishbones out of the salmon, only part of which having been irradiated. The recovered ESR signals are identical to those obtained with irradiated bones giving evidence of the presence of an irradiated ingredient containing bones. In this case, it was the ingredient salmon, detected even if it was only a part of the ingredient salmon.

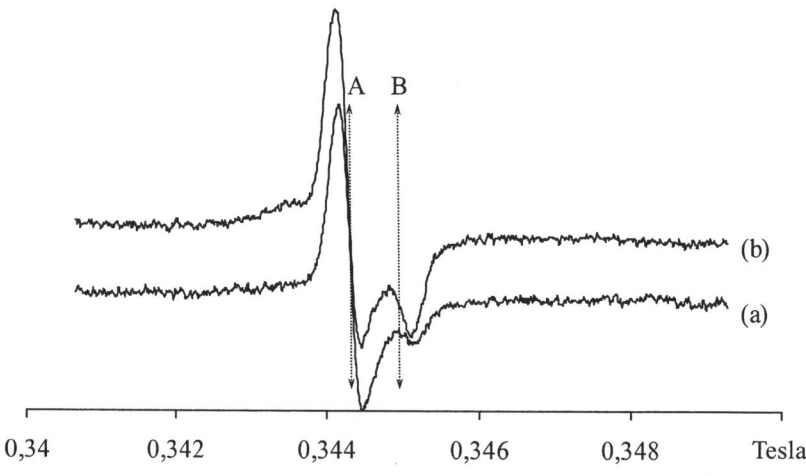

Figure 7 *ESR spectra obtained (a) with the CEN protocol applied on a piece of drumstick chicken bone and (b) with recovered bone fragments from a sample of irradiated MRM.*

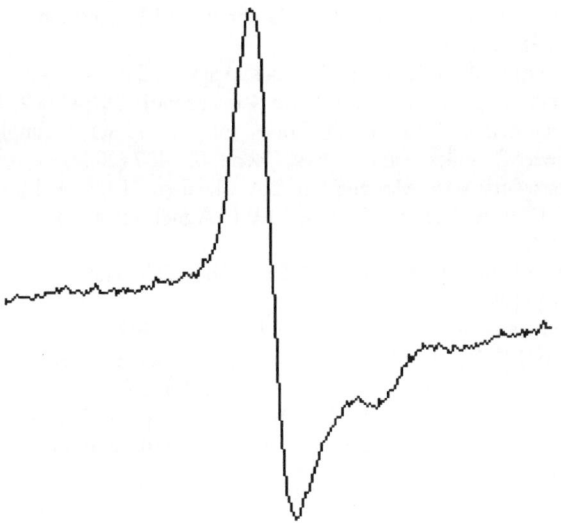

Figure 8 *ESR spectrum obtained with a sample of fish quenelle containing both 5 kGy irradiated salmon (7%; m:m) and non irradiated salmon (12%; m:m).*

References

1   J. Raffi, H. Delincée, E. Marchioni, C. Hasselmann, A.M. Sjöberg, M. Leonardi, M. Kent, K.W. Bögl, G. Schreiber, H. Stevenson, W. Meier,. Concerted action of the Community Bureau of Reference on the Methods of Identification of Irradiated Foods, EUR 15261 EN, Brussels (Belgium): BCR. 119 p. (1993).

2   McMurray CH, Stewart EM, Gray R, Pearce J.. Detection Methods for Irradiated Foods-Current Status. Cambridge (United Kingdom): The Royal Society of Chemistry. 431 p. (1996).

3   Anonymous. Foodstuffs. Detection of irradiated food containing bone. Method by ESR spectroscopy. EN 1786. Brussels (Belgium): European Committee for Standardization. 12 p. (1996).

4   Anonymous. Foodstuffs. Detection of irradiated food containing cellulose by ESR spectroscopy. EN 1787. Brussels (Belgium): European Committee for Standardization. 11 p. (1996).

5   Anonymous. Foodstuffs. Detection of irradiated food containing crystalline sugar by ESR spectroscopy. EN 13708. Brussels (Belgium): European Committee for Standardization. 9 p. (2000).

6   E. Marchioni, P. Horvatovich, B. Ndiaye, M. Miesch, C. Hasselmann, Radiat. Phys. Chem., 63, 447 (2003).

7   E. Marchioni, P. Horvatovich, H. Charon, F. Kuntz, J. Agric. Food Chem., 53, 3769 (2007).

8   G. Bacquet, V.Q. Truong, M. Vignoles, J.C. Trombe, G. Bonel, Calcif, Tissue Int., 33, 105 (1981).

9   J. Raffi, M.H. Stevenson, M. Kent, J.M. Thiery, J.J. Belliardo, Int. J. Food Sci. Technol., 27, 111 (1992).

10   N. Deighton, S.M. Glidewell, B.A. Goodmann, I.M. Morisson, Int. J. Food Sci. Technol., 28, 45 (1993).
11   E.F.O. Dejesus, A.M. Rossi, R.T. Lopes, Appl. Radiat. Isot., 47, 1647 (1996).
12   B.J. Tabner, V.A. Tabner, Int. J. Food Sci. Technol., 29,143 (1994).
13   E.F.O. Dejesus, A.M. Rossi, R.T. Lopes, Int. J. Food Sci. Technol., 34, 173 (1999).
14   H. Delincée, C. Soika, Radiat. Phys. Chem., 63, 437 (2002).
15   M.F. Desrosiers, W.L. McLaughlin, L.A. Sheahen, N.J.F. Dodd, J.S. Lea, J.C. Evans, C.C. Rowlands, J.J. Raffi and J.P.L. Agnel, Int. J. Food Sci. Technol., 25, 682 (1990).
16   M.F. Desrosiers, G.L. Wilson, C.R. Hunter and D.R. Hutton, Appl. Radiat. Isot., 42(7), 613 (1991).
17   M.F. Desrosiers, Appl. Radiat. Isot., 42(7), 617 (1991)
18   J.S. Lea, N.J.F. Dodd and A.J. Swallow, Int. J. Food Sci. Technol., 23, 625 (1988).
19   S.P. Chawla, P. Thomas and D.R. Bongirwar, Food Res. Int., 35, 467 (2002).
20   R. Gray and M.H. Stevenson, Int. J. Food Sci. Technol., 24, 447 (1989).
21   J. Raffi, M.H. Stevenson, M. Kent, J.M. Thiery and J.J. Belliardo, Int. J. Food Sci. Technol., 27, 111 (1992).

APPLICATIONS OF CW-EPR IN FOOD QUALITY CONTROL AND R&D

Andreas Kamlowski[1], David Barr[2], Hideyuki Hara[3]

[1]Bruker BioSpin, EPR/MicroSpin Division, 76287 Rheinstetten, Germany
[2]Bruker BioSpin Corp., EPR Division, 44 Manning Rd. Billerica, MA 01821, USA
[3]Bruker BioSpin K.K., ESR Division, Tsukuba 3050051, Japan

1 INTRODUCTION

Electron Paramagnetic Resonance (EPR) spectroscopy has a long tradition in food science, especially related to food irradiation control. Recently, EPR has also gained attention in other fields, related to antioxidant research as well as the characterization of dairy products, tea, wine and vegetable oil.

EPR, also referred to as Electron Spin Resonance (ESR) or Electron Magnetic Resonance (EMR), is the only direct method to detect free radicals (ROS, reactive oxygen species) or other paramagnetic species. Bruker BioSpin offers routine research systems as well as the e-scan, a dedicated table-top EPR scanner

2   CW-EPR INSTRUMENTATION

The method of choice for Quality Control (QC) with EPR is Continuous Wave (CW-) EPR at X-Band microwave frequencies, i.e. 9-10 GHz. X-Band is the most common operation frequency in research as well. It provides the best possible combination of sample size and inherent signal-to-noise to the task at hand.

For industrial and in particular QC purposes, an EPR system needs to fulfil a range of equally important criteria such as ease-of-use, turn-key operation as well as providing high sample throughput and automatic reporting of results. Bruker BioSpin's bench-top e-scan™ product line meets all of these demands. In the Food&Beverage industry samples for EPR are challenging since their high water content results in a high dielectric constant. However, special aqueous sample holder and probe head designs have been created to accommodate the problems caused by the dielectric loss from water containing samples.

3   FOOD IRRADIATION QUALITY CONTROL BY CW-EPR

European Union (EU) directives[1] clearly state that irradiated food as well as food containing irradiated ingredients (regardless of their percentage) must be labelled (i.e., with the international food irradiation symbol, the *radura*[2]). National authorities are responsible for issuing a clearance list of irradiated food (and packing materials for some countries).[3] For example, in Germany only spices, herbs, vegetable seasonings and, since recently, also frog legs are cleared for irradiation. Interestingly, in Japan the only product cleared for irradiation are potatoes. Thus, irradiation or even the import of irradiated poultry or shrimp is not allowed. National or federal bodies are obliged to control dairy and imported food according to international or national norms and to ensure that the corresponding laws are observed by the food industry and importers.

Food irradiation is used to reduce the health risk associated with food-borne pathogens such as *Salmonella* and to prolong shelf life (sprout inhibition, delay of ripening). In fact, ionizing radiation inhibits the division of microorganisms and creates radiolytic products as well as free radicals. In a dry environment these radicals are relatively stable. For example, irradiated poultry bones or dried spices may contain a substantial amount of stable radicals which can be easily detected by EPR spectroscopy. Extensive consultations and round-robin tests were conducted during the 1990s in order to set European-wide standards for sample preparation, measurement protocol and unequivocal identification of irradiated food via EPR.

Currently, three EU norms exist, defining food irradiation control via EPR spectroscopy.[4,5,6] With standard research EPR spectrometers such as Bruker's EMX and ELEXSYS series, food irradiation control can be conducted with superior sensitivity. However, experienced technicians or scientists are required to operate these more complex, general purpose spectrometers. For food irradiation control by EPR to become a real turn-key Quality Control device, a dedicated, easy-to-use bench-top EPR reader like the e-scan™ Food Analyzer is required. The e-scan™ Food Analyzer[7] is the improved successor of the EMS104 table-top EPR analyzer[8] with which many of the round-robin tests were conducted to established the EU norms.

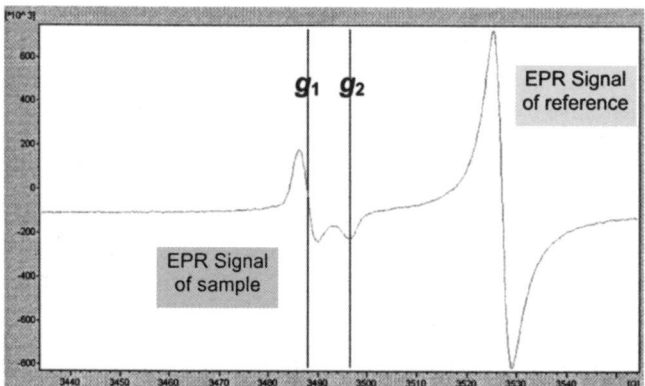

Figure 1: *EPR spectrum of an irradiated chicken bone recorded with the e-scan™ Food Analyzer. The spectrum of the sample (axially symmetric line shape at the center of the field sweep) is measured in parallel with the spectrum of the reference marker (isotropic signal line shape to higher field at the right) with a known g value (1.980). The horizontal axis gives the B_0 field values in Gauss. The two g-values are evaluated automatically.*

In case of bone-containing food, the EU norm requires careful evaluation of the g-factors of the EPR signal (see Figure 1).[4] To this end, the microwave resonance frequency needs to be known precisely and the magnetic field axis be calibrated. Both requirements are met with the e-scan™: The internal marker (see Figure 1, signal at high field) provides an in-situ tool for magnetic field calibration and validation. Two g-values are obtained automatically, providing the basis for the decision of whether the bone-containing food was subjected to irradiation (see [4] for details). As noted in passing, the g_2-value of the local minimum is evaluated, which is the physically correct position (axial g-tensor). The EU norm[4] instead specifies the inflection point low field from the local minimum.

4 FLAVOUR STABILITY IN BEER: QUALITY CONTROL BY CW-EPR

The "cardboard" like flavour that occurs in stale beer is thought to arise from the "free radical" mediated oxidation of various constituents in beer. The characteristic odour and taste are caused by decomposition products from the free radical process. Similar processes occur in many foods, but in beer, these "off-flavour" products can be detected by the consumer even at very low concentrations.

The environment that beer is stored in is critical for minimizing oxidative staling. If the beer is stored at cooler temperatures, the oxidation process will only occur very slowly; and of course, raising the temperature will increase the rate of oxidation. However, as production volumes and distribution distances increase, the ability to carefully control the storage environment for beer is compromised. Therefore, methods for measuring and controlling the oxidative stability of beer have become vital. All beers have a certain amount of naturally occurring antioxidants that protect their flavour by terminating or inhibiting oxidative free radical reactions. Beers with a higher "antioxidant activity" can resist the oxidation for longer periods, and thus, have better shelf life stability.

4.1 Spin Trapping EPR of Radicals

Spin trapping is a method that allows detection of very short-lived free radicals, such as the hydroxyl radical, the superoxide anion radical or carbon centred free radicals. It involves adding a "spin trap" to a sample (e.g., beer) that you suspect will contain free radicals. If free radicals are produced, the spin trap (which, itself, is not EPR detectable) will react with the free radical and form a stable chemical bond between the two. This complex of the radical and the spin trap is still a free radical, but is significantly more stable than the initial free radical. The complex, often called a "spin adduct" or a "radical adduct", is then detected by EPR. The general reaction scheme for a spin trapping experiment is shown in Figure 2.

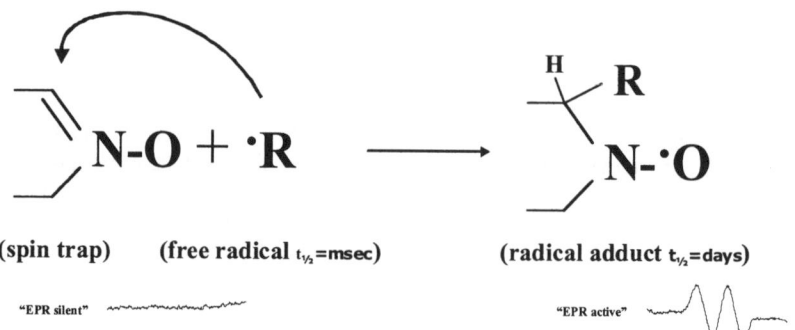

Figure 2: *Spin trapping of a radical, R, and part of CW-EPR spectra of the sample without (EPR silent) or with trapped radicals (EPR active), resp.*

4.2 The Automated EPR/lagtime Assay

The so called "lagtime" assay uses "forced oxidation" combined with EPR spin trapping to effectively measure the antioxidant activity of a beer and to even predict a beers shelf life.[9] It has been well established that the time in minutes (i.e., lagtime) before a dramatic EPR signal increase occurs, correlates with the time in days that are required for a sensory panel to detect the characteristic "cardboard" off-flavour. A parameter known as the T150 (the EPR intensity at time = 150 minutes) is another metric that is used to evaluate the resistance of the beer to oxidation. The T150 value is particularly used in ale-style (top-fermented) beers; which do not have a lagtime.

Although the information the lagtime assay provides is undoubtedly powerful, the assay itself has been hindered by a lack of automation and low throughput. Here, we demonstrate the use of an EPR system from Bruker that simplifies, automates and greatly increases throughput for the lagtime assay. Using an automatic sample changer the brewer can measure the EPR from several samples simultaneously in one time period. The data analysis step is also greatly aided by a software package that was specifically designed for calculating lagtimes.

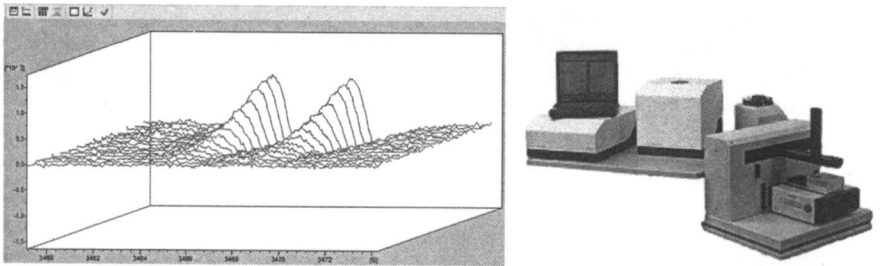

Figure 3: 2D CW-*EPR spectrum of a beer sample with spin trap (left). e-scan™ Beer Analyzer with autosampler (right).*

To study free radical formation in a beer sample a spin trapping agent is added to the beer and incubated at 60 °C. This forced aging accelerates the free radical oxidation process to a rate that is measurable within a relatively short time period (i.e., 1-3 hours). As free radicals form, they are trapped by the spin trap and "spin adducts" will begin to accumulate (cf. Figure 3, left).

In the automated assay, the samples are introduced by an autosampler system (see Figure 3, right) to the spectrometer at specific time intervals and the EPR spectrum of the spin adducts are measured. At the end of the assay, for each of the up to 20 beer samples a time course for free radical formation is obtained. It can be thought of as an "oxidation profile" for the beer, and is actually a measure of the beers resistance to oxidation. Lagtime and T150 values are used as the primary indicators of the beer's performance (cf. Figure 4).

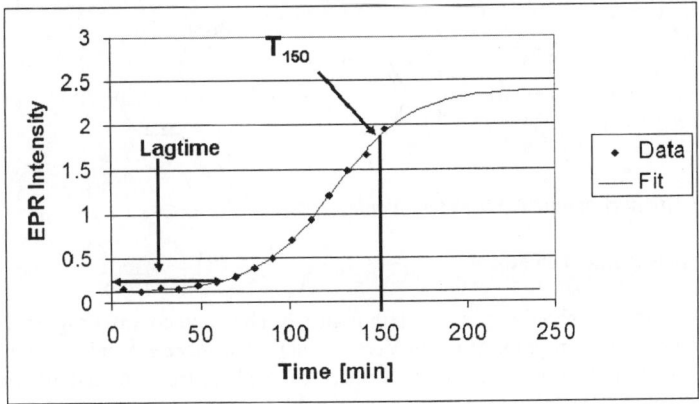

Figure 4: *EPR/lagtime time course of a beer sample (see text). Plotted is the peak-to-peak intensity of the spin adduct. The longer the lagtime value, the better the flavour stability. The converse is true for the T150 value.*

The information from the EPR oxidation assay is extremely useful for a brewery. It is not only used as a quality control check for beer as it is packaged, but can be used as a research tool to improve the shelf life of a beer. Several "process" changes can be made in a full scale or pilot brewing plant. The efficacy of these process changes can be rapidly monitored using the EPR technique.

5   ANTIOXIDANT & CW-EPR: THE DPPH-ASSAY

5.1 Antioxidants and EPR

Antioxidant molecules themselves are not free radicals and are not EPR-active. However, these compounds do react rapidly with reactive oxygen species, ROS, and other free radicals to render them harmless. The EPR-based DPPH assay described below can be used to provide quantitative information concerning the relative effectiveness of various dietary substances and their antioxidant capacities.

5.2 The DPPH-Assay

DPPH (1,1-diphenyl-2-picrylhydrazyl) (see Figure 5) exists in solution as a semi-stable free radical. Compounds with antioxidant activity may undergo a redox reaction with DPPH, leading to a time-dependent decrease of the DPPH radical concentration, which is easily detectable by EPR spectroscopy. A particular advantage of the EPR-based assay is that it is free of any background signal or interference and is, therefore, superior to the optical DPPH assay, which monitors the bleaching of the violet colour of the dissolved DPPH radical. For reproducible results, the timing of mixing of assay components needs to be precisely controlled.

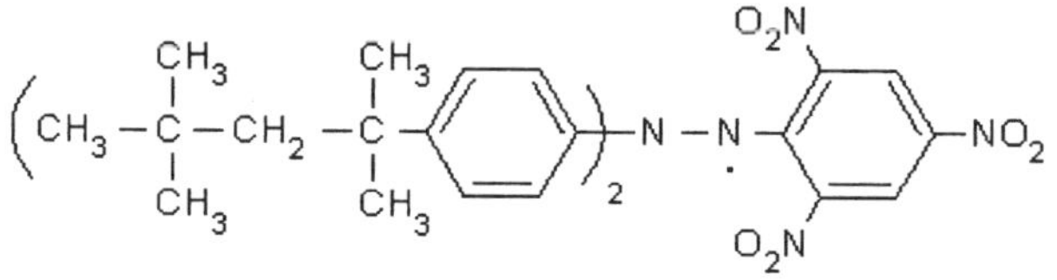

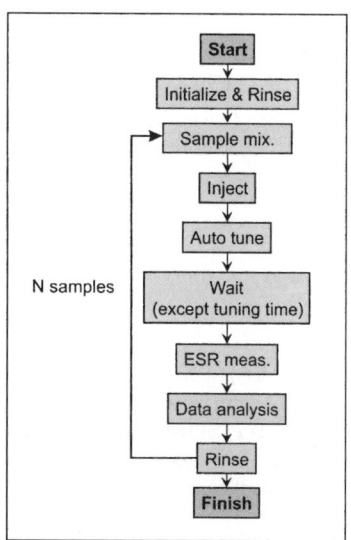

Figure 5: *The paramagnetic DPPH molecule.*

5.3 The Measurement Protocol

A flow chart for the procedures is shown in Figure 6. The main control program for the automated assay was programmed with Visual Basic®. The *e-scan*™ unit was remotely controlled through its proprietary ActiveX interface. An EXCEL® template file contained all relevant information about the sample and the desired automatic processing (number of samples, spectrum file names, peak position and line width). At the end of the assay, the measurement results and signal intensity plots are generated automatically.

Figure 6: *Flow chart for the DPPH assay procedure using the e-scan™ with LC autosampler.*

5.4 Results

In Figure 7, the results of a DPPH antioxidant assay at a single time point are shown for nine test samples with distilled water as control. EPR spectra were obtained 60 s after sample mixing (injection from autosampler). The control measurement with water shows a high EPR signal intensity, indicating little antioxidant effect. Ascorbic acid at 2.5 mM or higher concentrations leads to an essentially complete quenching of the DPPH signal in 60 s. The repeated measurement with water demonstrates the reproducibility of the control measurement, and the measurement with white wine indicates almost no antioxidant capacity for that test sample. In contrast, the red wine sample causes a reduction of the EPR signal to about 60% of control, corresponding to an antioxidant capacity equivalent to about 0.8 mM ascorbate solution.

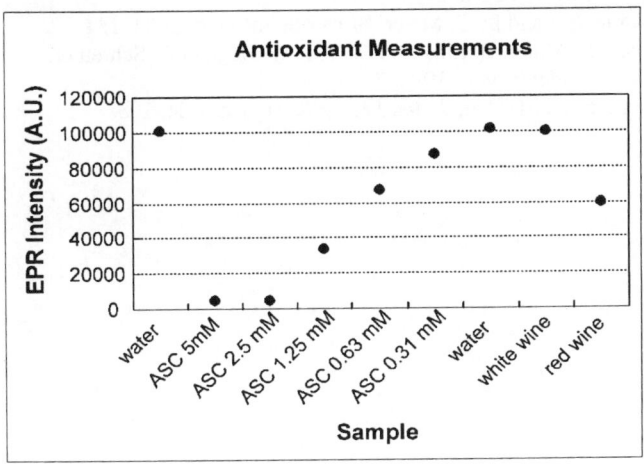

Figure 7: *Antioxidant capacity measurements obtained by the DPPH method. EPR signal intensities were determined at 60 s after mixing of the DPPH assay solution with each of the samples listed at the bottom of the automatically generated chart. From left to right, the test solutions were: distilled water (control), ascorbic acid at 5, 2.5, 1.25, 0.625, and 0.3125 mM, distilled water, white wine, red wine.*

6 SUMMARY AND CONCLUSION

Electron Paramagnetic Resonance (EPR, ESR) spectroscopy is a unique tool to answer specific questions evolving radical processes in both research and development (R&D) as well as quality control (QC). Key QC examples comprise the food irradiation control for customer care according to European Union (EU) norms and the flavour stability and shelf life assessment on fresh beer. For the first time, an automated DPPH-assay to measure the effectiveness of antioxidants by CW-EPR is reported.

References

1. CODEX General Standard for Irradiated Foods, CODEX STAN 106-1983, rev. 1-2003
2. http://en.wikipedia.org/wiki/Radura
3. http://nucleus.iaea.org/NUCLEUS/nucleus/Content/Applications/FICdb/BrowseDatabase.jsp
4. EN 1786:1996. Foodstuffs – Detection of irradiated food containing bone – Method by ESR spectroscopy
5. EN 1787:2000. Foodstuffs – Detection of irradiated food containing cellulose by ESR spectroscopy
6. EN 13708:2001. Foodstuffs – Detection of irradiated food containing crystalline sugar by ESR spectroscopy
7. A. Kamlowski and D. C. Maier, Bruker SpinReport, 2004, **154**, 32.
8. N. Helle, D. Maier, B. Linke, P. Such, K.W. Bögl, G.A. Schreiber, *Bundesgesundheitsblatt*, 1992, **7**, 331
9. M. Uchida and M. Ono, *J. Am. Brew. Chem.*, 1996, **54**, 198.

APPLICATION OF ELECTRON SPIN RESONANCE TO STUDY FOOD ANITOXIDATIVE AND PROOXIDATIVE ACTIVITIES

J.J. Yin[1,*] and P.P. Fu[2]

[1]Center for Food Safety and Applied Nutrition, U.S. Food and Drug Administration, College Park, MD 20740, USA
[2]National Center for Toxicological Research, U.S. Food and Drug Administration, Jefferson, AR 72079, USA

*Corresponding author. Tel: 301-436-1991, E-mail: junjie.yin@fda.hhs.gov

1 INTRODUCTION

Reactive oxygen species (ROS) and lipid peroxidation are associated with many human diseases including cancer, athereosclerosis, cardiovascular diseases, ischemia, inflammation, and liver injury.[1-3] Epidemiological evidence indicates that higher intake of antioxidant-rich foods, such as fruits, vegetables, and whole grains, is associated with lower disease risk. To date, a variety of phytochemicals present in the diet and neutraceuticals (e.g., functional foods and dietary supplements) have been identified that exhibit antioxidant or prooxidatant activity.[3] For human health protection, it is important to evaluate what food stuffs and phytochemicals possess anti- or pro-oxidative activity and determine their mechanisms of action.

Free radicals, including ROS, are very short-lived entities and, consequently, difficult to detect in biological systems. Electron spin resonance (ESR) has been the most powerful methodology used to detect free radicals generated chemically or formed in biological systems. ESR spin trapping and ESR spin labeling are the two principal ESR techniques used for the detection of free radical formation and the identification of the different types of free radicals formed.[3-12] The ESR spin trapping technique utilizes specific spin trapping agents to react with short-lived free radicals that result in the formation of adducts, which have half-lives sufficiently long enough for ESR spectroscopic measurement.[3-7] The ESR spin labeling technique involves the use of a stable paramagnetic spin label agent to interact with the target chemical, e.g. the oxygen molecule, or tissue *in vitro* and *in vivo*, and is a powerful tool to probe the structural and/or dynamical changes in a specific biological system.[8-12]

We have long been interested in development of new ESR techniques[13-19] and utilization of ESR techniques to study anti- and/or pro-oxidant activities in a variety of foods products, dietary supplements, functional foods, herbals, and phytochemicals. Our studies have included wheat,[20,21] herbal products,[22,23,24] dietary supplements,[21,24,25] conjugated linoleic acid,[5,12] strawberry,[26] flavonoids,[23] Aloe vera,[27] vitamin A,[3,4,6,7] and food contaminants, including fumonisin B1[10,11] and polycyclic aromatic hydrocarbons.[28-30] In this paper, we demonstrate how we utilize ESR spin-trapping techniques to detect ROS (including hydroxyl radical, superoxide radical anion, and singlet oxygen) generated from chemical, photochemical, and biochemical reactions, and to verify food extracts and phytochemicals that are capable of scavenging ROS and exhibiting antioxidative properties. We also show how the ESR oximetry technique is used to determine the effects of anti- and pro-oxidative phytochemicals on the modification of cell membrane leading to enhanced or suppressed lipid peroxidation.

2 DETECTION OF REACTIVE OXYGEN SPECIES (ROS) by ESR SPIN TRAPPING TECHNIQUE

2.1 Hydroxyl Radicals

Hydroxyl radicals can be generated by the classical Fenton reaction, by mixing freshly pre-pared $FeSO_4$ with H_2O_2,[4] or by Fenton-like reactions using $Fe(II)/H_2O_2$, $Cu(II)/H_2O_2$, or $Co(II)/H_2O_2$ as reagents.[20] DMPO (5,5-dimethyl-1-pyrroline N-oxide) is the most com-monly used spin trapping agent to identify the hydroxyl radical. The resulting DMPO-OH adduct exhibits two characteristic sets of four-line (1:2:2:1) ESR spectral profiles, with hyperfine splitting parameters $a^N = a^H = 14.9$ G. The standardized ESR profile obtained from the Fenton reaction in the presence of DMPO is shown in Figure 1A.

The antioxida-tive activity of cinnamon and whole grain soft wheat extracts was studied.[4] Comparison of their ESR profiles shown in Figure 1B and Figure 1C with the control (Figure 1A) clearly indicates that the intensities of the DMPO-OH ESR signals in Figure 1B and Figure 1C are significantly reduced. These results provide evidence that both cinnamon and whole grain soft wheat extracts exhibit anti-oxidative activities, capably of scavenging the hydroxyl radical.

2.2 Superoxide Radical Anions

The superoxide radical anion generated by reaction of xanthine and xanthene oxidase can be detected by ESR spin trapping technique. Besides being used for trapping the hydroxyl radical, DMPO is also used as a spin trapping agent to react with the superoxide radical anion, which forms the DMPO-·OOH adduct for ESR spectroscopic measurement.

However, the DMPO-·OOH adduct is unstable and easily decomposes to the corres-ponding hydroxyl DMPO-·OH adduct.[6] Consequently, it is often unclear whether the free radical generated is the superoxide radical anion, a hydroxyl radical, or a combination of both. To ascertain whether the superoxide radical anion is indeed formed, and not the hydroxyl radical, it is prudent to employ another spin trap agent, 5-*tert*-butoxycarbonyl 5-methyl-1-pyrroline N-oxide (BMPO). The advantage of BMPO is that, once BMPO-super-oxide adduct (i.e., BMPO-·OOH) is formed from the reaction of BMPO with the superoxide radical anion, the resulting adduct is stable and does not decompose into the corresponding hydroxyl adduct (i.e., BMPO-·OH).[6]

We have successfully employed the ESR spin trapping technique with BMPO as the spin trapping agent to determine whether the UVA irradiation of retinyl palmitate, the major ester form of vitamin A (retinol), generates the superoxide radical anion.[6] We showed that the photoirradiation of BMPO alone with UVA light did not generate ESR signals (Figure 2A). Similarly, no ESR signal was observed when retinyl palmitate was mixed with DMPO without UVA light irradiaion (Figure 2B). However, the irradiation of

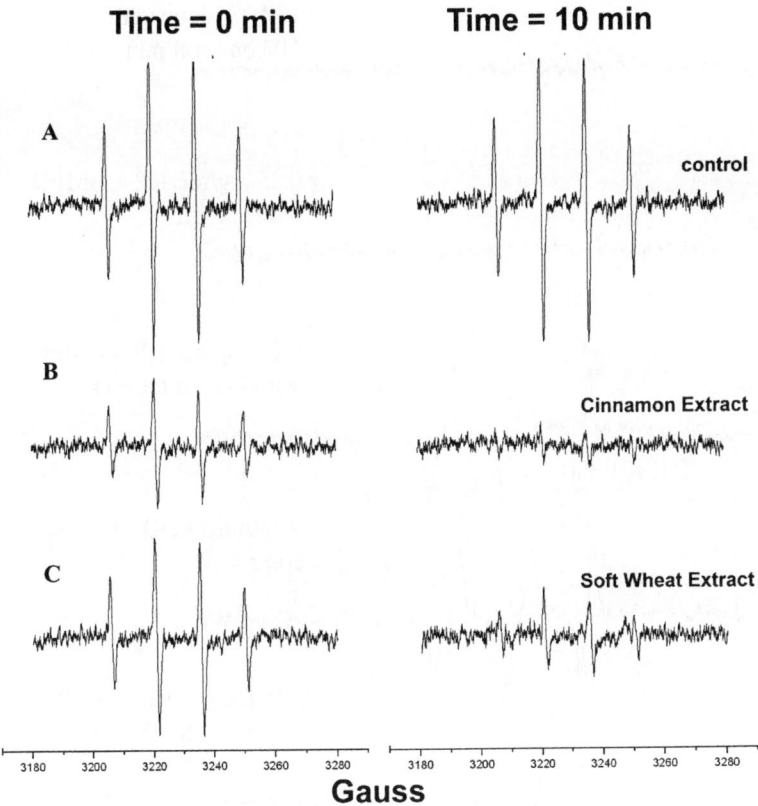

Figure 1 *ESR spin-trapping technique using DMPO as the spin trapping agent (A) to detect hydroxyl radicals generated by Fenton reaction and to examine scavenging activity of (B) cinnamon extract and (C) whole grain soft wheat extract.*[2]

retinyl palmitate with UVA light and in the presence of DMPO produced ESR spectral signals (Figure 2C). The ESR spectral profiles are identical to those produced from the reaction of xanthine and xanthine oxidase in the presence of BMPO (Figure 2D) and confirms that the generated ESR profile is identical to that of BMPO-superoxide adduct (i.e., BMPO-˙OOH). The ESR signal generated from the photoirradiation of retinyl palmitate and BMPO, as shown in Figure 2C, was quenched when superoxide dismutase (SOD) was incorporated into the reaction mixture (Figure 2E). The overall results provide clear direct evidence that the superoxide radical anion is generated from photoirradiation of retinyl palmitate with UVA light.

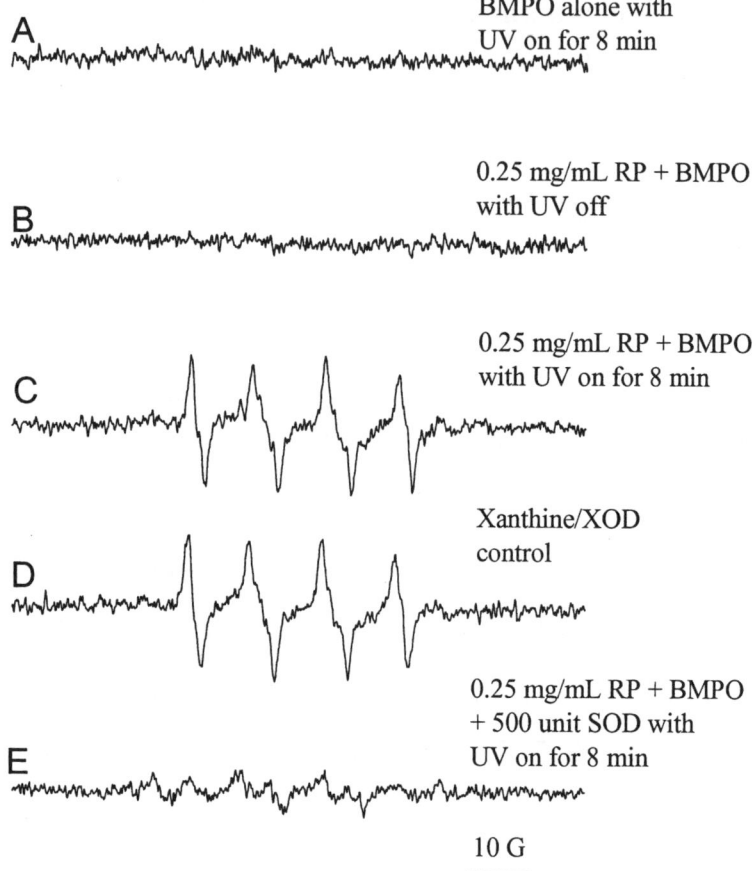

BMPO alone with
UV on for 8 min

0.25 mg/mL RP + BMPO
with UV off

0.25 mg/mL RP + BMPO
with UV on for 8 min

Xanthine/XOD
control

0.25 mg/mL RP + BMPO
+ 500 unit SOD with
UV on for 8 min

10 G

Figure 2 *ESR spectra formed from UVA (320 nm) irradiation of 0.25 mg/mL retinyl palmitate (RP) in 70% ethanol in water with 200 mM BMPO for 8 min and from reaction of xanthine and xanthine oxidase.*[6]

2.3 Singlet Oxygen

In 1976, Lion and co-workers found that TEMP (2,2,6,6-tetramethyl-piperidine) could be used as a specific probe for trapping singlet oxygen.[31] Upon reaction of singlet oxygen with TEMP, the resulting TEMPO, a stable nitroxide, can be detected by ESR spectroscopy.[6,31] Consequently, TEMP has been popularly used as an ESR spin trapping agent for the detection of singlet oxygen formation.

We employed TEMP to study the generation of singlet oxygen generated by the UVA irradiation of retinyl palmitate.[6] As expected, the UVA irradiation of the TEMP spin trap agent in the absence of retinyl palmitate did not result in an ESR signal (data not shown). Similarly, the reaction of retinyl palmitate (0.35 mg/mL) in 70% ethanol/water containing 200 mM TEMP did not produce ESR signals (Figure 3A). However, the concomitant exposure of retinyl palmitate to TEMP and UVA light (320 nm) for 15 min generated singlet oxygen production, as evidenced by an ESR spectral profile which is typical of TEMPO (Figure 3B).[32] The intensity of these ESR signals were progressively enhanced when the photoirradiation time was subsequently increased to 30, 45, and 60 min, respectively (Figure 3C-D). These results provide direct evidence that the photoirradiation of retinyl palmitate with UVA light generates singlet oxygen and that the quantity of singlet oxygen formed is dependent on the dose of administered light.

3 APPLICATION OF ESR SPIN LABEL FOR OXIMETRY STUDY

The ESR oximetry measurement is based on the bimolecular collision of O_2 with a spin probe (spin label) which is a stable free radical. Molecular oxygen (O_2) is paramagnetic. Collision of the spin probe with O_2 produces a spin exchange between O_2 and the spin probe, which results in shorter relaxation times and leads to the ESR signals of the spin probe with broader line widths (Figure 4A). Since the integrated area of the ESR signal over the scanning range is unaffected by these effects on the relaxation times, broadening of the spin probe's ESR signal is necessarily accompanied by a decrease in the peak height of the ESR signal (Figure 4B). The extent of spin exchange is dependent on molecular oxygen concentration. Thus, a decrease in oxygen concentration due to oxygen consumption, such as by lipid peroxidation, results in a decreased line width for the spin probe. A time dependent decrease in line width and increase in intensity of the ESR signal indicates continuous oxygen consumption, and this spectral change can be quantified (Figure 4B).[4] By repeated measurement of the spin probe's line widths, one can assess rate of lipid peroxidation in the sample.

Conjugated linoleic acids (CLAs) are a group of octadecadienoic acids (18:2) that are naturally present in food products and may have beneficial and/or adverse health effects. We measured the effects of CLA isomers on oxygen diffusion-concentration products in liposomes and phospholipid solutions using ESR spin-label oximetry methods.[5] Liposomes were prepared by mixing synthetic phosphatidylcholines (PCs) with c9,t11-CLA, t10,c12-CLA, and linoleic acid (LA) in the sn-2 position into natural PCs from soybean, egg yolk, rat brain, and rat heart at 5 mol %. It was determined that individual synthetic PCs, the phospholipid matrix, and the tested lipid systems all exhibited an influence on the oxygen diffusion-concentration products during lipid peroxidation. The oxygen consumption kinetics of liposomes prepared using soy PC during 30 min is presented in **Figure 5**. The order of oxygen consumption rates is as follows: 5 mol % PC(c9,t11-CLA) in soy PC >

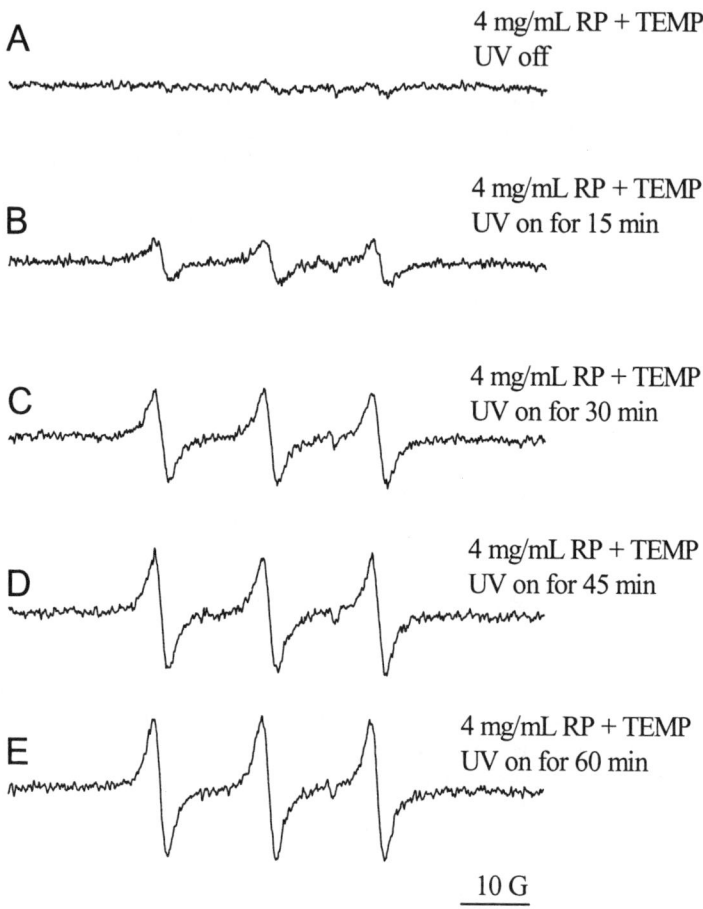

A 4 mg/mL RP + TEMP
UV off

B 4 mg/mL RP + TEMP
UV on for 15 min

C 4 mg/mL RP + TEMP
UV on for 30 min

D 4 mg/mL RP + TEMP
UV on for 45 min

E 4 mg/mL RP + TEMP
UV on for 60 min

10 G

Figure 3 *ESR spectra of TEMPO formed from UVA (320 nm) irradiation of 4 mg/mL retinyl palmitate (RP) in 70% ethanol in water with 200 mM spin trap TEMP.*[6]

soy PC alone > 5 mol % PC(*t*10,*c*12-CLA) in soy PC > 5 mol % PC(LA) (Figure 5). These results provide evidence that CLA isomers affect the permeability of membranes to oxygen and lipid peroxidation.[5] The results are consistent with our earlier finding that the perturbation of membrane structure and the increased CLA-induced incorporation of the relative oxygen diffusion-concentration products into membrane lipids affect oxidative stress.[12]

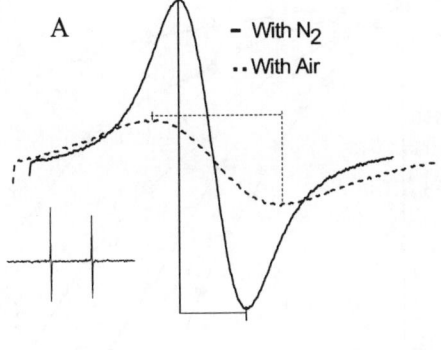

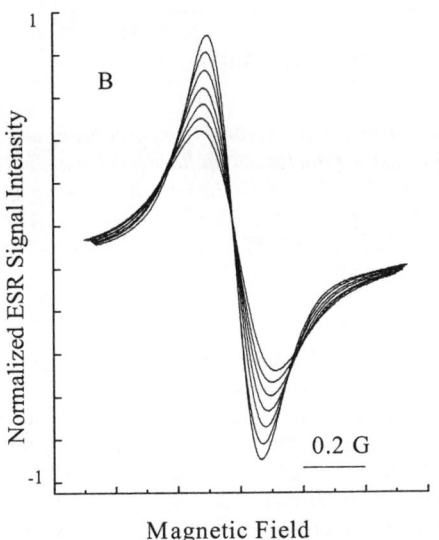

Figure 4 *Measurement of lipid peroxidation by ESR oximetry measured in a closed chamber Panel. A shows ESR spectra of the scans of the low field line of ^{15}N-Tempone (spin label) in a nitrogen atmosphere (solid line) or air-saturated (broken line) aqueous solutions. Panel B indicates the presence of oxygen results in a broader and less intense ESR signal for the spin probe. Sample contained liposomes (30 mg/mL liver phosphatidylcholine) and 0.1 mM ^{15}N-Tempone and lipid peroxidation was initiated by adding 25 mM AAPH. The progressive increases in peak to peak signal intensity (and accompanying progressive narrowing of line width) in each panel are due to time-dependant oxygen consumption resulting from lipid peroxidation.[4]*

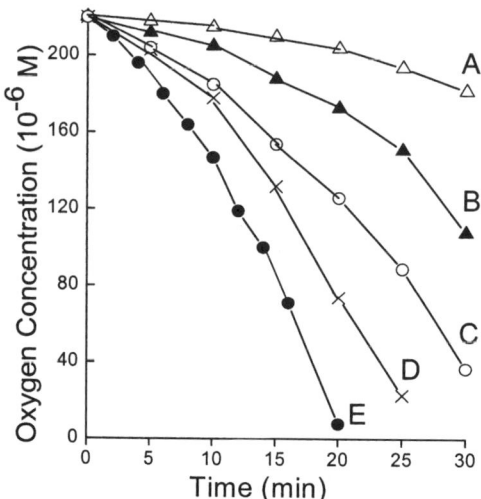

Figure 5 *Oxygen consumption kinetics of liposomes prepared using soy PC. Oxygen consumption was measured in closed capillary tubes with individual test systems that included (A) 5 mol % PC(c9,t11-CLA) in l-R-dimyristoylphosphatidylcholine, (B) 5 mol % PC (LA) in soy PC, (C) 5 mol % PC (t10,c12-CLA) in soy PC, (D) soy PC alone, and (E) 5 mol % PC (c9,t11-CLA) in soy PC.[5]*

4 CONCLUSION

Because most of the free radicals, including hydroxyl radical, superoxide radical anion, and singlet oxygen, are short-lived and generated in very low concentrations in biological systems, detection of free radicals is difficult. In this paper we selected examples from our research to illustrate the potential application of ESR spectroscopy to study the anti- and pro-oxidative activities of food extracts, phytochemical constituents, and dietary supplements, and to elucidate the mechanisms underlying the antioxidant and prooxidant capabilities via the gene-ration and/or reduction of lipid peroxidation and damage or protection of cellular membrane.

Acknowledgement

We thank Dr. Mary D. Boudreau for critical review of this manuscript.

References

1   E.R. Stadtman and B.S. Berlett, *Chem. Res. Toxicol.*, 1997, **10**, 485.
2   J. Moore, J.J., Yin and L. Yu, *J. Agric. Food Chem.*, 2006, **54**, 617.
3   P.P. Fu, Q. Xia, J.J. Yin, S.-H.Cherng, J. Yan, N. Mei, T. Chen, M.D, Boudreau, P.C. Howard and W.G. Warner, *Photochem. Photobiol.*, 2007, **83**, 409.
4   J.J. Yin, L. Fang, P.P. Fu, W.G. Warner, Y. Zhao, G. Xing, B. Sun, X. Li, P.C. Wang, C. Chen and X.-J. Liang, *Mol.* Pharma., 2008, in press.
5   J.J. Yin, J.K.G. Kramer, M.P. Yurawecz, A.R. Eynard, M.M. Mossoba and L.Yu, *J. Agric. Food Chem.*, 2006, **54**, 7287.

6   Q. Xia, J.J. Yin, S.-H. Cherng, W.G. Wamer, M. Boudreau, P.C. Howard, and P.P. Fu, *Toxicol. Lett.*, 2006, **163**, 30.

7   W.H. Tolleson, S.H. Cherng, Q. Xia, J.J. Yin, **W.G.** Wamer, P.C. Howard, H. Yu and P.P. Fu, *Int. J. Environ. Res. Public Health*, 2005, **2**, 147.

8   J.J. Yin and J. S. Hyde, *Z. Phys. Chem.*, 1987, **153**, 541.

9   J.J. Yin, M.J., Smith, R.M. Eppley, S.W. Page and J.A. Sphon, *Biochem. Bioph. Res. Co.* 1996, **225**, 250.

10 J.J. Yin, M.J. Smith, R.M. Eppley, A.L. Troy, S.W. Page and J.A. Sphon, *Arch. Biochem. Biophys.*, 1996, **335**, 13.

11 J.J. Yin, M.J. Smith, R.M. Eppley, S.W. Page and J.A. Sphon, *Biochem. Biophys. Acta-Biomembranes,* 1998, **1371**, 134.

12 J.J. Yin, M.M. Mossoba, J.K.G. Kramer, M.P. Yurawecz, K. Eulitz, K.M. Morehouse and Y. Ku, *Lipids,* 1999, **34**, 1017.

13 J.S. Hyde, J.J. Yin, W. Froncisz and J.B. Feix, *J. Magn. Reson.*, 1985, **63**, 142.

14 C.-S. Lai, M.D.Wirt, J.J. Yin, W. Froncisz, J.B. Feix, T.J. Kunick and J.H. Hyde, *Biophys. J.*, 1986, **50**, 503.

15 J.J. Yin, M. Pasenkiewicz-Gierula and J.S. Hyde, *Proc. Natl. Acad. Sci. USA.*, 1987, **84**, 964.

16 J.B. Feix, J.J. Yin and J.S. Hyde, *Biochemistry*, 1987, **26**, 3850.

17 J.J. Yin, J.B. Feix and J.S. Hyde, *Biophys. J.*, 1987, **52**, 1031.

18 J.J. Yin and J.S. Hyde, *J. Magn. Reson.*, 1987, **74**, 82-93.

19 J.J. Yin, J.B. Feix and J.S. Hyde, *Biophys. J.*, 1988, **53**, 525.

20 J. Moore, J.J. Yin and L. Yu, *J. Agric. Food Chem.*, 2006, **54**, 617.

21 K. Zhou, J.J. Yin and L. Yu, *Food Chem.*, 2006, **95**, 446.

22 J. Xie, Z.H. Shao, T.L. Vanden Hoek, W.T. Chang, J. Li, S. Mehendale, C.Z. Wang, C.W. Hsu, L.B. Becker, J.J. Yin and C.-S. Yuan, *Eur. J. Pharmacol.,* 2006, **532**, 201.

23 W.T. Chang, Z.H. Shao, J.J. Yin, S. Mehendale, Y. Qin, W.J. Chen and C.-S. Yuan, *Eur. J. Pharmacol.*, 2007, **566**, 58.

24 Z.H. Shao, C.Q. Li, T.L. Vanden Hoek, P.T. Schumacker, L.B. Becker, J.J. Yin and C.S. Yuan, *Am. J. Chin. Med.*, 2004, **32**, 89.

25 H.N. Yu, J.J. Yin and S.R. Shen, *J. Agric. Food Chem.*, 2..4, 52, 462.

26 C.Y. Wang, S.Y. Wang, J.J. Yin, J. Parry and L.L. Yu, *J. Agric. Food Chem.*, 2007, **55**, 6527.

27 Q. Xia, J.J. Yin, F.A. Beland, M.D. Boudreau and P.P. Fu, *Toxicol. Lett.*, 2007, **168**,165.

28 Q. Xia, M.W. Chou, J.J. Yin, P.C. Howard, H. Yu and P.P. Fu, *Toxicol. Ind. Health*, 2006, **22**, 147.

29 J.J. Yin, Q. Xia and P.P. Fu,. UVA photoirradiation of anhydroretinol – formation of singlet oxygen and superoxide. *Toxicol. Ind. Health*, 2007, **23**, 625.

30 J.J. Yin, Q. Xia, S.-H. Cherng, I-W. Tang, P.P. Fu, G. Lin, H. Yu and D. Herreño Sáenz, *Int. J. Environ. Res. Public Health,* 2008, **5**, 26.

31 Y. Lion, M. Delmelle and A. Van De Vorst, *Nature*, 1976, **263**, 442.

32 S. Rinalducci, J.Z. Pedersen and L. Zolla, *Biochim. Biophy. Acta.*, 2004, **1608**, 63.

WATER/BIOPOLYMER INTERACTIONS: COMPARISON OF NMR WITH OTHER TECHNIQUES

A. Almutawah, S. A. Barker and P. S. Belton

School of Chemical Sciences and Pharmacy, University of East Anglia, Norwich, NR4 7TJ, UK

1 INTRODUCTION

The interaction of water with food materials is a critical one in the determination of food texture and safety. Many studies using a variety of techniques have been published. However there have been relatively few which have attempted to compare results from different techniques in order to obtain a consistent model of water behaviour. When data obtained using different methods are compared there often appears to be a conflict, leading uncertainty as to which, if any, of the techniques is truly reporting on the interactions of interest. However developments in the theoretical understanding of the origins of some of the apparent conflicts have recently emerged. Key to these has been the work of Halle [1], Steinhauser [2] and Weingartner [3] who have developed sophisticated understandings of the interpretation of NMR, dielectric and other data. In this paper these results will be considered along with some other thermodynamic and dielectric results.

2 RESULTS FROM NMR

The current interpretation of NMR data from water biopolymer systems has been summarised in reviews [1,4] and books [5]. It is not intended to re-review the subject here, merely to summarise the conclusions.

These are:

1 All proton and deuteron NMR relaxation times are strongly affected by water/ biopolymer proton exchange and/or dipolar interactions.

2 Multiple exponential proton relaxation times may reflect multiple sites, but the numbers of components and their relative intensities may or may not represent the number and relative populations of the sites

3 As result of these conclusions proton NMR measurements cannot be said in general to represent the "state" of water in a food system.

4 ^{17}O NMR results are a more reliable guide to the behaviour of water because there are no exchange effects.

5 In dilute biopolymer solutions water at the biopolymer/water interface has its rotational motion slowed by a factor of about 4

6 One or two water molecules may be structurally integrated into the biopolymer or be trapped in clefts and exchange with bulk water on a time scale of about 1 microsecond.

7 Above about 30% water content neither ^{1}H nor ^{17}O NMR results offer any evidence for "bound" water or any significant amount of water with greatly slowed correlation times for motion.

8 At low water contents the NMR results are much more problematic. Relatively little has been published. Results are complicated by line width and exchange problems.

Figure 1 represents the current state of knowledge about water dynamics in dilute protein solutions

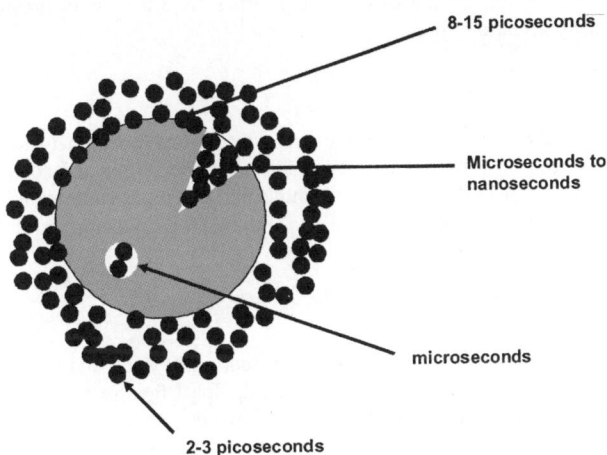

Figure 1   *Water dynamics in the presence of proteins. The time scales attached to the arrows represent the rotational correlation times for water in various locations*

3   DIELECTRICS

What might be termed the classical approach to the interpretation of high frequency dielectric spectra of water protein solutions is summarised in Table 1. Most reported results involve a number of dispersions (for a summary see Nandi, Battacharyya and Bagchi [6]) interpreted as resulting from protein, bound water and bulk water reorientations.
An interpretation based on Weingartner's [3] approach is also included in Table 1

Table 1 *Origins of correlation times obtained in water protein systems by dielectric measurements*

Designation	Correlation time[a] range/ns	Classical interpretation	New interpretation[3]	comments
β	20-100	Protein tumbling	Protein tumbling	N, B, B[6] question this interpretation
δ1	2-10	Bound water	Cross terms	N, B, B[6] suggest exchange effects
δ2	20-100 x 10⁻³	Bound water	Water at the protein interface	Values are consistent with NMR[b]
γ	2-20 x10⁻³	Bulk water	Bulk water	Values are consistent with NMR[b]

a. These ranges are indicative and are not exclusive. Some authors report and additional δ dispersion
b. Typically dielectric rotational correlation times are 3 to 4 times greater than NMR rotational correlation times

Many previous authors have ignored the cross terms that occur between the various dielectric components in a biopolymer/water system. Steinhauser and co-workers [2] have analysed the situation theoretically and with molecular dynamics simulations. Their model may be summarised by the following equation for the total polarisation, φ, of the system.

$$\varphi = \sum_{p,w} M_{p,w} = M_{ww} + M_{pp} + M_{pw} + M_{wp}$$

Where M is the dipole moment of component I, w represents water and p protein. However the dipole moment of the protein has an effect on the local field experienced by the water and vice versa. Hence cross terms of the sort M_{wp} have to be included. Steinhauser's work showed that these terms were responsible for some of the dispersions attributed to bound water. In detailed study of Ribonuclease A [3]. It was found that the dispersion obtained could be adequately explained by this model and that the δ2 and γ correlation times were consistent with those obtained by NMR. In fact it is possible to write a similar expression for NMR relaxation taking the form:

$$\frac{dM}{dt} = \frac{dM_w}{dt} + \frac{dM_p}{dt} + \frac{dM_{w,p}}{dt} + \frac{dM_{p,w}}{dt}$$

In this case M is the total magnetisation, the subscripts w and p have the same meaning as above and the cross terms refer interactions between protein and water, the subscript w,p refers to the effect of protein on water and p,w to the effect of water on protein. These terms may involve a number of sub terms such as chemical exchange terms which themselves could involve both frequency and relaxation time effects. In the case of proton and deuteron relaxation the term in w,p predominates. It can be ignored for ^{17}O.

Low frequency dielectrics (i.e. in the range 10^{-3} to 10^{6} Hz) respond in rather more complex way to changes in dynamics. However in low water content regions useful data may be obtained. A study[7] of wheat gluten/water interactions showed a sharp change in water dynamics as the water content changed from 0 to 60% by weight. This is illustrated below in Figure 2

It is clear from the diagram that a transition in behaviour occurs at around 30% water content. This transition is also observed in changes in the behaviour of the amide II band of the infrared spectra, shown in figure 3. A consistent interpretation of these results is that full hydration of the amide groups only occurs at about 30% water content and after that water behaves essentially as bulk water[7].

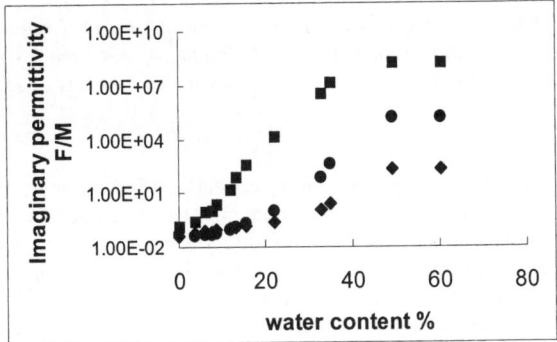

Figure 2 *The variation of the imaginary permittivity of a gluten water system at 288K with water content. Squares: 10^{-2} Hz; triangles: 10^{3} Hz; diamonds: 10^{6} Hz*

4   THERMODYNAMIC MEASUREMENTS

The phenomenon of non-freezing water is well documented and is often considered as a measure of bound water. Typically the amount of non-freezing water is a strong function of initial water content[8]. This appears inconsistent with the bound water concept as it would be expected that, above the bound water threshold, the amount of bound water, hence non-freezing water, would remain constant. An alternative model is one in which that non-freezing water arises because the biopolymer/water system is not at thermodynamic equilibrium. As the system is cooled water crystallises out and, if it were an equilibrium system, at sufficiently low temperatures biopolymer would crystallise out in a eutectic.

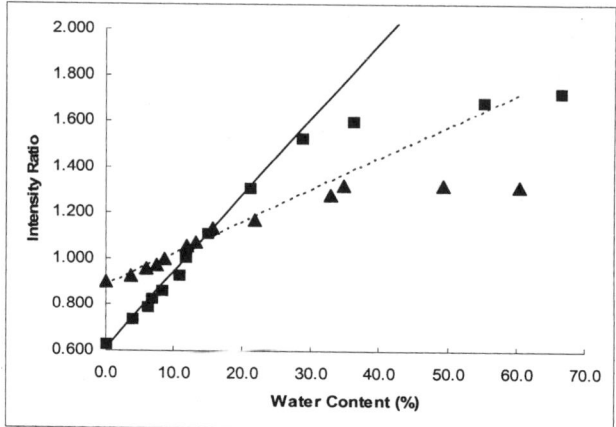

Figure 3   *Changes in ratios of band intensities in the amide II region of gluten with water content. Squares: ratio of intensities at 1548 cm⁻¹ to 1515 cm⁻¹; triangles: ratio of intensities 1533 cm⁻¹ to 1515 cm⁻¹. Straight line fits to the two sets of low water content data are shown.*

However the biopolymer does not easily form crystals, so ice continues to form until the remaining liquid water associated with the biopolymer has the same vapour pressure as the ice. At this point crystallisation ceases and no further change in the ice water ratio occurs. The amount of ice formed will be a function of the starting conditions of the system and thus will vary from sample to sample. More details are given elsewhere[4,8]. Typical values for non-freezing water are around 0.3 to 0.5 g water per g protein and ice formation is not observed below about 0.3 g water per g protein. The process is summarised in the pseudo-phase diagram shown in Figure 4. At temperatures above 273 K the biopolymer can exist either as a hydrated solid or, if the water content is high enough, in solution. Below 273 K, depending on the polymer content, there may be ice formation. Ice will continue to form and concentrate the polymer in contact with the liquid phase until a critical temperature, usually around 263 K where no more ice forms and the polymer liquid water ratio remains constant.

5 HYDRODYNAMICS

The hydrodynamic properties of protein solutions have consistently suggested that [9] the apparent dimensions of the protein are larger than expected. They can be explained by assuming that there is a hydration shell of water that moves with the protein. The exact value of the size of the shell varies but calculations which take account of protein structure and electrostatic properties [9] suggest values of around 0.4 g of water per g of protein are involved. These figures are remarkably similar to the amounts of non-freezing water observed and therefore are suggestive of a real effect due to some kind of water with different properties.

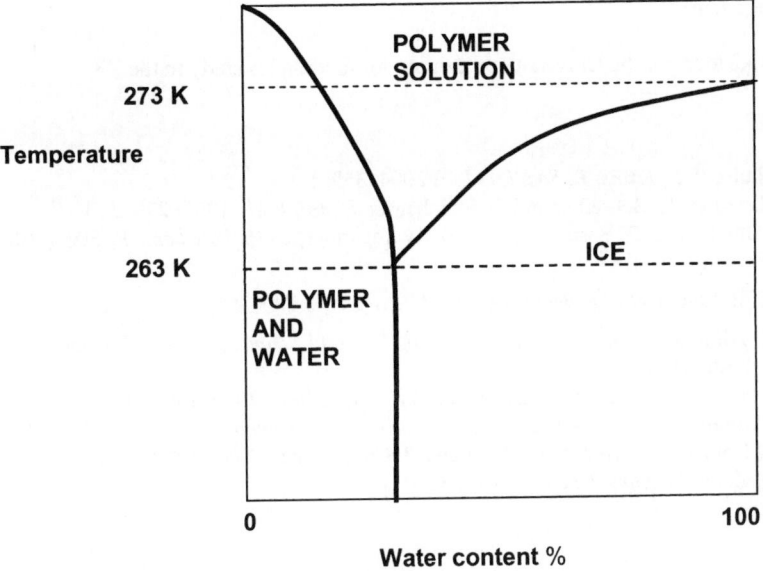

Figure 3   *The pseudo phase diagram indicating the behaviour of a cooled biopolymer water system*

However at first sight results from NMR and dielectrics suggest that there is no evidence in solution for any amount of water that has sufficiently different dynamic properties to explain this. Halle [1] has argued that water near to the protein interface, which has its motion slowed by a factor of about 4, will have an associated increase in viscosity. This can account four the slowing of the rotational diffusion of proteins. It has not been demonstrated that this is true for translational diffusion or sedimentation

6   CONCLUSIONS

Bound water in dilute biopolymer solutions does not exist in any but very small amounts. Results in low water content systems, from freezing experiments and hydrodynamics, suggest that hydration levels of the order of 0.3 to 0.5 g water per g protein are in some way critical. If it is assumed that this level of water is required to hydrate polar groups then the measurements may be reconciled. Below this level hydration of polar groups will be incomplete and ice formation will not be possible, as there is no bulk water available. Currently hydrodynamic calculations make the assumption, implicit or explicit, that the protein surface is smooth. This is not the case. There will be side chains and amide groups which stick out of the surface, there may also be water filled clefts. These will make a significant difference to hydrodynamic properties of the molecule and will be related to the available surface area for hydration. They may therefore be interpreted as an apparent hydration sphere. It might be expected that these extensions are where the water molecules will hydrate initially and that they represent a hydrodynamic volume not dissimilar to that of the hydrating water.

Acknowledgements

AA would like to thank the University of Bahrain for funding his study in the UK

References

1   B Halle, *Phil. Trans. R. Soc. Lond.,* B 2004, **359**, 1207
2   G. Loeffler, H. Schreiber and O. Steinhauser *J. Mol. Biol.*, 1997, **270**, 520
3   A. Oleinikova, P. Sasisanker and H. Weingartner, *J. Phys. Chem. B*, 2004, **108**, 8467
4   P. S. Belton, *Cellular and Molecular Life Sciences* 2000, **57**, 993
5   B.P. Hills, *Magnetic Resonance Imaging in Food Science*, John Wiley and Sons, Chichester, 1998
6   N. Nandi, K. Battacharyya and B. Bagchi, *Chem. Rev.*, 2000, **100**, 2013
7   A. Almutawah, S. A. Barker and P.S. Belton, *Biomacromolecules*, 2007, **8**, 1601
8   A.D Roman-Gutierrez, S. Guilbert and B. Cuq, *Cereal Chem.* 2002, **79**, 471
9   H-X Zhou, *Biophys. Chem.,* 2001, **93**, 171

Fish and meat

LOW FIELD NMR STUDY ON WILD AND FARMED ATLANTIC COD (*GADUS MORHUA*)

M. Guðjónsdóttir[1], V.N. Gunnlaugsson[1], G.A. Finnbogadóttir[1], K. Sveinsdóttir[1], H. Magnússon[1], S. Arason[1,2]

[1]Matís ohf. Food research, innovation and safety. Process development. Skúlagata 4, 101 Reykjavík, Iceland. E-mail: mariag@matis.is
[2]University of Iceland, Department of food science, Suðurgata, 107 Reykjavík, Iceland.

1 INTRODUCTION

1.1 Physical and chemical differences between farmed and wild cod

The production of farmed cod has increased rapidly in the last few years, both due to an increased market demand of cod products as well as to a simultaneous decrease of available wild cod in the oceans. Recent studies have shown that farmed cod can not be used in the production of all kinds of fish products, in which wild cod is used today. To meet this growing demand it is therefore important to find the variances between the physical and chemical properties of farmed and wild cod and how these parameters can be affected by various processing methods, such as salting, cooling, drying etc.

Low field NMR has become a valuable tool in the research of muscle structure of meat and fish and has given valuable information about the water behaviour in such biological systems. The aim of this paper was to study the differences in farmed and wild cod muscle as indicated by low field relaxation measurements and how these results can be related to more traditional measurements of physical, chemical and sensory analysis and how different processing, such as filleting pre or post rigor mortis, salting and superchilling affected these parameters.

2 MATERIALS AND METHODS

2.1 Experimental design

Wild Atlantic cod (*Gadus morhua*), caught with a shrimp warp during November 3rd-6th 2006 and farmed Atlantic cod, caught as spawn in September to October 2003 and fed on dry feed until slaughtering on November 6th 2006 were used in the experiment. The mean weight of the wild cod was 2.78 kg at time of catch and the mean weight of the farmed cod was 2.84 kg at the time of slaughter. The fish was processed and filleted at two occasions, i.e. firstly pre rigor mortis on the day of slaughter or catch and secondly three days after slaughter to show the effect of processing and filleting after rigor mortis. All fillets were pre-chilled before packaging in a superchilled 1.5% salt brine and then packed in 10 kg

Styrofoam boxes. The boxes were then placed in a chilled storage with a temperature of 2°C or a superchilled storage with an average temperature of -2°C. A part of the samples stored in the superchilled storage were also injected with a 1.5% salt brine before packed and placed in the superchilled storage. Sampling for all measurements took place on day 1, 3, 6, 9, 13 and 15 of storage.

2.2 Yield measurements

The yield was assessed through the brining of the samples in a pre-cooling slurry ice before packaging and through brine injection by comparing the sample weight before and after each processing step. Drip was evaluated through the storage by measuring the sample weight before packaging and on each sampling day.

2.3 Chemical measurements

The water content of each fillet was measured by drying 5 grams of minced muscle in a ceramic bowl mixed with sand for 4 hours at 103 ± 2 °C. The water content was based on the weight differences before and after the drying of three replicates for each sample.[1] Salt content was measured with the Volhard Titrino method according to AOAC ed. 17.[2] The water holding capacity (WHC) was found with the centrifugal method described by Eide et al. (1982).[3] Total Volatile bases (TVB-N) and trimethylamine (TMA) was measured by steam distillation (Struer TVN distillatory, STRUERS, Copenhagen, Denmark) and titration after extracting the fish muscle with 7.5% aqueous trichloracetic acid solution.[4] TMA was measured in trichloroacetic acid (TCA) extract by adding 20 ml of 35% formaldehyde. TVB-N and TMA measurements were done in duplicate. pH measurements were performed with a pH electrode (SE 104 Mettler Toledo GmbH, Greifensee, die Schweiz) connected to a Knick pH meter (Portames 913 pH, Knick, Berlin, Germany). The electrode was immersed in the minced samples at 20 ± 2 °C.

2.4 Microbial analysis

Total viable psychrotrophic counts (TVC) and counts of H_2S-producing bacteria were evaluated on iron agar (IA) as described by Gram et al. (1987) with the exception that 1% NaCl was used instead of 0.5% with no overlay.[5] Surface-placing was used for all counts. Plates were incubated at 15°C for 4-5 days. Bacteria forming black colonies on IA produce H_2S from sodium thiosulphate and/or cysteine. In all experiments cooled Maximum Recovery Diluent (MRD, Oxoid) was used for dilutions. All samples were analyzed in duplicate and results presented as an average.

2.5 Sensory analysis

Quantitative Descriptive Analysis (QDA), introduced by Stone and Sidel (1985), was used to assess the nine different groups.[6] Eleven panellists all trained according to international standards (ISO 1993); including detection and recognition of tastes and odours, trained in the scales and in the development and use of descriptors, participated in the evaluation. The sensory attributes were 30, including various odour, appearance, flavour and texture attributes. Samples weighing approximately 40g were cut from the loins and placed in aluminium boxes coded with three-digit random numbers. The samples were cooked for 6 minutes in a pre-warmed oven (Convotherm Elektrogeräte GmbH, Eglfing, Germany) at 95-100°C with air circulation and steam before served to the panel. Each panellist evaluated duplicates of each sample in a random order in each session. A computerized system (FIZZ, Version 2.0, 1994-2000, Biosystemes) was used for data recording. The QDA data was corrected for level effects by the method of Thybo and Martens (2000) and analysis of variance (ANOVA) was performed on this data in the statistical program NCSS 2000 (NCSS, Utah, USA).[7] The program calculates multiple comparisons using Duncan's multiple comparison test. The significance level was set at 5%, if not stated elsewhere.

2.6 Microscopic measurements

Samples were collected from anaesthetized wild and farmed fish before slaughtering as well as after slaughter, both before and after rigor mortis to see the differences in the amount of extracellular water between the experimental groups. The samples were placed in oval plastic glasses along with Tissue-Tec freezing glue (Tissue-Tek, OCT, Sakura, USA) where they were frozen in liquid nitrogen for 50 seconds. The samples were then stored at -83°C until they were cryosectioned. 10 μm thick slices were cut from the samples at -27°C with a Leica CM 1800 cryosectioner after 20 minutes of thawing. The samples were stored on sampling glasses (SuperFrost/Plus, 25 x 75 x 1,0 mm from Menzel-Gläser, Germany) at -83 °C until died with Orange G and Methyl Blue. The samples were covered with a covering glass with MOUNTEX (Histolab, Sweden) after drying at 20°C and then analysed in the microscope and photographed with a Leica DC 300F digital camera. The mean extracellular space was calculated as a ratio of the total area of the sample.

2.7 Low field NMR measurements

A 20 MHz Bruker Minispec benchtop instrument (Bruker Optics GmbH, Am Silberstreifen D-76287 Rheinstetten, Germany) was used for the NMR measurements in the study. Samples were minced and placed into 10 mm sample vials. Four replicates were made from each sample group and all measurements were perfomed at 2 °C. Longitudinal relaxation times, T_1, were measured with an inversion recovery (IR) pulse sequence and fitted with a mono-exponential curve. Transverse relaxation times, T_2, were measured with a Carr-Purcell-Meiboom-Gill (CPMG) pulse sequence with an interpulse spacing τ of 500 μs and fitted with a bi-exponential curve.

2.8 Data handling

Multivariate analysis were made on all data in Latentix 1.0 (Mathworks Inc, Natick, MA, USA). Principal Component Analsysis (PCA) was made for unsalted samples separately as well as for all sample groups. The results were plotted in a bi-plot showing the distribution and correlation of samples as well as characteristic measurements.

3 RESULTS

3.1 Yield measurements

The yield gained during brining and brine injection of the farmed and wild cod was assessed by comparing the weight of the samples before and after each step. The results can be viewed in Table 1. The post rigor farmed cod showed to much gaping and was of too low quality to be kept in the study. This group was therefore not tested further in this

Table 1: *Yield results of wild and farmed cod through brining and brine injection.*

Description	Brining yield [%]	Brine injection yield [%]
Farmed pre rigor	4,6	7,2
Wild pre rigor	4,9	11,3
Wild post rigor	8,9	16,5

evaluation. According to Table 1 brining of the fish in the pre-cooling slurry ice with a salt content of 1.5% had very little effect on the pre rigor mortis farmed and wild cod, resulting in a low brining yield of 4.6% and 4.9% respectively. The wild post rigor cod gained on the other hand 8.9% during the brining, showing an effective salt uptake. After brine injection a significantly higher yield was gained in the wild cod than in the farmed cod and especially in the wild post rigor mortis cod. This indicates that the salt uptake of the pre rigor muscle in both farmed and wild cod is not as good as in the traditional handling of wild post rigor mortis muscle.

Drip measurements showed that the drip was within 7% for all groups during storage which was found satisfactory. The chilled samples (0-4°C) had the fastest drip loss and especially in the farmed pre rigor cod. By injecting the pre rigor fish with salt the drip loss could be slowed down through the storage, but the lowest drip was found in the unsalted superchilled pre rigor samples, which only had a drip loss of 4%. In the post rigor wild cod the highest drip loss was found in the salt injected samples, while the lowest drip was found in the superchilled samples.

Yield measurements could therefore be summarized by the facts that the best salt uptake was gained by salt injection and that salt addition was most effective in wild post rigor mortis cod muscle. Drip measurements also showed that superchilling had a positive effect on the yield, in correlation to slower spoilage mechanisms than in the chilled samples. Salt injections also showed some positive effect on the drip loss, especially in the pre rigor fish.

3.2 Chemical measurements

Salt measurements on the samples showed that the samples had a salt content between 0.3 and 0.4%. When salt was added by salt injection the salt content of the wild post rigor group increased to 0.6% NaCl, but no significant change was found in the salt content for salt injected farmed or wild pre rigor samples. This indicated that only the wild post rigor muscle had an effective salt uptake in the study.

The water content was significantly higher in the wild cod (81.6 to 83.9%) than in the farmed cod (78.8-80.5%) and higher water content was gained in the salt injected samples than in the unsalted samples for both wild and farmed cod. This can be explained by increased binding of water with added salt and thus higher water binding properties of the muscle. It should also be mentioned that most of the brine (98.5%) injected into the muscle was water. The water content was not significantly different between pre and post rigor samples in the wild fish, nor with different storage temperatures throughout the storage. Changes in the water content during the storage time were then also insignificant.

The water holding capacity (WHC) of the wild cod was significantly higher in the wild cod (87-97 %) than in the farmed cod (72-95%). Measured pH values were significantly higher for the wild cod (pH 6.7-7.1) than the farmed cod (pH 6.0-6.4). pH is generally lower in farmed cod than in wild cod due to a higher glycogen content in the farmed cod muscle. This glycogen can react to form greater amounts of lactic acid, which in turn lowers the pH. A low muscle pH has also been connected to the degredation of collagen in connective tissue, which leads to increased gaping and drip and hence affects the water binding properties of the muscle. No significant difference was found in either WHC nor pH between pre and post rigor muscle, salt injection or through the storage time.

The chemical spoilage indicators of total volatile base nitrogen (TVB-N) and trimethylamine (TMA) first became evident in the chilled samples on storage days 9-13 and not until on day 15 for superchilled pre rigor muscle (both salted and non salted). It could therefore be seen that superchilling had a positive effect on the slowing down the spoilage of the fish.

3.3 Microbial analysis

It is worth mentioning when interpreting the bacterial results that three days were between the processing and filleting of the pre rigor fish and the post rigor mortis fish. The post rigor mortis fish was well chilled on flake ice while it went through rigor mortis.

On the first storage day the total bacterial count (TVC) was slightly higher in the farmed pre rigor cod than in the wild pre rigor cod. No difference was found between these groups on other days through the storage at chilling temperatures (0-4°C). In the superchilled samples little or no difference was found between the sample groups until on storage day 15. The salt injected samples did though generally show slightly higher bacterial counts than the unsalted groups, especially in the post rigor wild fish. Lowering the temperature had a positive effect on the bacterial counts, but after 13 days of storage the TVC of fillets stored at superchilling was in the range of log 6-7.5/g and of fillets stored at chilled conditions (0-4°C) in the range of log 8-9/g.

On the first storage day no H_2S-producing bacteria could be detected. It is though worth mentioning that the detection threshold was log 1.3/g with the dilutions used in the study. Little or no difference was found in the groups under superchilled conditions, but as with the total bacterial count slightly higher values were obtained for the salt injected samples than the unsalted ones. On the first storage day counts of H_2S-producing bacteria were in the range of log 2-2.5/g in the superchilled samples, which was slightly higher than the counts in the chilled samples. This can be explained by the fact that there are more competition between different bacterial species at chilling conditions than at superchilling. It is possible that *Shewanella putrefaciens*, which is a very cold temperature resistant and is H_2S-producing, has got a head start on other bacterial species on the first storage day and is therefore dominant in the superchilling samples. The advantages of superchilling on the growth rate of H_2S-producing bacteria was also evident through the storage time.

3.4 Sensory analysis

Principal Component Analysis (PCA) of the Quantitative Descriptive Analysis (QDA) sensory data showed that 88% of the variation in the data could be explained with the first two principal components (Figure 1). The first principal component (65%) mostly described differences in texture, but also partly differences in freshness attributes in taste and smell. Samples grouped into two distinguished groups, where wild cod samples (right grouping on Figure 1) were connected to higher scores in texture attributes such as softness, juiciness and a more mushy texture, while the farmed cod (left grouping on

Figure 1) showed more meaty, rubbery and clammy texture as well as more meat smell and taste. Freshness attributes, such as sweet and metallic smell and taste, were also more characteristic for the farmed cod early during the storage period, while the wild cod showed a darker appearance. The second component (23%) describes freshness and spoilage characteristics of the samples through the storage. At the beginning of storage the samples were described with characteristic freshness smell and taste attributes such as sweet smell and taste, shellfish smell and metallic taste. At the end of the storage period samples were described with attributes such as trimethylamine (TMA) smell and taste, sour taste and smell, sulphur smell and bitter taste. The analysis showed that the wild cod showed spoilage characteristics faster than the farmed cod. The superchilling slowed down the spoilage mechanism compared to the chilled storage but the salt injection had small affect on the storage life.

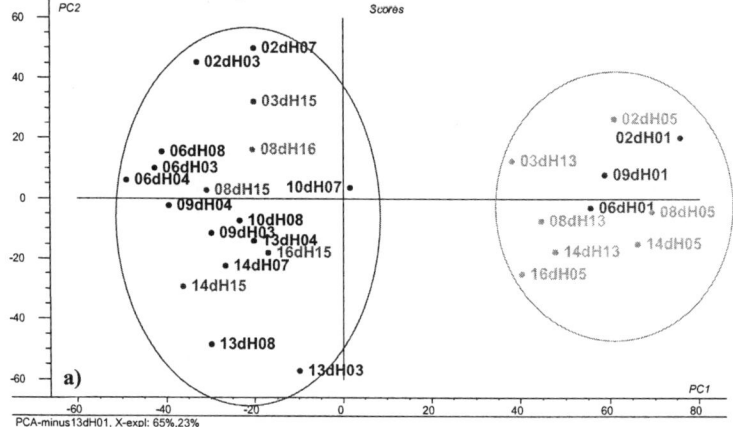

Figure 1: Principal *Component Analysis from Quantitative Descriptive Analysis of sensory data. Description of samples: 02dH03: 02d = day 2 of storage, H03 = group 3. (H01 = farmed pre rigor, H03 = wild pre rigor, H04 wild post rigor, H05 = farmed pre rigor superchilled, H07 = wild pre rigor superchilled, H08 = wild post rigor superchilled, H13 = farmed pre rigor salted, H14 = wild pre rigor salted, H15 = wild post rigor salted).*

3.5 Microscopic Imaging

Microscopic imaging was performed on samples from both alive wild and farmed cod, which had been anaesthetized with benzocain, as well as pre and post rigor mortis wild and farmed cod. Sample pictures from these measurements can be seen in Figure 2 and 3. According to Figure 2 the amount of extracellular water was higher in the farmed cod (7.56 ± 3.8%) than in the wild cod (4.86 ± 2.9%) before slaughtering. This relationship changed post mortem, but more extracellular water could be found in the chilled post rigor wild cod (16.6 ± 3.2%) than in the pre rigor wild cod (10.6 ± 2.2%) and the farmed cod (10.3 ± 2.3%) (Figure 3). No significant difference was between the wild and farmed cod in pre rigor samples. Similar results were found in the superchilled samples. This increase in extracellular water post mortem in the wild cod can be associated with the sensory results of a juicier and softer texture, while the appearance of the farmed cod is dryer and more meaty.

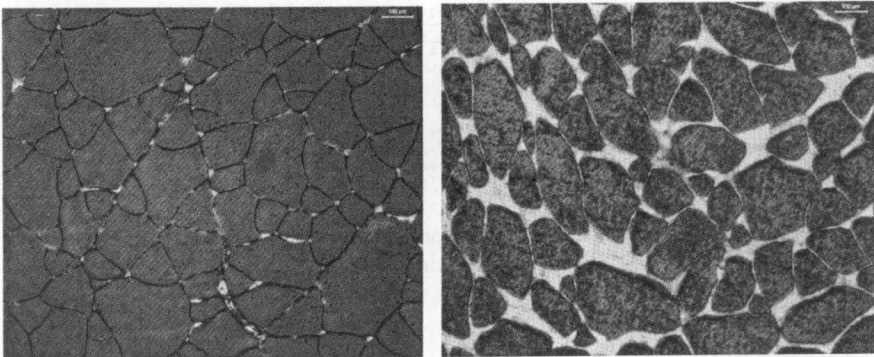

Figure 2: *Microscopic imaging of alive, anaesthetized wild cod (left) and farmed cod (right) before slaughter.*

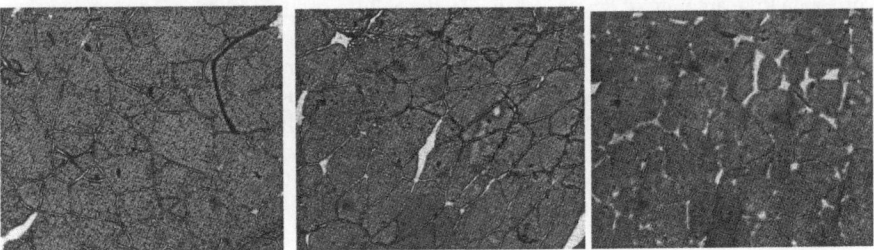

Figure 3. *Microscopic images of chilled cod muscle after slaughter/catch of farmed pre rigor mortis muscle (left), wild pre rigor mortis muscle (middle) and wild post rigor mortis muscle (right).*

3.6 Low field Nuclear Magnetic Resonance

The longitudinal and transverse relaxation times of the samples were measured with a 20 MHz Bruker Minispec Benchtop Instrument on all sampling days. The longitudinal relaxation times fluctuated sligthly during the storage time but no decreasing or increasing trends were observed. These fluctuations in the relaxation time measured are corresponding to the fluctuations found in the water content and water holding capacity through the storage. In the transverse relaxation times, T_{21} and T_{22} an increasing trend was on the other hand observed with storage time, after an initial decrease from day 1 to day 3. The results obtained on storage day 6 can be seen in Table 2.

According to Table 2 it could be seen that the storage temperature (chilled or superchilled) did not have a major effect on the relaxation times measured. Longer relaxation times were on the other hand obtained in the wild cod samples than in the farmed cod samples, especially in the post rigor samples regardless of processing method in correlation to the water content of the samples. Salting also increased the relaxation times in the wild cod but had no or little effect on the farmed cod. This is in correlation to the less salt uptake observed in the farmed cod.

Table 2: *Low field NMR relaxation time results from storage day 6.*

Description	T1 [ms]	dT1 [ms]	T21 [ms]	dT21 [ms]	T22 [ms]	dT22 [ms]	App. pop A1 [%]	App. pop A2 [%]
Farmed pre rigor chilled	458,3	3,3	18,3	2,7	63,5	0,9	4,9	95,1
Wild pre rigor chilled	499,5	10,1	14,4	1,5	65,6	0,5	3,0	97,0
Wild post rigor chilled	530,8	4,6	15,6	3,4	68,8	0,5	2,6	97,4
Farmed pre rigor superchilled	444,0	11,6	15,0	1,9	62,2	0,4	4,4	95,6
Wild pre rigor superchilled	513,3	8,3	15,7	2,4	66,5	0,7	2,9	97,1
Wild post rigor superchilled	546,3	6,1	16,9	3,3	73,6	1,2	2,9	97,1
Farmed pre rigor salted	454,0	4,1	12,5	2,0	65,1	0,3	3,0	97,0
Wild pre rigor salted	582,3	8,2	19,3	2,4	80,5	1,3	3,7	96,3
Wild post rigor salted	619,0	5,8	17,6	3,0	81,4	0,5	2,9	97,1

A multiple component analysis was made of all data by performing a principal component analysis in Latentix 1.0. The resulting bi-plots of sample groupings and characteristic measurements can be seen in Figure 4 and 5.

According to Figure 4 the first principal component (PC1) described 47.8% of the variation between the samples. The difference between farmed (left) and wild cod (middle (pre rigor) and right (post rigor)) was obvious from the analysis. The PCA also showed that the farmed cod showed higher values for the shorter transverse relaxation time, T_{21} and corresponding amount of water A_1, while the wild cod showed higher values for the longitudinal relaxation time, T_1, the longer transverse relaxation time, T_{22} and relative amount of extracellular water A_2 from the NMR measurements. This was in correlation to the amount of extracellular water as measured by microscopic imaging. The wild cod also had higher water content, water holding capacity and pH which the T_1 and T_{22} relaxation times correlated to according to the PCA. The second principal component (PC2), which described 28.6% of the sample variation, describe the effect of the storage time and spoilage on the samples, thus showing high counts of bacteria (TVC and H_2S producing) as well as high values for the chemical spoilage indicators of TVB-N and TMA. Since an increasing trend in the transverse relaxation times were found with storage time, this can also be seen in the figure.

When adding the salt injected samples the figure became slightly more complicated and the groupings were no longer as evident (Figure 5). The first principal component (PC1) described 37.0% of the sample variation and still a clear difference could be found between the farmed cod (left) and wild cod (right). By adding salt to the samples a right-shift could be seen in the PCA towards higher values of water and salt content, water holding capacity, higher injection yield and more extracellular water as well as higher values for T_1 and T_{22}.

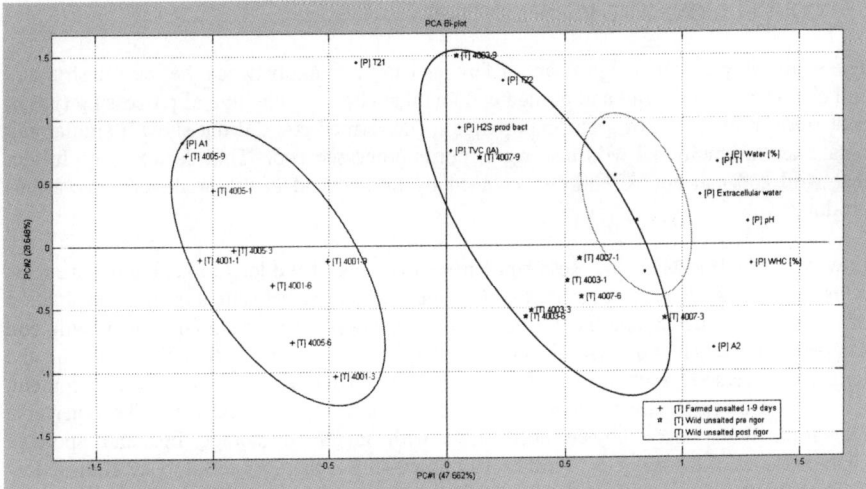

Figure 4: *Principal Component Analysis (PCA) bi-plot results for unsalted samples, showing groupings of samples and characteristic measurements. The farmed pre rigor cod samples grouped on the left side of the plot, while wild cod grouped on the right side of the plot (pre rigor – middle, post rigor – right).*

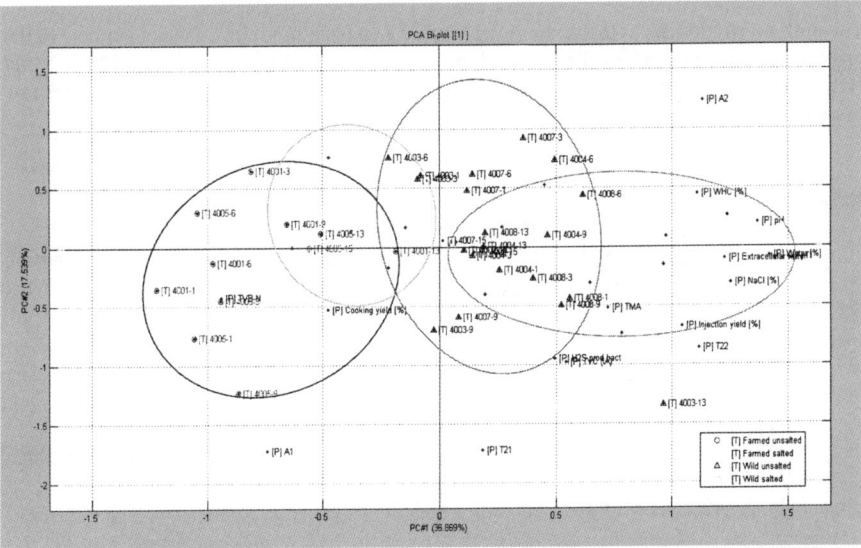

Figure 5: *Principal Component Analysis (PCA) bi-plot results for all samples (both salted and non salted), showing groupings of samples and characteristic measurements. The farmed pre rigor cod samples grouped on the left side of the plot, while wild cod grouped on the right side of the plot. Salt injection caused a right-shift of the groupings.*

5   CONCLUSIONS AND PERSPECTIVES

The study showed that a significant difference can be found between the muscle structure and characteristics of wild and farmed cod and that choosing the time of processing (pre or post rigor mortis) was of great importance. The farmed cod did not show a similar salt uptake as the traditional wild post rigor mortis processed cod. This showes even further that wild cod can not be entirely replaced by farmed cod in the processing of various products.

Low field Nuclear Magnetic Resonance measurements of the longitudinal and transverse relaxation times of the samples showed that the method is a valuable tool in the evaluation of cod muscle structure and that a clear distinction was made between farmed and wild cod samples. The evaluation also showed that the longitudinal relaxation time, T_1, and the longer transverse relaxation time, T_{22}, could be correlated to high water and salt content, the amount of extracellular water, high water holding capacity and pH. The transverse relaxation times also showed correlation with increased storage life and spoilage characteristics of the samples. Further data processing is though needed to see if these low field NMR measurements can be used to predict these chemical and physical parameters with good approximations. The results though indicate that the technique can be used for traceability, with the purpose of destinguishing between wild and farmed cod.

Acknowledgements

The authors would like to thank the AVS Marine Research fund for financial support to the project "Processing and quality control of farmed cod" (Project number R026-06) and Hraðfrystihúsið Gunnvör (HG) cod farming company for provision of facilities and raw material.

References

1   ISO The Intl. Organization for Standardization. Animal feeding stuffs - Determination of moisture and other volatile matter content. 6496. Genf, Switzerland, 1999.
2   AOAC Assn. Official Analytical Chemists. Official methods of analysis. 17th ed. Washington, D.C. (no. 976.18), 2000.
3   Eide O, Børresen T, Ström T. *J Food Sci., 1982, 47*:347-354.
4   Malle P, Tao SH. *J. Food Prot.* 1987, **50**: 756-760.
5   Gram L, Trolle G, Huss HH. *Int. J. Food Microbiol.* 1987, **4**: 65-72.
6   Stone H, Sidel JL. *Sensory evaluation practices.* Orlando, Fla.: Academic press, Inc., 1985, 311p.
7   Thybo A, Martens M. *Food Qual and Pref., 2000,* **11**(4):283-288.

A LOOK AT NMR RELAXOMETRY APPLICATIONS IN MEAT SCIENCE – RECENT ADVANCES IN COUPLING NMR RELAXOMETRY WITH SPECTROSCOPIC, THERMODYNAMIC, MICROSCOPIC AND SENSORY MEASUREMENTS

H.C. Bertram[1], R.L. Meyer[2] & H.J. Andersen[2,3]

[1]University of Aarhus, Faculty of Agric. Sciences, Dept. Food Science, Research Centre Aarslev, Kirstinebjergvej 10, DK-5792 Aarslev, Denmark. E-mail: HanneC.Bertram@agrsci.dk
[2]Interdisciplinary Nanoscience Center (iNANO), University of Aarhus, Faculty of Natural Science, Building 1521, Ny Munkegade, DK-8000 Århus C, Denmark
[3]Arla Foods amba, Skanderborgvej 277, DK-8260, Viby J, Denmark

1 INTRODUCTION

1.1 NMR in meat science

NMR has become a powerful tool in meat science. Especially the use of proton NMR relaxation measurement has experienced considerable success in meat science because of its potential for characterising water mobility and compartmentalisation. Recently, investigations combining proton NMR relaxation with other techniques have explored additional unique relationships between water characteristics and specific biophysical and structural features of the meat. This chapter aims at unfolding present status of proton NMR relaxation applications in meat science with focus on latest studies combining proton NMR relaxation with spectroscopic, microscopic and sensory measurements.

1.2 Water-holding capacity

Water-holding capacity (WHC) is one of the most important quality traits within the meat industry, as it has both vast economical consequences during production of meats and is critical for how consumers perceive the quality of the meats. WHC is a measure of the ability of fresh meat to retain inherent water. Traditionally WHC was determined by gravimetric methods.[1] However, within the past two decades low-field proton NMR relaxometry has been established as a superior tool in the determination of WHC of meat. Renou et al.[2] were the first to show that proton NMR relaxation times measured in pork correlate with the so-called imbibition time, which is an indirect measure of WHC in fresh meat. Subsequently, Tornberg et al.[3] reported that proton T_2 relaxation differed significantly between specific pork qualities spanning from low WHC to high WHC. More recently, it has been demonstrated that proton NMR T_2 relaxation directly correlates with WHC as determined by gravimetric methods over a wide range of WHC in pork.[4-7] The expansion of this area has been given in several extensive reviews.[8-10]

1.3 Interpretation of the relationship between NMR relaxation and WHC

The biophysical rationale for a correlation between proton NMR T_2 relaxation times and WHC is owed to the fact that NMR relaxation contains information on the interaction between water protons and the surrounding environment including distances to potential relaxation sinks, which makes it possible to translate relaxation to water compartmentalisation and distribution in the meat. Distributed exponential fitting analysis[11] on NMR T_2 relaxation decays acquired on meat shows the presence of three water populations in meat (Figure 1). In order to obtain evidence how to interpret these populations, studies including manipulation of the micro- and macrostructure of meat[12], exchange of muscle water (myowater) with deuterium oxide[10] and use of Gd-DTPA as an extra-cellular marker[13] have been carried out. Interpretation of these investigations suggests that the three populations are attributed to water closely associated with macromolecules (T_{2B}), myofibrillar water (T_{21}) and extra-myofibrillar water (T_{22}). However, the assignment of the fast-relaxing T_{2B} population is still uncertain, and a recent study has suggested that it reflects macromolecular protons located in water-plasticized structures.[14] In contrast, the assignment of the slow-relaxing T_{22} population to extra-myofibrillar water is supported by the fact that a strong correlation between the size of T_{22} population and potential drip loss, as well as a direct correlation between and amount of water loss created by centrifugation and the amount of this T_{22} water population, have been demonstrated.[7] Accordingly, proton NMR relaxation is a powerful tool in the determination of WHC, as it gives a direct measure of the proportion of myowater in the meat that is susceptible to be lost as drip.

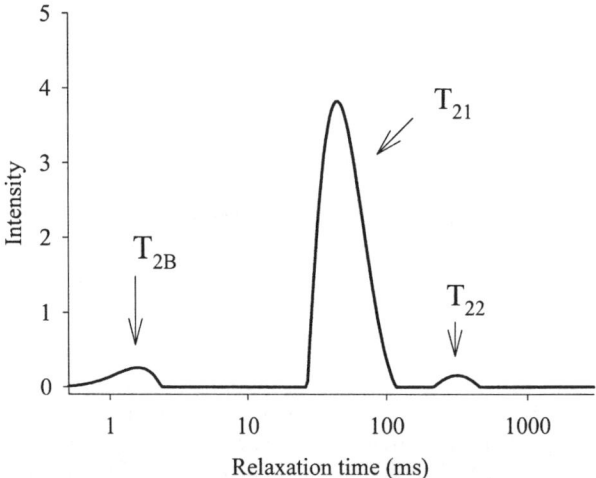

Figure 1. *Distributed proton NMR T_2 relaxation times measured in a pork sample at 25 ℃. Three populations at approx. 1-3 ms, 40-80 ms and 200-400 ms assigned T_{2B}, T_{21}, and T_{22}, respectively, are detected.*

2 NMR COUPLED WITH DIFFERENTIAL SCANNING CALORIMETRY

Differential scanning calometriy (DSC) can be used to explore protein denaturation as transitions in the obtained thermograms directly reflect temperature-induced changes in protein structures.[15] DSC studies on meat have demonstrated three denaturation steps during cooking. These denaturation steps heating that have been attributed to mainly myosin denaturation (~40-60 °C), denaturation of sarcoplasmic proteins and collagen denaturation (~60-70 °C temp) and primary actin denaturation (~80 °C).[16-17] As proton NMR relaxometry monitors the interactions between water protons and exchangeable protons in proteins, an association between proton relaxation as function of heating temperature and the heat-induced denaturation of meat proteins could be expected. Consequently, the relationship between changes in water characteristics and protein denaturation during cooking of meat was explored by low-field proton NMR T_2 relaxation measurements combined with parallel DSC measurements on pork samples heated to various temperatures in the region between 25 and 75 °C.[18] Principal component analysis (PCA) was carried out on the distributed proton NMR T_2 relaxation data, which revealed that the major changes in myowater characteristics during heating occurred between 40 and 50 °C. This major change is probably initiated by denaturation of myosin heads, which however, could not be confirmed by the performed DSC measuerments. In contrast, the DSC thermograms revealed exothermic transitions at 54, 65 and 77 °C, which reflect denaturation of myosin (rods and light chain), sarcoplasmic proteins together with collagen and actin, respectively. Simultaneous modelling of DSC and NMR data by partial least squares regression (PLSR) showed a correlation between denaturation of myosin rods and light chains at ~53-58 °C and heat-induced changes in myofibrillar water characteristics (T_2 relaxation time ~ 10-60 ms) as well as between actin denaturation at ~80-82 °C and expulsion of myowater from the meat. In conclusion, the study displayed a direct relationship between thermal denaturation of specific proteins/protein structures and heat-induced changes in myowater mobility during heating of pork.

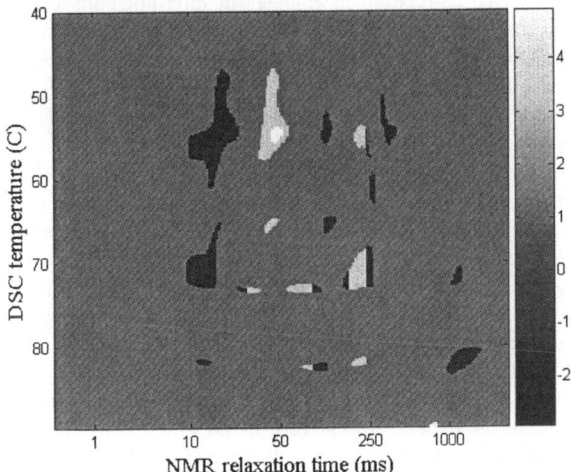

Figure 2. *Correlation between DSC amplitude and distributed NMR T_2 relaxation times measured during step-wise heating of pork from 25 °C to 75°C. Reported values are Jack-knife based significance levels presented as number of standard deviations away from zero (presenting non-significance).*

Concomitant use of NMR and DSC have also been employed in a study with focus on water activity in meat.[14] Water dynamics in freeze-dried, rehydrated chicken was elucidated, which revealed the presence of mobile water at a water content of 30-40% using DSC, while mobile water was detectable using NMR relaxometry at a water content of even 17%. These findings were suggested to be a result of different effects of present solutes for the two analytical tools.[14] Accordingly, the study demonstrated that NMR provides a supplementary measure of water activity. Prospected studies relating water activity probed by NMR to microbial activity should establish the potential of using NMR for probing shelf-life of meat.

3 NMR COUPLED WITH FT-IR SPECTROSCOPY

Infrared (IR) spectroscopy measures fundamental molecular vibrations, which facilitate studies of the morphology and structure of proteins at the molecular level.[19] Accordingly, the combination of FT-IR spectroscopy and low-field ^1H NMR T_2 relaxometry is appealing, as it facilitates a simultaneous characterization of protein structures and myowater properties. Thus, FT-IR spectroscopy and low-field ^1H NMR T_2 relaxometry has recently been combined to characterize changes in myofibrillar protein structures and the chemical-physical properties of myowater upon heating of meat.[20] Pronounced changes in both T_2 relaxation data and FT-IR spectroscopic data were observed during cooking, revealing severe concomitant changes in the water properties and structural organization of proteins during heating of meat. NMR and FT-IR spectra were compared by partial least square regression (PLS). This showed a correlation between the FT-IR peaks reflecting β-sheet and α-helix structures and the NMR T_{2B} and T_{21} relaxation populations and also expelled 'bulk' water (T_2 relaxation times >1000 ms). More specifically it was seen that the T_{2B} population had a positive correlation to changes in intramolecular antiparallel β-sheets and α-helical structures. Myofibrillar-entrapped water as depicted in the T_{21} population had a positive correlation to changes in intermolecular extended β-structures (1695 cm^{-1}, 1628 cm^{-1}, 1619 cm^{-1}). Thus, water trapped within the highly dense and structured protein network was positively correlated to intramolecular antiparallel β-sheets and α-helical structures, and it was demonstrated that definite structural changes in proteins during cooking of meat are associated with simultaneous alterations in the chemical-physical properties of the water within the meat.[20]

Having established that usefulness of combining FT-IR spectroscopy and low-field ^1H NMR T_2 relaxometry is the study of affiliated heat-induced changes in protein structures and myowater inspired to a similar approach in the elucidation of the effect of heating rate on water characteristics and protein structural changes in meat.[21] Thus, the effect of a slow and a fast heating rate to an internal centre temperature of 65 °C was investigated on three pork qualities (low, intermediate and high WHC). NMR data acquired on meat samples upon the heating procedure revealed that fast heating broadened the T_{21} distribution and decreased the relaxation times of the T_{21} population for all three meat qualities. FT-IR spectroscopy showed that fast heating caused a higher gain of random structures and aggregated β-sheets at the expense of native α-helixes, and that these changes facilitated the fast heating-induced broadening of T_{21} distribution and reduction in T_{21} relaxation times. In summary, this study clearly demonstrated that the changes in T_2 relaxation times of myowater protons as function of heating rate and raw meat quality are well related to simultaneous changes in the protein secondary structure as probed by FT-IR spectroscopy.[21]

Curing is one of the most commonly used preservation procedures in meat industry. Recently, a study investigating the multi-factorial effect of meat ageing, meat curing and cooking on water characteristics and concomitant changes in protein structures was performed using proton NMR relaxometry coupled with FT-IR spectroscopy.[22] Distributed NMR T_2 relaxation times revealed significant effects of heating and salting and aging on water distribution and associated protein structural changes in meat. Heating caused a decrease in the T_{21} relaxation times with a concomitant broadening of T_2 distribution, while the salting increased the T_{21} relaxation times because of salt-induced swelling of meat myofibrils. The width of major T_2 population for cooked meat decreased with increasing salt concentration up to 6%, while further increased salt concentration induced an opposite effect on the width of T_{21} population, indicating that myowater is more heterogeneously distributed in cooked and highly salted meat. FT-IR absorption bands intensities showed that both heating and salting caused a pronounced decrease in the native α-helical structure of myofibrillar protein. However, the mechanism of heat- and salt-induced denaturation of α-helical structure seems to differ, as these changes in α-helical structure gave rise to opposite roles in relation to the ability to retain water in the meat product.[22]

5 NMR COUPLED WITH ANALYSIS OF SPECIFIC PROTEIN DEGRADATION

Another way to examine protein features in meat is to follow specific protein degradation using western blot analyses. It has been hypothesized that the degradation of cytoskeleton proteins plays a key role for WHC of pork, as such a degradation allows the myofibrils to swell and hereby to restrain more myowater.[23] More specifically, it has been proposed that integrin, which attaches the cytoskeleton to the extracellular matrix, should have impact on the formation of drip channels in pork.[24] In order to investigate a potential relationship between integrin content in the pork and water mobility and distribution, low-field proton NMR relaxation measurements on pork were coupled with western blot analyses and confocal laser scanning microscopic investigations of integrin.[25] Even though this study did not reveal a strong link between integrin degradation and water distribution, the study indicated that integrin degradation had impact on the succeeding development in water mobility and distribution during ageing.

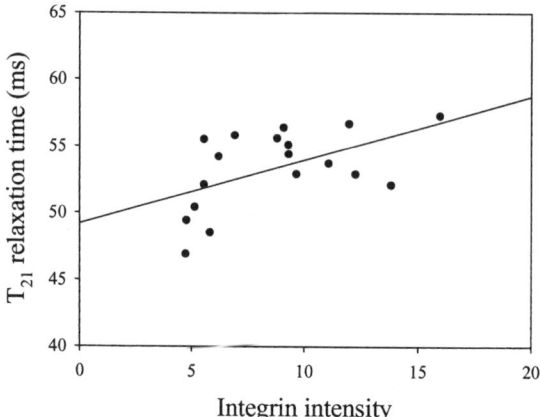

Figure 3.   *Relationship between integrin intensity determined in muscle samples 24 h post mortem and the relaxation time of T_{21} population measured in muscle samples after 7 days of ageing.*

6   NMR COUPLED WITH MICROSCOPIC MEASUREMNTS FROM MICRO TO NANO LEVEL

Microscopy is a unique technique to image the microstructure of biological specimens. Current advances in microscopy such as laser scanning confocal microscopy and atomic force microscopy, although once considered complicated methods, are now frequently utilized in laboratories around the world for analysis of cellular structure and molecular dynamics. Microstructure of muscle foods is decisive for their technological and textural attributes, and an understanding of the relationship between microstructure and the characteristics of water and/or added water in meat products during cooling, storage, and/or processing is essential for optimising the WHC and water-binding capacity of meats. A limited number of studies have combined the use of different microscopic techniques with concomitant NMR measurements to elucidate the relationship between structural features and water characteristics in meats.[3,26-27] These studies have found obvious relationships between water characteristics and structural changes in the muscle protein matrix in meat and meat products, however, limitations in the analysis of microscopic images have not yet made it possible to give a quantitative understanding of the structure of meats during cooling, processing and/or storage and the water characteristics at the given point of measure. Recently, proton NMR relaxation measurements have for the first time been combined with confocal laser scanning microscopy (CLSM) and atomic force microscopy (AFM) in an investigation of the influence of different marinating agents on water-binding capacity in enhanced pork samples.[28] The study showed that inclusion of sodium chloride, phosphate or bicarbonate in the marinade gave rise to distinct technological properties of the meat as a result of effects of the marinating agents on water mobility and distribution measured by low-field proton NMR relaxometry. The CLSM- and AFM-based examinations revealed that the marinating-induced differences in water mobility and distribution could be ascribed to effects on the structure both at the micro and nano level (Figure 4). Accordingly, a link between micro-/nanostructures and water mobility and distribution were demonstrated in relation to technological properties of the different marinated pork samples. Development of suitable techniques to quantitatively analyse structural features of microscopic images will undoubtedly transfigure our understanding of the WHC or water-binding capacity of meats when combined with NMR relaxometry measurements, and therefore this area needs much further attention.

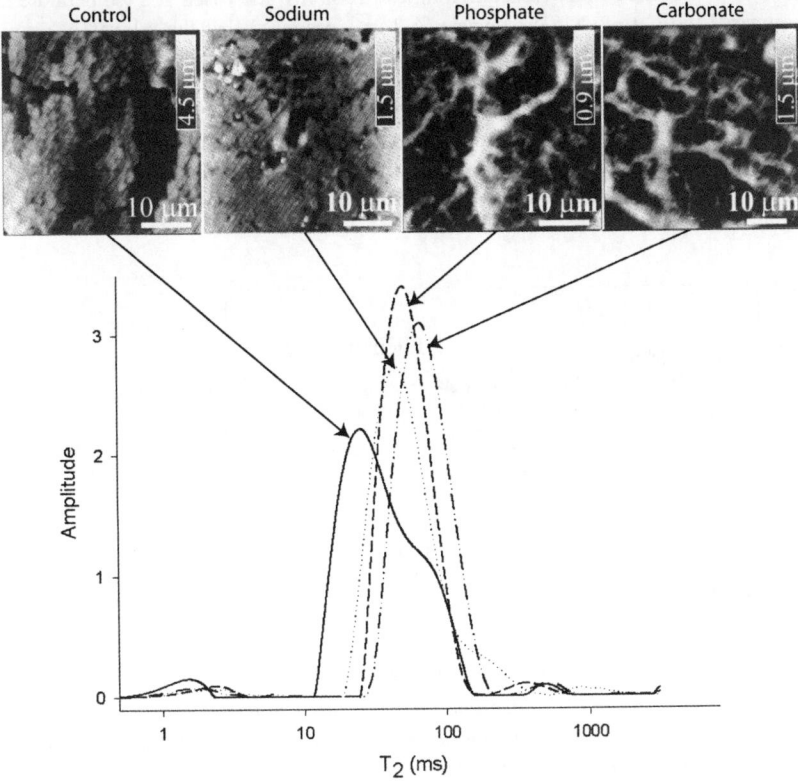

Figure 4. *Atomic force microscopy images and corresponding NMR T_2 distribution of pork treated with different curing agents.*

7   NMR COUPLED WITH SENSORY ANALYSIS

The way meat is perceived at the time of consumption is of utmost importance, as it is decisive for the consumers' view of a given meat quality.[29] Accordingly, there is an interest in being able to predict the sensory properties of meat. Since proton NMR relaxation probes water mobility and distribution, a relationship between proton NMR relaxation data and sensory properties related to water properties, i.e. juiciness can be hypothesised. Accordingly, a study has been conducted where a variation in the juiciness of pork was introduced through heating to different end temperatures, which is known to affect the perceived juiciness of meat.[30-32] The study showed that heating of pork samples to a core temperature of either 62 °C or 75 °C resulted in pronounced effects on the proton NMR relaxation profile, which was correlated with significant effects on juiciness assessed by a sensory panel.[33] The effect of end temperature on the proton NMR relaxation profile was depicted as a lower relaxation time of T_{21} in meat samples heated to 75 °C compared with meat samples heated to 62 °C. In addition, an effect of temperature on the slower-relaxing part of the distributed relaxation profile was also observed, and it was clear that the amount of expelled water (relaxation times >1000 ms) increased at a temperature of 75 °C compared with at 62 °C. Brownstein & Tarr[34] proposed that the relaxation time of a water molecule depends on the probability of the water protons to meet a surface and relax, implying that the relaxation time depends on the size of the pores confining the water protons. In accordance with this idea, the relaxation time of T_{21} has been shown to reflect the structural organization of myofibrils and myofilamentous lattice spacing.[35] Consequently, the study demonstrated that the reduction in juiciness at 75 °C can be ascribed to changes in the size of the pores confining the myofibrillar water together with an expulsion of water.

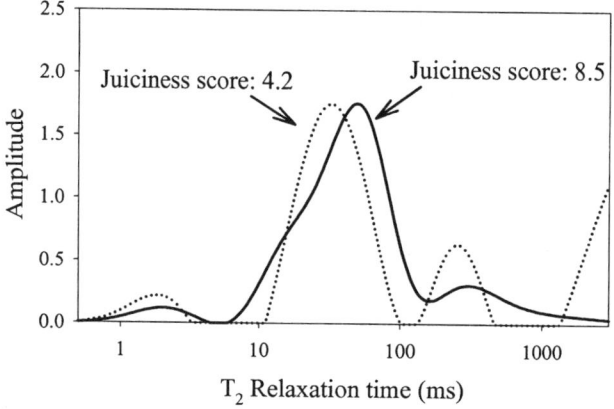

Figure 5. *Distributed T2 relaxation times measured in pork samples heated to 62 °C (full line) and 75 °C (dotted line). Sensory analysis on the same loins revealed that pork heated to 62 °C was significantly juicier than pork heated to 75 °C.*

8   CONCLUSIONS AND PERSPECTIVES

The number of NMR applications in meat science has increased markedly in recent years. These applications have established that NMR can be considered an extremely capable method, which productively has added important basic knowledge within essential areas of importance for the continuous elucidation of critical challenges within meat science. Especially low-field proton NMR relaxometry, which is dealt with here, has proven to be an excellent tool for determination of WHC in fresh meat. In addition, the combination of low-field proton NMR relaxometry with other spectroscopic, biophysical and sensory measurements has shown that low-field proton NMR relaxometry can provide important information with regard to many aspects of meat quality involving water characteristics in meats. This includes processing of meat, where proton NMR relaxometry has demonstrated its capability to register heat-induced and curing-induced changes in water mobility and distribution. Through coupling with complementary measurements of protein characteristics by DSC and FT-IR, it has been possible to demonstrate a link between process-induced changes in protein structure and water properties probed by NMR relaxometry. Likewise, coupling NMR with other techniques in multi-analytical approaches can be expected to prevail in future meat research, and there is no doubt that NMR will be an important technique in eras governed by omics-strategies.

Acknowledgements

The authors wish to thank The Danish Ministry of Food, Agriculture and Fisheries for financial support through the project entitled "Integrated characterization of quality and microbial safety of foods" (3304-FVFP-07-784-01) and the Danish Research Council FTP for financial support through the project entitled "NMR-based metabonomics in tissues and biofluids" (274-05-0339).

References

1. K. O. Honikel, *Meat Sci.,* 1998, **49**, 447.
2. J. P. Renou, G. Monin and P. Sellier, *Meat Sci.*, 1985, **15**, 225.
3. E. Tornberg, A. Andersson, Å. Göransson and G. von Seth, In: Pork Quality: Genetic and Metabolic Factors. Ed. E. Puolanne & D.I. Demeyer (with M. Ruusunen & S. Ellis) (Eds). Wallingford, Oxon: CAB International. (pp. 239).
4. J. Brøndum, L. Munck, P. Henckel, A. Karlsson, E. Tornberg and S.B. Engelsen, *Meat Sci.*, 2000, **55,** 177.
5. R. J. S. Brown, F. Capozzi, C. Cavani, M. A. Cremonini, M. Petracci, G. Placucci, *J. Magn. Resonan.*, 2000, **147**, 89.
6. H. C. Bertram H. J. Andersen and A. H. Karlsson, *Meat Sci.*, 2001, **57**, 125.
7. H. C. Bertram, S. Dønstrup, A. H. Karlsson and H. J. Andersen, *Meat Sci.*, 2002, **60**, 279.
8. H. C. Bertram and H. J. Andersen, *Ann. Rep. NMR Spectr.*, 2004, **53**, 157.
9. H. C. Bertram, H. J. & Andersen, *J. Anim. Breed. Genet.*, 2007, **124**, 35.
10. H. C. Bertram and H.J. Andersen, In: Modern Magnetic Resonance, Ed. G.A. Webb. Springer (pp. 1707).
11. R. S. Menon, M. S. Rusinko and P. S. Allen, *Magn. Reson. Med.*, 1991, **20**, 196.
12. H. C. Bertram, A. H. Karlsson, M. Rasmussen, S. Dønstrup, O. D. Petersen and H. J. Andersen, *J. Agric. Food Chem.*, 2001, **49**, 3092.
13. H. C. Bertram, J. Stagsted, J. F. Young and H. J. Andersen, *J. Agric. Food Chem.*, 2004, **52**, 6320.
14. L. Venturi, P. Rocculi, C. Cavani, G. Placucci, M.D. Rosa and M.A. Cremonini, *J. Agric. Food Chem.*, 2007, **55**, 10572.
15. G. Bruylants, J. Wouters and C. Michaux, *Curr. Med. Chem.*, 2005, **12**, 2011.
16. D. J. Wright, *J. Sci. Food Agric.*, 1978, **29**, 1088.
17. E. Stabursvik and H. Martens, *J. Sci. Food Agric.*, 1980, **31**, 1034.
18. H. C. Bertram, Z. Wu, F. van den Berg and H. J. Andersen, *Meat Sci.*, 2006, **74**, 684.
19. J. Kong and S. Yu. *Acta Biochim. Biophys. Sinica*, 2007, **39**, 549.
20. H. C. Bertram, A. Kohler, U. Böcker, R. Ofstad and H. J. Andersen, *J. Agric. Food Chem.*, 2006, **54**, 1740.
21. Z. Wu, H. C. Bertram, U. Böcker, R. Ofstad and A. Kohler, *J. Agric. Food Chem.*, 2007, **55**, 3390.

Magnetic Resonance in Food Science

22. Z. Wu, H. C. Bertram, A. Kohler, U. Böcker, R. Ofstad and H. J. Andersen, *J. Agric. Food Chem.*, 2006, **54**, 8589.
23. L. Kristensen and P. P. Purslow, *Meat Sci.*, 2001, **58**, 17.
24. M. Lawson, *Meat Sci.*, 2004, **68**, 559.
25. I. K. Straadt, M. Rasmussen, J. F. Young and H.C.Bertram, *Meat Sci.*, 2008, in press. doi:10.1016/j.meatsci.2008.03.012
26. M. Mortensen, H.J. Andersen, S.B. Engelsen and H.C. Bertram, *Meat Sci.*, 2006, **72**, 34.
27. H.C. Bertram, Z. Wu, I.K. Straadt, M. Aagaard and M.D. Aaslyng, *J. Agric. Food Chem.*, 2006, **54**, 9912.
28. H.C. Bertram, Z. Wu, R.L. Meyer, X. Zhou and H.J. Andersen, *J. Agric. Food Chem*, 2008, in press. DOI: 10.1021/jf8007426
29. W. Verbeke, M.J. Van Oeckel, N. Warnants, J. Viaene and C.V. Boucque. *Meat Sci.*, 1999, **53**, 77.
30. J. A. Bowers, J.A. Craig, D. H. Kropf, and T. J. Tucker, *J. Food Sci.*, 1987, **52**, 533.
31. J. K. Joseph, B. Awosanya, A. T. Adeniran and U. M. Otagba, *Food Qual. Preference*, 1997, **8**, 57.
32. C. Bejerholm and M. D. Aaslyng, *Food Qual. Preference*, 2003, **15**, 19.
33. H. C. Bertram, M. D. Aaslyng and H. J. Andersen, *Meat Sci.*, 2005, **70**, 75.
34. K. R. Brownstein and C. E. Tarr, *Phys. Reviews A*, 1979, **19**, 2446.
35. H. C. Bertram, P. P. Purslow and H. J. Andersen, *J. Agric. Food Chem.*, 2002, **50**, 824.

SODIUM MRI AS A TOOL FOR OPTIMIZATION OF SALTING PROCESSES

Emil Veliyulin[1], Ida G. Aursand[1,2], Ulf Erikson[1] and Bruce J. Balcom[3]

[1] SINTEF Fisheries and Aquaculture, N-7465 Trondheim, Norway
[2] Department of Biotechnology, NTNU, N-7491, Trondheim, Norway
[3] MRI Centre, Department of Physics, University of New Brunswick, Fredericton, NB, Canada E3B 5A3

1   INTRODUCTION

Magnetic Resonance Imaging (MRI) is a powerful imaging modality that can produce high quality cross-section images of biological systems and thus can be used for studying the chemical and physical properties, anatomical structure and dynamic processes in foods[1,2]. MRI has significant advantages such as being non-invasive and non-destructive technique. In most cases no special sample preparation is needed prior to analysis. But due to the large footprint of MRI instrumentation, high investment costs and the necessary related infrastructure, MRI cannot presently be considered as a standard analytical tool in food processing industry. Using MRI one can obtain basic insight into a number of issues related to anatomical studies, composition and structure of tissues, fat, water and salt distributions. Moreover, theoretical transport models can in turn be used to interpret the images. For example, for the aquaculture industry, MRI studies may for be used to study the effect of diet and different feeding regimes on fat contents and distribution in fish. In fish processing, MRI can be helpful for optimization of various unit operations such as freezing, thawing, salting, smoking, etc. A detailed review of various MRI applications in food science is written by Hills[3].

Salting of muscle foods is an important food preservation technique, which affects its sensory properties, water holding capacity, microbiological stability, color and other properties. Industrial salting procedures is still often associated with experience and tradition with minimal use of scientific approaches to understand details of the process and improve the final product. Conventional chemical methods are typically used to study fish and meat salting[4-6], but an increasing interest for non-invasive and non-destructive methods such as nuclear magnetic resonance (NMR) is obvious. Achieving high sensory quality, reduced salt content, homogeneous salt distribution, long shelf life, high retail price and low production costs are some of the most important goals in the salted food production.

Performing MRI on salted muscle food can provide with a better understanding of the transformations during the freezing, thawing, salting, curing, drying and rehydration stages. Spin-echo MRI modality always produces images that are T_2 weighted to a certain extent. A known effect of partial MRI salt "invisibility"[7,8] often makes quantification of the salt content impossible. Single point imaging[9] (SPI) and SPRITE[10] are pure phase encoding imaging methods which use the free induction decay (FID) for NMR signal detection, hence minimizing loss of the observed sodium NMR signal due to the relaxation. Still many commercial MRI systems use spin-echo based ^{23}Na imaging sequences and understanding of the partial sodium "invisibility" is still relevant.

2   DYNAMIC SALT MRI STUDIES

One-dimensional salt diffusion into cod fillet pieces was studied using a specially designed diffusion cell (Figure 1) allowing exposure of the sample to brine from the open side only. A closed chamber with reference brine (20%) was placed in the bottom of the cell. ^{23}Na NMR images were taken on a 2.34 T Bruker Avance DBX100 instrument (Bruker Optics GmbH, Germany) using an in-house made 60 mm ^{23}Na RF volume probe and using spin-echo protocol with the following parameters: echo time (TE) = 3.0 ms, relaxation delay (RD) = 500 ms, number of scans (NS) = 128, field of view (FOV) = 8 cm, image matrix size (MS) = 64 x 64 pixels giving a pixel resolution of 1.25 mm. Figure 1 shows images of the system taken at 4 different salting times, clearly demonstrating the progress of the salt penetration into muscle. The figure also shows one dimensional profiles extracted from the MRI images that can further be used for mathematical modeling of the salting process under given conditions. Similar MRI salting experiment can also be performed in real-time by placing a fresh muscle sample inside a closed chamber filled with brine and taking MRI images of the whole system during the salting process. Details of a real-time salting MRI experiment are described by Erikson et.al.[11].

3   EFFECT OF THE SALTING METHOD ON SALT DISTRIBUTION

Salt uptake rate in muscle foods depends on many technological parameters, one of them being the availability of the water in a muscle which is a function of the quality of the raw material. Another important parameter is availability and local concentration of the sodium ions at the interface with the muscle. ^{23}Na MRI can be used to visualize variations in the sodium distribution of similar raw materials salted or processed by different methods, and thus serve as a tool for the process optimization.

Two groups of salmon fillets (3 fillets in each group) were salted using either brine salting (7,4% NaCl, 4°C, 3 days, fish to brine weight ratio 1:10) or dry salting (10°C, 24 h, without drainage, fine salt blended with 28% sugar) methods, simulating corresponding industrial processes. The smoking (4h, 27-28°C, beech chippings) was done after drying in air (24 h, 10°C). Sodium imaging was performed on Avance AV300 instrument (Bruker Optics GmbH, Germany) with a magnetic field strength of 7 T and horizontal bore opening using a commercial ^{1}H/^{23}Na double tuned 72 mm Birdcage probe. A multi-echo imaging protocol with the following parameters was used: one slice with 10 mm slice thickness was acquired, MS = 64 × 64, NS = 128, FOV = 6.4 cm and RD = 240 ms. Five successive echoes with TE = 5.6 ms were acquired. In addition T_1 weighted ^{1}H MRI of 9 corresponding slices was done to visualize fat distribution for comparison with the sodium images. Parameters for the ^{1}H MRI were: MS = 128 × 128, NS = 1, FOV = 6.4 cm, TE = 10 ms and RD = 500 ms.

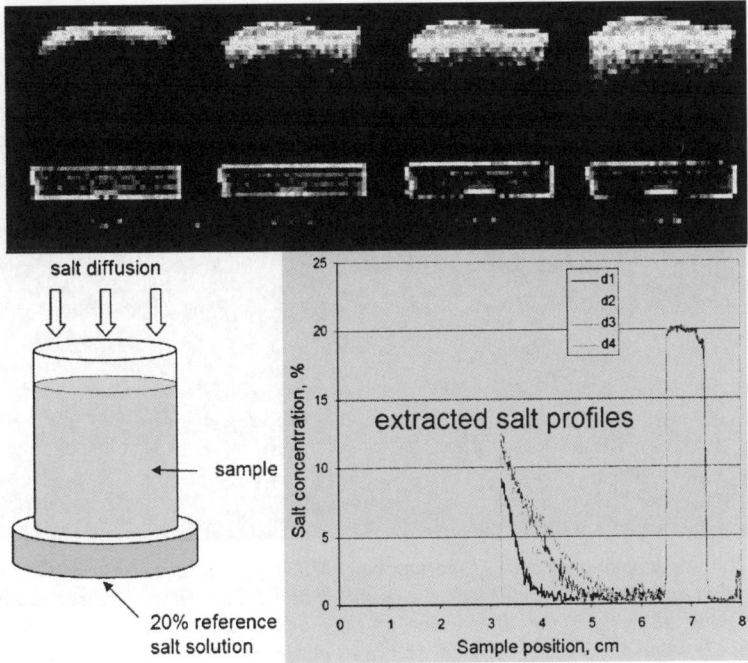

Figure 1. *One dimensional salt diffusion MRI study in cod muscle: MR images taken at 4 salting times, diffusion cell used in the experiment and the corresponding extracted salt profiles.*

Representative sodium and proton images of each fish group are shown in Figure 2. After the MRI experiment all samples were analyzed chemically for total salt content and the average salt contents in the sample groups were: 1.40 ± 0.28% (brine salted), 1.82 ± 0.39% (dry salted and smoked) indicating differences in the salt uptake of salmon by the two salting methods. After smoking the salt distribution in the dry salted became much more homogeneous. The fat layer inside the skin is known to act as a barrier for salt to diffuse into the muscle[12]. Figure 2 shows that this effect is most pronounced in the dry salted fillets compared to brine salted fillets. However, after the drying period and smoking, an evenly salt distribution was observed in the dry salted fillets. This indicates that salting method might not affect the salt distribution in the finished smoked product.

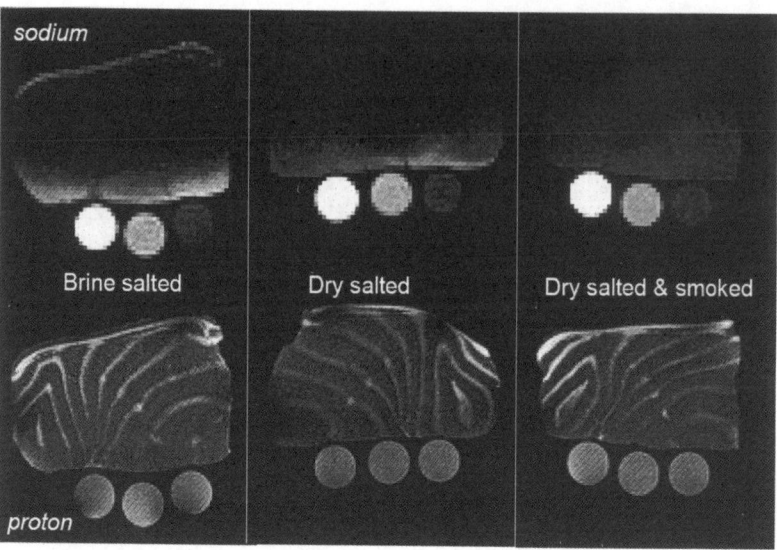

Figure 2. *^{23}Na (upper row) and ^{1}H (bottom row) MR images of three salmon fillet pieces processed in different ways: brine salted, dry salted and dry salted with subsequent drying and smoking showing clear differences in the sodium distribution. Three reference solutions (2, 4 and 6% NaCl brines) are shown as circles in the middle of each image: The circle with highest light intensity in the ^{23}Na MR images reflects the cylinder containing 6% brine solution, and the circle with lowest light intensity reflects the cylinder containing 2% brine solution.*

4   PARTIAL "INVISIBILITY" AND QUANTIFICATION OF SODIUM

In a muscle matrix, quadrupolar interactions of the sodium bound to proteins and other macromolecules result in, so called, partial sodium NMR "invisibility"[13] due to presence of very short NMR relaxation components with T_2 of about few milliseconds. ^{23}Na nuclei with so short relaxation times can not be detected in most spin-echo MRI experiments and which is often referred to as "loss" of sodium signal. At physiological concentrations (< 0.4 % NaCl) the NMR visibility of the sodium was reported to be around 40 % for many muscle types[7]. Sodium MRI visibility factors can be estimated as ratios of the MRI calculated salt contents in the samples to those determined chemically in percents. In our recent MRI study[14] of cod and salmon fillet pieces homogeneously salted at several predefined brine concentrations (0 to 25 wt.% NaCl) it was found that sodium visibility reaches its maximum for both cod and salmon at the brine concentration about 10%, indicating that the amount of the "structured" sodium at this concentration is minimal. This correlates well with swelling of the fish muscle salted in brine of similar concentration[6]. Comparing cod and salmon salted at the same brine concentrations it was observed that all salmon samples had lower sodium visibility than the corresponding cod samples which is apparently a result of higher structural inhomogeneity of salmon muscle compared to cod due to the presence of the micro-network of fat[15]. Figure 3 demonstrates linear increase of the sodium visibility factors with the water content for all samples ($R^2 = 0.87$). This was expected since the MRI "visible" free part of the sodium must be dissolved in the muscle water. Numbers at the plot symbols in Figure 3 indicate brine concentration at which the corresponding sample was salted.

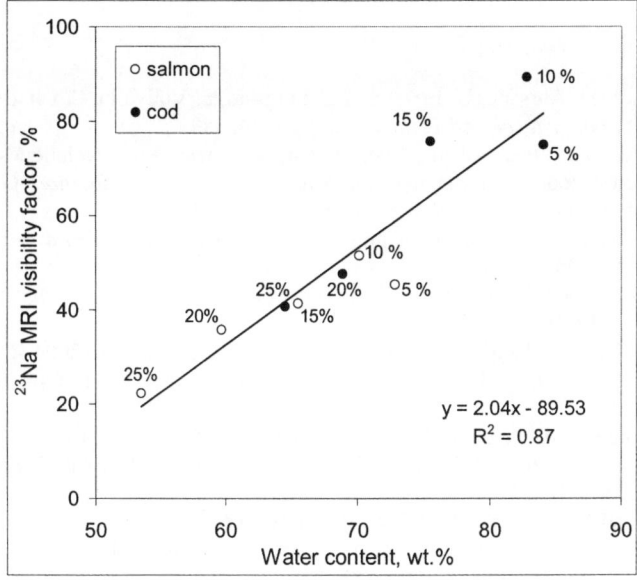

Figure 3. Sodium visibility factors for salmon and cod pieces salted at different brine concentrations as a function of the water content in the samples.

Possibility to correctly quantify sodium from the MR images implies full recovery of the [23]Na signal and should therefore be based on imaging techniques with shortest possible delay between the excitation pulse and the signal detection. SPI[9] and SPRITE[10] techniques are pure phase encoding imaging methods using the free induction decay (FID) for the signal detection, hence minimizing loss of the observed sodium NMR signal due to the relaxation. SPRITE MRI has been successfully used by Romanzetti et.al.[16] to produce [23]Na spin density maps of human brain. Recently a comparative MRI study was carried out by our group in which a number of model food samples were imaged by both spin-echo and SPRITE MRI methods, demonstrating the possibility of accurate sodium quantification by the SPRITE approach. It was also shown that both bulk [23]Na FID and CPMG relaxation measurements can be used for accurate measurement of the total salt content in foods.

5 CONCLUSIONS

Although not always quantitative, [23]Na imaging based on spin-echo detection gives an opportunity to obtain multi-slice [23]Na images of salted foods for comparative purposes or to follow a real-time salting process without special requirements to hardware. Sodium MRI is an excellent tool to study influence of the raw material quality on the salt uptake rate, optimizing industrial salting regimes, choosing the best salting method to achieve target product with predefined sensory and commercial qualities as well as for development of new salting processes. The degree of partial sodium "visibility" by MRI spin-echo techniques is approximately linearly correlated with the water content in the fish muscle, reaching its maximum for fillet pieces salted at 10% brines for both salmon and cod. The partial sodium MRI "invisibility" appears to be a purely methodological problem that can be avoided by applying SPI and SPRITE imaging techniques.

References

1   I. Martinez, M. Aursand, U. Erikson, T.E. Singstad, E. Veliyulin, C. van der Zwaag C. *Trends in Food Science and Technology,* 2003, **14**, 489
2   R.R. Ruan and P.L. Chen, Nuclear magnetic resonance techniques and their application in food quality analysis, in *Nondestructive food evaluation;* Gunasekaran, S. Ed.; Marcel Dekker Inc.: New York, 2001, pp.165-216.
3   B. Hills, "Magnetic resonance imaging in food science", John Wiley & Sons Inc., New York, 1998, p.96
4   K.A. Thorarinsdottir, S. Arason, S.G. Bogason, K. Kristbergsson, *Int. J. Food Sci. Tech.* 2004, **39**, 79
5   A.M. Rørå, R. Furuhaug, S.O. Fjæra, P.O. Skjervold, *Aquaculture* 2004, **232**, 255
6   L. Gallart-Jornet, J.M. Barat, T. Rustad, U. Erikson, I. Escriche, P. Fito, *J. Food Eng.* 2007, **79**, 261
7   C.S. Springer, *Ann. Rev. Biophys. Chem.,* 1987, **16**, 375
8   E.M. Shapiro, A. Borthakur, R. Dandora, A. Kriss, J.S. Leigh and R. Reddy, *J. Magn. Reson.,* 2000, **142**, 24
9   S. Emid and J.H.N. Creyghton, *Physica B,* 1985, **128**, 81
10 B.J. Balcom, R.P. Macgregor, S.D. Beyea, D.P. Green, R.L. Armstrong, T.W. Bremner, *J. Magn. Reson. A,* 1996, **123**, 131
11 U. Erikson, E. Veliyulin, T. Singstad and M. Aursand, *J. Food. Sci.,* 2004, **69**, 107
12 Gallart-Jornet, L.; Barat, J. M.; Rustad, T.; Erikson, U.; Escriche, I.; Fito, P. *J. Food Eng.* 2007, **79**, 261–270
13 J.P Renou, S. Benderbous, G. Bielicki, L. Foucat, J.P. Donnat, *Magn. Reson. Imaging* 1994, **12**, 131
14 E. Veliyulin and I.G. Aursand, *J. Sci. Food. Agric.,* 2007, **87**, 2676
15 M. Aursand, B. Bleivik, J.R. Rainuzzo, L. Jørgensen, V. Mohr, *J. Sci. Food Agric.,* 1994, **64**, 239
16 S. Romanzetti, M. Halse, J. Kaffanke, K. Zilles, B.J. Balcom, N.J. Shah, *J. Magn. Reson.,* 2006, **179**, 56

Subject Index